AF545206

PERSPECTIVES IN ANIMAL ECOLOGY AND REPRODUCTION

PERSPECTIVES IN ANIMAL ECOLOGY AND REPRODUCTION

VOL. II

V.K. GUPTA
ANIL K. VERMA

DAYA PUBLISHING HOUSE
NEW DELHI
2004

ISBN 81-7035-354-8

Published by : **Daya Publishing House**
1123/74, Deva Ram Park
Tri Nagar, Delhi - 110 035
Phone: 27383999
Fax: (011) 23244987
e-mail: dayabooks@vsnl.com
website: www.dayabooks.com

Showroom : 4762-63/23, Ansari Road, Darya Ganj,
New Delhi - 110 002
Phone: 23245578, 23244987

Laser Typesetting : **APCON**
New Delhi - 110 064

Printed at : **Chawla Offset Printers**
Delhi - 110 052

PRINTED IN INDIA

Foreword

In the present perspectives, animal ecology and reproduction is not only confined to academic pursuits and research in various institutions but has become the subject matter of general public awareness. There are, however, only a few authoritative books dealing with animal ecology and reproduction. The present book is a genuine endeavour in that direction. I believe this volume will meet the requirement of researchers and animal knowledge seekers to a great extent. The book covers a wide range of organisms from insects, arthopods to the highly evolved mammals. The information provided in this volume would be of inherent interest and immense value to the students, teacher and researchers interested in this field. The exhaustive information provided in this book would provide the benchmark data useful to a number of organizations and individuals deeply interested and committed to understand the reproductive strategies adapted by the animals and the ecological problems being faced by them. This book is, therefore, extremely welcome.

I am glad that editors have done a commendable job in compiling a set of original and significant papers/review articles on animal ecology and their reproductive patterns, which can very well be used as a ready-reckoner for the ecologists, reproductive biologists and ethologists in particular and the zoologists in general. Editors themselves have made important contributions in the area and a good deal of enthusiasm on their part is reflected in this book. Various research articles contained in this volume have been authored by eminent researchers and educationists of the country and they fulfill the objectives of sensitizing the scientific community and common masses towards crucial issues in the animal world. The excellent illustrations given in the book extend the range of usefulness of this effort.

I congratulate the editors for their efforts in bringing out such an excellent book which will give all round readers an exciting and serious reading material and also for continuing the outstanding work for which Regional Research Laboratory, Jammu has been well known. I am very glad to write the foreword for this book and I wish the authors all success in this endeavour.

Dr. G. N. QAZI
Director
Regional Research Laboratory
Jammu (INDIA)

Preface

It is a matter of pleasure and satisfaction for us that the first volume of this book *"Perspectives in Animal Ecology and Reproduction"* published in 2002 has been liked and warmly accepted by exports and workers in these fields throughout the country. Encouraged by the success of the publication of volume I and the good response, we once again requested the distinguished scientists/research workers who have themselves made pioneering and significant contribution to their respective fields of study to contribute their original research papers for publication in the volume II of this book series. The response to our invitation for this book has been overwhelming and the present book is the outcome of a collaborative effort of the experts in these fields. We are therefore, pleased to place before our generous readers the volume II of this boo series.

Vast strides have been made in the fields of animal ecology and reproduction in recent years and it is a gigantic task to compile all this information in a single volume. Knowledge of animal ecology and reproduction, their life-cycle strategies and effective conservation of the threatened forms is essential to everyone, directly or indirectly connected with the aquacultural teaching, research and extension. The present volume is the outcome of our extensive teaching, research and extension experience. Well written in simple and lucid manner, comprehensive and up-to-date coupled with clear and suitable photographs and drawings. This volume has been bifurcated into two sections and contains nineteen chapters and each prepared by expert scientists. We are grateful to the contributors for their diligence in preparing their comprehensive and insightful accounts. We are also greatly indebted to our readers for the warm and cordial reception given to the previous volume. Suggestions for further improvement of the book will be thankfully acknowledged in the future edition.

Jammu, India

V.K. GUPTA
ANIL K. VERMA

CONTENTS

1

Bio-Attributes of Predaceous Coccinellids – A Review

Omkar*, Ahmad Pervez, Shefali Srivastava and Barish E. James

Ladybird Research Laboratory, Department of Zoology, University of Lucknow, Lucknow-226007, India

ABSTRACT

Predaceous coccinellids are important tools of integrated pest management programmes, as they prey on various phytophagous pests. A perusal of literature gives an insight into their various ecological aspects, viz. distribution, population dynamics, predation, cannibalism and intraguild interactions, feeding-, foraging- and reproductive behaviour, factors influencing growth, development and reproduction. Empirical data on their certain biological attributes, such as oviposition, fecundity, hatchability, emergence, longevity and mortality, have also been reviewed.

Key words: Coccinellids, ladybeetles, aphids, coccids, predation, reproduction

Introduction

Coccinellids are potent predators of several phytophagous pests, possessing the potential of being utilized in integrated pest management programmes. They prey upon aphids and coccids with certain species being predaceous on both (Omkar and Pervez, 2000a). In addition, certain species feed on adelgids (Pope, 1973), aleyrodids (Heinz and Zalom, 1996), chrysomelids (Ghorpade, 1979), pentatomids (Afroze and Shujauddin, 1998), phylloxera (Pope, 1973), psyllids (Anonymous, 1995) and mites (Espinha and Torres, 1995). There is also a report of coccinellid preying on larvae of mycophagous ladybirds (Camargo, 1937). Such vast

* Corresponding author.

prey-ranges of coccinellids and their prey interactions have attracted intensive investigations. Different coccinellidologists have reviewed their ecology and emphasized their pragmatic utilization in bio-control (Hodek and Honek, 1996; Majerus, 1994; Obrycki and Kring, 1998; Dixon, 2000). Recently, certain aphidophagous coccinellids have been studied and/or reviewed, *viz. Cheilomenes sexmaculata* (Bind, 1998; Agarwala and Yasuda, 2000), *Coccinella septempunctata* (Srivastava, 2000; Omkar and Pervez, 2002a) and *Coccinella trnsversalis* (James, 2001). Owing to their importance, certain ecological aspects of predaceous coccinellids have been reviewed to present a brief account of their distribution, population dynamics, predation, foraging, reproduction and influence of certain factors on their biological attributes.

Distribution

About 90% of the approximately 4200-coccinellid species are predaceous (Iperti and Paoletti, 1999). They occur in different habitats including high altitudes, agricultural and horticultural fields, etc. (Rice, 1992; Brown and Miller, 1998; Easterbrook, 1998; Singh and Singh, 1998; Zhou and Carter, 1992; Omkar and Pervez, 1999; 2000b). Their record from various regions is as presented in Table 1.1.

Coccinella sp. dominates in Canada and is found in abundance in maize fields of Quebec (Coderre *et. al.*, 1987). Japanese coccinellid fauna was dominated by *Coccinella septempunctata brucki*, *Harmonia axyridis* and *Propylea japonica* (Takahashi and Naito, 1984). *C. sexmaculata and Coccinella* sp. are most common amongst Indian coccinellids (Omkar and Pervez, 2000b). Abundance of *C. septempunctata*, *Propylea quatuordecimpunctata* and *Coccinella quinquepunctata* was reported in Czech Republic (Honek, 1989) and the former two species were the main coccinellids of winter wheat stands in Germany (Bode, 1980), Russia (KuliKova, 1984) and USA (Elliott and Kieckhefor, 1990). The individuals of ladybeetle, *Adalia bipunctata* were observed in abundance in the apple orchards of Poland constituting 22% coccinellid fauna (Olszak and Niemczyk, 1987).

Ladybeetle species vary from stenotopic to eurytopic. Stenotopic species are restricted to specific habitats, *e.g. Bothrocalvia lewisi* and *Bothrocalvia albolineata*, existing in mountainous habitats of India (Shantibala and Singh, 1991). They are highly adapted to specific habitats and slight changes in environmental conditions may prove detrimental. For instance, *Spiladepha barovskii kiritschenkoi* and *Coccinella reitteri* live at high altitude of Kazakhistan and rarely fly due to poorly developed wings. Flight is an unfavourable activity and causes hindrance in survival (Savoiskaya, 1970). Eurytopic species are cosmopolitan owing to their high plasticity and are found frequently and abundantly, *e.g. C. septempunctata, C. sexmaculata, A. bipunctata, H. axyridis* and *C. transversalis.*

Table 1.1: Coccinellids from different regions

Name of the region	Reference
Bermuda	Gordon and Hilburn (1990)
Brazil	Arioli (1985)
Bulgaria	Jordanova (1979)
East Africa	Fursch (2001)
England	Majerus and Kearns (1989)
France	Iperti (1986)
Germany	Freier and Gruel (1993)
India	Agarwala and Ghosh (1988)
	Omkar and Pervez (2000d)
Jordan	Allawi (1989)
New Guinea	Chazeau (1989)
North America	Gordon (1985)
North Korea	Bielawski (1980)
Nepal	Canepari (1997)
Paraguay	Michel (1992)
Russia	Kuznetsov (1981)
South America	Gordon (1996, 1999)
Thailand	Chunram and Sasaji (1980)
Taiwan	Yu and Pang (1997)
West Indies	Gordon (1978)
Yemen	Raimundo and Harten (2000)

Coccinellids have seasonal occurrence and migratory tendency (Boraei *et al.*, 1993), owing to the ephemeral nature of prey. The prominent habitat preference is remarkable (Honek, 1985). Survey of various habitats, which differ in prey density, isolation and host plants, revealed that *A. bipunctata, Adonia variegata* and *C. septempunctata* prefer high aphid density, while *C. quiquepunctata* easily exists in low aphid density. *C. quinquepunctata, C. septempunctata, A. bipunctata* and *A. variegata* prefer sunny, whilst *Adalia decimpunctata and Calvia quatuordecimpunctata* like shady conditions. *C. septempunctata* and *A. variegata* prefer herbs, while *A. decimpunctata* and *C. quatuordecimpunctata* lived on trees (Honek, 1985). Prey scarcity makes them leave the habitat and search elsewhere. Three abundant species of coccinellids, *viz. C. septempunctata, A. bipunctata* and *P. quatuordecimpunctata,* produce their first brood on nettles infested with aphid, *Macrolophium evansi,* considered to be one of the most suitable habitats (Hodek and Honek, 1996). Overwintered adults and offsprings move to bean plots infested with *Aphis fabae* and breed there. Adults often move to weeds and to humid and moist places in search of aphids (Hodek *et al.*, 1966). They are also found on trees and shrubs at the edges of forest in search of their food (Lusis, 1961). Thus, they migrate from prey-poor to prey-rich habitat (Savoiskaya, 1965).

Population Dynamics

Coccinellid populations are directly dependent on the prey availability, in particular, that of the aphids. A hectare of bean field can produce about 4 billion aphids, *A. fabae* and about 0.28 million individuals of *C. septempunctata* per hectare were estimated predating on them (Dixon, 1998). The study of coccinellid relationships with abiotic (temperature, rainfall, sunshine and humidity) and biotic (pathogens and pests) factors in the field has pragmatic importance in bio-control. Seasonal occurrence of coccinellids and aphids indicates a seasonal relationship amongst them. Successful synchronization of coccinellids with aphids promotes progeny production in coccinellids (Agarwala and Bhattacharya, 1995). Environmental conditions, such as temperature, R.H., sunshine and rainfall, strongly affect coccinellid population (Thalji, 1994). Feeding potential of coccinellids increases with increase in temperature (Freier and Triltsch, 1996). R. H. and sunshine have also been positively correlated with coccinellid population size (Debraj *et al.*, 1997; Singh *et al.*, 1994). But, rainfall has a negative correlation with coccinellids (Prasana *et al.*, 1997). Biotic factors such as aphids have a initial positive correlation with ladybeetles (Sharma and Yadav, 1994; Chitradevi *et al.*, 1997; Manzar *et al.*, 1998). However, increase in ladybeetle population reduces the aphid population, thereby establishing a negative correlation (Sharma *et al.*, 1997).

Predation

(a) Foraging

Foraging, is a highly investigated aspect of predaceous coccinellids, characterized by both random and non-random searching by larvae and adults. After an initial random (extensive and non-systematic) search for prey, coccinellids switch over to non-random (intensive/area restricted) search once prey are located (Carter and Dixon, 1984a; Ferran *et al.*, 1994). The switching over is elicited after consuming suitable prey (Kalushkov, 1999). Prey-odours are key stimulators of coccinellids (Obata, 1986). Olfactometric experiments indicate that coccinellids, *Chilocorus nigrita, C. septempunctata and Cryptolaemus montrouzieri* are attracted by the prey-odours (Hattingh and Samways, 1995; Ponsonby and Copland, 1995). Mealy-bug wax is an arrestant for coccinellids, *Diomus* sp. and *Exochomus* sp. (van der Meiracker *et al.*, 1990). Visual stimulus secondarily brings the ladybeetls closer to their preys. Certain ladybeetles, *e.g. C. nigritus* are attracted to prominently silhouetted features, such as trees (Hattingh and Samways, 1995). Shape of the leaves also attracts them, *e.g.*, *C. nigritus* usually encounters plants having ovate, elliptical or oblong leaves (Hattingh and Samways, 1995). Prey colour sometimes affects the feeding propensity as *C. septempunctata* largely consumed red and green coloured pea aphids, whereas *H. axyridis* consumed more of red morphs (Harmon *et al.*, 1998).

Mostly, coccinellids are positively photo-tactic and negatively geo-tactic and exploit prominent plant structures to aid searching (Ponsonby and Copland, 1995). Plant structures (twigs, pronounced leaf veins, leaf edges and needles) and architecture widely influence the foraging in coccinellids (Vohland, 1996). Hairy and waxy plant surfaces hinder larval movement and thus reduce the foraging success (Shah, 1982). First instar walked only on the leaf edge and on protruding veins, avoiding interveinal areas. Since, aphids usually reside in interveinal areas, they are less vulnerable to the attack of first instar. Fourth instars are capable of walking across interveinal areas, improving searching successes over that of first instar (Shah, 1982). The spatial diverse plant structures create relatively more difficulties in prey finding than a homogenous environment and mostly the herbivore-pests feed on the outer and upper parts of the plants (Ferran and Deconchat, 1992).

Hunger level and temperature also affect the foraging in ladybeetles. A hungry first instar of *Hippodamia convergens* spends more time in searching the alfalfa crop for aphids than *H. axyridis* (Obata, 1986). There exists a circadian rhythm in the foraging pattern with peak activity during daytime (Nakamuta, 1987). Intensity of height also alters and orientates the movement of coccinellids towards the light source. Aphid density and its age distribution also affect coccinellid prey searching, as older aphids are much more difficult to capture than early aphid instars. The first coccinellid instar have the lowest survival rate, since it is difficult for them to attack and capture the older instars of aphids (Frazer *et al.*, 1981). Coccinellid-aphid encounters are grouped into three outcomes, i.e. (i) the aphids are consumed, (ii) merely contacted by the larvae or (iii) aphids permanently leave the plants. In the latter case, aphids may fall off the plant to avoid larva, but again climb back on the plant. The encounter frequencies vary according to the age and species of both prey and predator. Older aphids are captured more frequently than younger ones by fourth coccinellid instar, because it often fails to detect the first instar nymphs (Frazer *et al.*, 1981). Besides predating, the coccinellids also dislodge aphids from the substrata. Adults possess relatively more capacity than the larvae to dislodge the aphids (McConnel and Kring, 1990). Dislodgment reduces the damaging potential of the aphid.

The speed and track width of third instar of *C. septempunctata* was lesser than fourth instar and adults (Carter and Dixon, 1984b). Analysis of searching track reveals that prey consumption causes a significant change in foraging approach (Rowlands and Chapin, 1978). The mean searching rate of first instar along the ventral surface of leaf varied from 0.27 to 0.41 cm/s with *C. sexmaculata* having lower speed (0.27 cm/s), than those of *C. transversalis* and *Spilocaria bissellata*, i.e. 0.30 cm/s and

0.41 cm/s, respectively (Mondal and Kar, 1983). The turning threshold, i.e. the minimum prey density required to divert the attention of a searching larva, enabling it to switch over from slow to rapid turning determines whether it exploits or abandons its prey. Foraging is effective when threshold matches the physiological and energetic requirement (Grunbawn, 1998).

(b) Predation

Food relationship is a widely studied aspect of coccinellid ecology. Coccinellids are polyphagous having wide food-range, which include homopterans, lepidopterans, coleopterans, hymenopterans, phytophagous mites, fungi, etc. along with certain natural products, such as, pollen and honeydew. Majority are aphidophagous or coccidophagous with females being more voracious than males (Omkar and Srivastava, 2001a). Mean prey consumption by male and female, *C. septempunctata* was 62.24 and 69.42 aphid, *Lipaphis erysimi* per day, respectively (Sharma *et al.*, 1997). Lifetime consumption of *L. erysimi* by adult male and female *C. septempunctata* was 5180 and 6151, respectively (Omkar and Srivastava, 2001a). Male consumes about 88.70% female's diet (Lucas *et al.*, 1997a).

Coccinellid food is classified into preferred, essential, alternative and rejected foods. Preferred food is the ladybeetle's choice when it is subjected to multiple foods. Essential food supports both development of immature stages and ovariole maturation. Alternative food promotes immature survival and development but restrains oviposition. Rejected food is toxic and causes mortality. A different food-range concept was suggested, in which, 'included' was applied to the prey species that can be exploited successfully, 'peripheral' for those that are eaten but resulted in a significant decline in their fitness and 'excluded' for those preys that may prove fatal and thus cannot be exploited by the predator (Malcolm, 1992). Immature survival, development and reproduction are the criteria for prey-suitability. High growth index, low mortality and high oviposition can be obtained when predator stages feed on essential foods. Coccinellids have specialized food requirements based upon the biochemical composition of the prey and their host plants (Atwal and Sethi, 1963). Several workers have noticed a high degree of specificity of some ladybeetles for certain prey. *C. septempunctata* prefers

L. erysimi (Srivastava, 2000). Biochemical analysis of this aphid ascribed the preference to its high protein content (Atwal and Sethi, 1963). Aphids, *Aphis gossypii* (Singh and Devi, 1996) and *Hydaphis coriandri* (Omkar *et al.*, 1997) were also preferred by *C. septempunctata, C. sexmaculta* preferred *A. gossypii* over aphids, *Uroleucon compositae, A. craccivora, Brevicoryne brassicae, L. erysimi* and *A. nerii* (Omkar and Bind, 1996). *A. nerii* was also reported to be suitable prey of *C. sexmaculata* on account of high fecundity and short development time (Choudhury *et. al.*, 1983). Development time of *C. sexmaculata* was shorter when fed on *A. craccivora* (18.0 days) and *M. persicae* (18.7 days) than *A. gossypii* (20.9 days) (Sugiura and Takada, 1998). Oviposition in first ten days was higher while feeding on *A. craccivora* (172 eggs) than *A. gossypii* (99 eggs) (Sugiura and Takada, 1998). *Coccinella transversalis* prefer *A. craccivora* and *A. gossypii* (Omkar *et. al.*, 1999; James, 2001). Aphid, *Aphis pomi* was more suitable for development of *H. axyridis* than aphids, *B. brassicae* and *Hyalopterus arundinis* (Hukusima and Kamei, 1970). Ladybeetle, *Angleis cardoni* preferred *B. brassicae* (Afroze, 1999) while *Micraspis discolor* preferred *A. gossypii* and *A. craccivora* (Omkar and Pervez, 2001). Adults along with third and fourth instars of *P. japonica* preferred older instars of aphids, whereas first and second instars preferred young aphid instars (Zhou *et al.*, 1994).

C. septempunctata developed faster and had higher fitness on aphid, *Acyrthosiphon pisum* than on aphid, *Rhopalosiphum maidis* (Obrycki and Orr, 1990). Adults of *Hippodamia parenthesis* reared on *A. pisum* were also apparently larger and heavier than those reared on aphid, *Schizaphis graminum* (Obrycki and Orr, 1990). Aphids, such as, *A nerii* possess high alkaloid contents, which are toxic to certain coccinellids with exception of *A. variegata* (Iperti, 1966). *A. nerii* is unsuitable prey for *S. undecimpunctata, P. quatuordecimpunctata* and *A. bipunctata* (Iperti, 1966). It is also toxic to *H. axyridis* along with *C. septempunctata* and *C. transversalis* (Tao and Chiu, 1971). It was however not found to be toxic to *C. septempunctata* and is considered as an essential prey (Omkar and Srivastava, 2002c). High content of alkaloid, quenlizedine in aphid, *Macrosiphum albifrons* appears to be a reason for it being an excluded food for *C. septempunctata* (Hodek and Honek, 1996). Certain coccinellids, *e.g.*, *A. bipunctata*, survive well for long periods on aphids having alkaloids (Gruppe and Roemer, 1988).

The prey specificity might be ascribed to the differences in biochemical constituents of host plants. Conspecific aphids existing on different host plants also affect the growth, development and survival of coccinellid predators. Biochemical contents of host plants, entering the body of coccinellids via aphids are attributed to the variation in the effects on the biological attributes of the coccinellids (Kumar *et al.*, 1999).

Prey scarcity reduces the fitness of coccinellids larvae, which may develop even when the prey supply is reduced to 40% or even less (Baumgaertner *et al.*, 1981). The minimal food requirement for a developing larva could be as low as one-seventh of its normal requirement. Immature stages of *C. septempunctata* may even develop on one large aphid, *A. pisum* per day (Obrycki *et al.*, 1997). Larvae of *M. discolor* required a minimum of 35.2 aphids (*i.e.* 6, 4, 11.2, 14 aphids for I, II, III, IV instars, respectively) to undergo pupation (Agarwala *et al.*, 1988). The mean daily feeding of larvae was 1.99 aphids for 17.6 days (Agarwala *et. al.*, 1988). Prey scarcity may induce premature moulting and the adults may switch on the secondary prey or even to honeydews, pollens and nectars (Mills, 1981).

(c) Cannibalism

Starvation, hunger and search for the food are key factors responsible for the cannibalism among ladybeetles (Agarwala and Dixon, 1991). Usually eggs and just moulted individuals, are preferred and more susceptible (Agarwala *et al.*, 1997). Egg-cannibalism is frequent, since eggs are always vulnerable to cannibalism by predatory stages (Agarwala and Dixon, 1993a; Veeravel and Baskaran, 1995), especially by first instars (Agarwala, 1991). Egg cannibalism is frequent in laboratory but incidental in fields (Mills, 1982). Egg cannibalism is minimized if the neonate instars are in the vicinity of aphids (Hemptinne *et al.*, 1992). Cluster laying of eggs exhibits aposematic effect on neonate instars and is thereby an effective means of defence against cannibalism (Agarwala and Dixon, 1993b.) In aphid scarcity, cannibalism is the only mode of survival to overcome the pressure of immature development. Hungry larvae may indulge in egg cannibalism in the presence of aphids but the intensity is high in aphid scarcity. All the four instars survived more when fed on eggs than aphids of equal dry weight, which indicate that eggs offer higher nutrition than aphids. The instars consume conspecific eggs more efficiently than aphids, as less time as taken on capturing and consuming the eggs than aphids (Agarwala and Dixon, 1993a).

Kin-recognition is prominent in coccinellids, with females being reluctant in consuming their own eggs. Even a starved female, which had consumed all alien eggs, was reluctant to eat her own eggs. Conversely, the hungry males did not hesitate in consuming the eggs they had fertilized (Agarwala and Dixon, 1993a). The hungry second instars recognize their sibling eggs and prefer to eat alien ones (Agarwala and Dixon, 1993a). *C. septempunctata* (Koide, 1962) and *A. bipunctata* (Dimetry, 1974) were successfully augmented on conspecific and heterospecific eggs, but *C. septempunctata brucki,* failed to be reared on egg diet, as most of the first and second instars moulted to the next stages and then died (Takahashi, 1987b). Relatively, lesser dry weight of eggs is required for achieving an identical growth rate by fourth instars as compared to aphids (Agarwala, 1991). The hatched larvae stay for about 16-24 hours on their batches and than intensively search for food nearby (Brown, 1972). Cannibalism, i.e. feeding on eggs enhances the foraging potential of instars (Brown, 1972). The random and asynchronized hatching of eggs, and the occurrence of infertile eggs in an egg batch encourages egg sibling cannibalism (Mills, 1982). The cannibalism of conspecifics is known as non-sibling cannibalism (Osawa, 1989).

(d) Heterospecific Predation

Complex predator-predator interaction leads to intraguild predation (Polis and Holt, 1992). This results in instant energy gain to the starving predators and also reduces the potential competitors (Phoofolo and Obrycki, 1998). Coccinellid larvae depend on the surface chemistry of the prey due to their poorly developed eyes. This involves a risk of being poisoned by its victim, due to the presence of toxic chemicals (Hemptinne *et al.*, 2000). The eggs of *A. bipunctata* and *C. septempunctata* are toxic to heterospecifics (Hemptinne *et al.*, 2000). Both species prefer to eat the conspecific eggs than those of heterospecifics, supporting that toxic contents protect eggs from intraguild predation but do not prevent cannibalism. *C. septempunctata* is considered to be a successful member of the guild because of its establishment in the North America (Wheeler and Hoebeke, 1995). Presence of conspecifics of *C. maculata* and *C. septempunctata* did not significantly affect their survival. Introduction of heterospecifics led to the intraguild predation with reduced survival of former (Obrycki *et al.,* 1998), whereas, survival fitness of C. *septempunctata* increased in the presence of heterospecifics (Obrycki *et al.*, 1998). The reason for this asymmetry was the high rate of prey capturing by C. *septempunctata.* When two indigenous species of Japan, *i.e.*, *H. axyridis* and *C. septempunctata,* and an exotic, *Adalia bipunctata* were kept together, the latter was treated as an intraguild prey by the two indigenous species (Kajita *et al.*, 2000). C. *septempunctata,* has been dominated by *H. axyridis,* the second most established exotic predator in North America (Yasuda and Katsuhiro, 1997). The larvae of *H. axyridis* appeared to be a poor food to C. *septempunctata,* as the survival of latter reduced after feeding on the former (Yasuda and Ohnuma, 1999).

Reproduction

Reproductive behaviour comprises of various activities surrounding reproduction including, mate location and recognition, courtship and mating, and oviposition. Sex-pheromones are primarily responsible for mate location and recognition, which attract the males and bring them to the females' vicinity (Omkar and Pervez, 2002b). Visual cues secondarily bring the two sexes closer. Sex-odours are so strong in certain coccinellids that males are attracted towards females still in puparium, as in *Lepthothea galbula.* Males remain there and guard female pupae in order to copulate with them when they emerge (Richards, 1980).

(a) Courtship and Mating

The urge to mate is the greatest driving instinct in insects and prominent in coccinellids when they attain maturity. Duration of maturation was 10 days in *H. axyridis* (Obata, 1987) and 11 days in *Synia melanaria* (Afroze and Shujauddin, 1998). The males tend to mature earlier than the females; the males of C. *septempunctata,* became sexually mature within 8.80±0.29 days, while the female took 11.20±0.39 days (Omkar and Srivastava, 2002b). Mating amongst coccinellids is a frequent and easily observable phenomenon (Obata, 1987; Omkar and Srivastava, 2002b).

Courtship occurred prior to mating in five steps (i) ***Approach*** - when male proceeds and approaches towards female (ii) ***Watch***- when it pauses at a distance (iii) ***Examine*** - when it touches the female (iv) ***Mount*** - when it ultimately mounts over female's elytra, and (v) ***Attempt*** - when it inserts aedeagus in female genitalia (Obata, 1987; Omkar and Srivastava, 2002b). Mating durations in different coccinellids ranged from 10 min to 4-5 hours depending upon species, their sexual status and eco-conditions. Mating duration in *H. axyridis, C. septempunctata, P. japonico* and *C. sexmaculata,* was 119.08±16.82, 85.40±13.90, 215.60±34.43 and 37.67±10.04 min, respectively (Obata and Johki, 1991). Latent period or the static duration after establishing genital connection between two sexes in *H. axyridis* and *C. septempunctata,* was 34.55±2.43 and 48.00±5.25 min, respectively (Obata and Johki, 1991). Mating in *H. axyridis,* characterized with body shakings or bouts and the number of bouts in unmated adults (ranges 233-168 bouts) was relatively more than copulated male and virgin female (range 82-177 bouts) (Obata, 1987). A spermatophore containing sperms is injected inside the female's bursa copulatrix, during mating (Obata and Hidaka, 1987). After satiation the mating is terminated as female kicks off the male and starts walking (Omkar and Srivastava, 2002b). Less than two-day-old females of some Malayan species reject the attempts of the males to copulate. They move away rapidly from the male and kick the extended aedeagus with their legs (Ming, 1971).

(b) Preferential Mating

Polymorphism is a wide occurrence in coccinellids and usually two morphs have been reported, *viz.* melanic and non-melanic morphs (Osawa and Nishida, 1992; Holloway *et al.*, 1995). The preference of a type or variety of males or females by the opposite sex for mating is referred to as preferential mating or non-random mating (Kearn *et al.*, 1992; O'Donald and Majerus, 1992). There is a general non-random mating advantage in favour of melanic males and the females have a strong genetic preference to mate with the melanic forms of males (Kearn *et al.*, 1992; Tomlinson, 1996). Furthermore, mated females have a higher preference for the melanic males (O'Donald and Majerus, 1992; Rhamhalinghan, 1998). It appears that the female is gaining some compensatory selective advantage by choosing melanics at the cost of non-melanic males (Tomlinson, 1989). This mating advantage of the melanics is crucial in maintaining polymorphism between melanic and non-melanics in many populations. Conversely, no genetic preference towards the melanic females was exhibited by the males (Tomlinson, 1996).

Fluctuating asymmetry (random deviation from a perfectly symmetrical body of an individual resulting in the inability of individuals to undergo precisely and previously determined development path) in coccinellids also affected the male mating success (Ueno, 1994). It has negative correlation with fitness components, such as fecundity and growth rates (Leary and Allendorf, 1989). In *H. axyridis,* a negative correlation was established between the male longevity and male mating success with fluctuating asymmetry, which revealed that females preferred symmetrical males and rejected asymmetrical ones (Ueno, 1994).

Biological Attributes

(a) Oviposition and Fecundity

Females that mate frequently lay a high proportion of fertile eggs (Omkar and Srivastava, 2001b.) There is a decline in the fertility and viability of females that have mated once. Conversely, high fertility and improved quantity of viable eggs was observed in multiple mated females, inferring promiscuity as a crucial factor for high fertility (Majerus, 1994). The mean number of eggs laid/female/day increased curvilinearly with increase in aphid density (Agarwala and Bardhanroy, 1997; Omkar and Pervez, unpubl.). Ladybeetles sometimes abstain from oviposition at optimal food density (Kindlmann and Dixon, 1993). There is always a search for a convenient place to lay eggs and it should be laid in the vicinity of prey. Females have a specific choice and preference for the substrata to lay eggs. *A variegata, C. septempunctata,* and *Semiadalia undecimpunctata* lay eggs on plant surfaces, while *A. bipunctata* and *P. quatuordecimpunctata* prefer to oviposit on the lid of jar (Iperti, 1966). *C. sexmaculata,* prefers to lay eggs on the host plant twigs as compared to smooth glass surface of beaker, muslin cloth and filter paper (Bind, 1998). *C. septempunctata,* laid maximum number of eggs when recently enclosed and preferred oviposition on host plant twigs and yellow colour substrata (Gautam, 1990a;

Omkar and Srivastava, 2001b). *C. transversalis* also preferred host plant twigs with green and yellow colour substrata (James, 2001). *A decempunctata* and *Adalia tetraspilota* deposited most of their eggs in humid microcosms on foliages (Lusis, 1973). *A bipunctata* and *C. septempunctata,* also oviposited preferably on *Berberis vulgaris* leaves than on any other plant species (Shah, 1983). *C. septempunctata brucki* preferred to lay eggs on inanimate objects, such as rocks, mainly in April, whereas overwintered *H. axyridis* laid eggs subsequently on plants in the early May (Takahashi, 1987a). *C. sexmaculata,* exhibited no substrata preference when fed on mealy bug, *Ferrisa virgata,* but preferred to oviposit on cotton twigs when fed on *L. erysimi* (Gautam, 1990b). *A bipunctata* preferred red colour to green, yellow and blue; and curved surfaces for oviposition than plane surfaces (Iperti and Prudent, 1986). The exposure to the ethanol extracts of the surface of plants increases the incidence of oviposition in *C. maculata* (Smith and Williams, 1976).

Pseudoscymnus kurohimae covers its eggs with a secretion consisting of undigested remnants of prey (Arakaki, 1988), whereas *Scymnus hoffmani* covers the eggs with the exuviae of it's aphid prey (Kawauchi, 1985). Gravid females of *A. bipunctata* do not oviposit on vial, which had previously contained conspecific larvae of its own species. They were not deterred by the previous presence of conspecific adults or the instars of two heterospecifics, *A. decempunctata* and *C. septempunctata,* in the vial (Doumbia *et al.,* 1998). This shows that larvae of ladybeetles produce a specific Oviposition Deterring Pheromone (ODP). In response to ODP, females reduce the risk of their eggs being cannibalized by deterring oviposition on such places (Doumbia *et al.,* 1998; Kindlmann and Dixon, 1993). Cluster egg laying, protecting them with secretion and prey exuviae and deterring oviposition where sensing ODP are remarkable examples of parental care by females.

Usually females lay eggs in batches, whose sizes may be anatomically determined by the number of ovarioles, as eggs per batch is equal to half the number of ovarioles (Stewart *et al.,* 1991). Thus, half the number of ovarioles participates in the egg output while the other half prepares the next batch (Stewart *et al.,* 1991). Ovarioles have a seasonal variation with their numbers increasing during winter and decreasing during summer and monsoon (Rhamhalinghan, 1987a). The optimum ovariole-range found in *C. septempunctata,* was 41-50 ovarioles per ovary (Rhamhalinghan, 1986). When aphid are scarce, ovarioles ranged between 31-40 and in contrast high ranges of 51-60 and 61-70 ovarioles per ovary have been observed at adequate and high prey supply (Rhamhalinghan, 1986). Thus, fecundity is positively correlated with food consumption (Rhamhalinghan, 1987b; Veeravel and Baskaran, 1997). Table 1.2 provides information on the fecundity of certain ladybeetles.

(b) Hatchability

Hatchability or viability of eggs is an important constituent of the reproductive success of a ladybeetle and has been well explored by various workers (Table 1.3). R.H. affects hatchability with percent viability of 45.5, 46.9, 49.7, 61.4, 81.7 and 88.6% at R.H. 0, 20, 40, 60, 80 and 100%, respectively in *C. septempunctata,* (Singh and Malhotra, 1979). The larvae hatch out the best at optimum temperature and further increase in temperature has a negative correlation with hatchability. Normally, egg hatchability in *M. discolor* is 50%, which reduces to 18.18% in June-July (Choudhuri and Agarwala, 1990).

(c) Emergence

Table 1.4 reveals percent emergence in predaceous coccinellids. Temperature has a major impact on the emergence of *C. montrouzieri,* as it was significantly delayed for 28 days when 1 and 3 days old pupae were chilled at 10°C (Babu and Azam, 1988). In addition, the emerging adults laid fewer eggs on maturation (Babu and Azam, 1988). Adult emergence of *C. sexmaculata,* declined from 65.7% to 6.1%, when temperature increased from 30°C to 35°C (Ali Khan and Yousuf, 1986). Parasitization also causes reduction in adult emergence (Kadona *et al.,* 1988). Emergence is also affected, though relatively less, by the prey quality as seen in the case of *B. suturalis* where it was found to be 100%, 85.71%, 100% and 100%, when the larval stages were fed on mealy-bug, *Ferrisa virgata, Planococcus pacificus, F. virgata + P. pacificus* and *P. operculella,* respectively (Gautam, 1990c).

(d) Longevity

Longevity in coccinellids varies from one month to approximately three years (Milevoj, 1997). Certain ladybeetles, such as *Calvia quinquedecimgutta* have a long inactive period due to which their longevity extend beyond a year. Longevity in *C. septempunctata* varied from 25 days to 840 days when fed on artificial diets (Milevoj, 1997). Longevity of *C. sexmaculata* decreased in four successive generations in the laboratory; 89 to 50 days in males and 111 to 50 days in females (Hukusima and Kouyama, 1974). Table 1.5 provides data on the longevity of male and female ladybeetles.

(e) Mortality

Parasitism is a major mortality factor, apart from the normal deaths in various stages of a ladybeetle (Triltsch, 1996; Kadona *et al.*, 1998) the parasitoid, *Dinocampus (=Perilitus) coccinellae parasitizes* adult, pupal and larval stages of *C. septempunctata*, (Triltsch, 1996). It has a strong preference to oviposit and parasitize the adult stage, in particular, the adult females (kadona *et al.*, 1998) rather than larvae and pupae (Geogheghan *et al.*, 1998). Parasitoids, *Tetrastichus coccinellae, Tetrastichus* sp. (Omkar and Srivastava, 2001c), *Phalacrotophora fasciata* (Fallen), *Oomyzus scaposus* (Thompson) (Triltsch, 1996), *Homalotylus flaminius* and *Phalacrotophora* sp. (Won *et al.*, 1996) also attack *C. septempunctata*. Four hymenopterans, one dipteran and a pathogenic fungi, *Beauveria bassiana*, were recorded as mortality factors of *C. septempunctata*. The overall rate of parasitism was relatively low but in certain seasons, high pupal parasitism was noted (Thalji, 1997). *B. bassiana* and fungus, *Paecilomyces fumosorseus* infect whiteflies and when the latter have been devoured by the ladybeetles, *e.g., S. parcesetosum*, they may be infected (Poprawski *et al.*, 1998). The susceptibility test of *B. bassiana* inferred that all ladybeetles were susceptible to infection, in particular, *Coccinella* sp. (Haseeb and Murad, 1998). Prepupae of *C. septempunctata*, are also vulnerable to parasitism by Phoridae (Diptera) (Hurst *et al.*, 1998) along with the larvae, *Allothrombium triticium* (Acari : Trombidiidae) (Zhang and Saboori, 1996). The phorids prefer older prepupae over new prepupae (Hurst *et al.*, 1998).

Maternally inherited bacteria that kills the male but not the female hosts during embryogenesis occur widely in aphidophagous coccinellids (Majerus and Hurst, 1997). Certain ladybeetles, such as *A. variegata,* bear maternally inherited flavobacteria (Hurst *et al.,* 1999a) that prove to be fatal for male embryos. In *A bipunctata,* at least four different bacteria are known to cause male-killing: a *Rickettsia* (Werren *et al.,* 1994; Hurst *et al.,* 1999b), a *Spiroplasma* (Hurst *et al.,* 1999b) and two strains of *Wolbachia* (Hurst *et al.,* 1999b).

Table 1.2: The fecundity of ladybeetles

Name of the ladybeetle	Fecundity	Reference
Adalia bipunctata	1011.00	Kalushkov (1994)
Cheilomenes sexmaculata	1096.00	Bind (1998)
	1068.00 to 1391.00	Saha (1987)
Coccinella septempunctata	1660.00	Kawauchi (1985)
	809.16	Gautam (1994)
	466.00	Devi *et al.* (1996)
	617.20	Singh and Malhotra (1979)
	634.75	Singh and Singh (1993)
Coccinella transversalis	650.00	Gautam (1994)
Coccinella undecimpunctata	370.50	El-Hag and Zaitoon (1996)
Chilocorus bijugus	423.00	Jaio *et al.* (1997)
Cryptolaemus montrouzieri	150.00	Gautam (1994)
Harmonia eucharis	618.00	Chakraborty *et al.* (1995)
Hippodamia convergens	344.00	Rodriguez and Miller (1995)
Micraspis discolor	665.16	Agarwala *et al.* (1988)
	270.50	Prodhan *et al.* (1995)
Nephus regularis	110.00	Gautam (1994)
Oenopia quadripunctata	367.00	Singh *et al.* (1993)
Propylea japonica	1481.00	Kawauchi (1985)
Scymnus brunnescens	200.00	Gautam (1994)
Scymnus nubilus	700.00	Gautam (1994)
Scymnus quadrillum	132.00	Gautam (1994)
Scymnus pyrocheilus	520.00	Gautam (1994)
Synia melanaria	1401.2	Afroze and Shujauddin (1998)

(f) Sexual dimorphism and sex-ratio

In majority of ladybeetles, both sexes are morphologically similar. Usually females are slightly larger than males. However, in certain ladybeetles, certain distinctive features on head and pronotum enable easy discrimination, as in *Propylea dissecta* (Omkar and Pervez, 2000b). There is a serrated crown on the head of males and the females possess a pair of triangular spots on their heads (Omkar and Pervez, 2000b). The antenna of male *C. sexmaculata* (516.67 μm) is smaller than the female (633.33 μm) (Bind and Omkar, 2000). The two sexes of *C. nigritus* can be identified by the patches present on their sternite (abdominal plates). The lateral portions of 4th and 7th sternite of males have dark patches, which are absent in females (Samways and Tate, 1984).

Table 1.3: Percent hatchability of eggs in ladybeetles

Name of the ladybeetle	Tempe-rature (°C)	R.H. in %	% Hatch-ability	Reference
Adalia bipunctata	—	—	69.0	Majerus (1994)
Adonia variegata	25±2	60±5	81.8	El-Hag and Zaitoon (1996)
Cheilomenes sexmaculata	27±1.5	60±5	48.94	Saha (1987)
	25±2	65±5	59.5	Bind (1998)
Coccinella septempunctata	23.6	—	87.5	Singh *et al.* (1994)
	30±1	70-80	88.0	Singh and Malhotra (1979)
	25±2	65	87.08	Singh and Singh (1993)
Coccinella transversalis	25±2	65±5	85.7	Veeravel and Baskaran (1995)
Coccinella undecimpunctata	25±2	60±5	74.5	El-Hag and Zatioon (1996)
Harmonia eucharis	20.8±3.3	—	73.6	Chakraborti *et al.* (1995)
Micraspis discolor	26	60-80	54.31	Agarwala *et al.* (1988)
Scymnus coccivora	27.1	—	89.86	Padmaja *et al.* (1995)
Synia melanaria	25±5	80±50	97	Afroze and Shujauddin (1998)

The number of male and female, *C. septempunctata* was 536 and 964, respectively, in a field survey, maintaining a sex ratio of 1:1.8 (Singh and Malhotra, 1979). The sex-ratio of *C. nigritus* (male:female) varied from 1:1.44 to 1:3.55 in the orchard infested with scale, *Aspidiotus nerii* (Erichsen *et al.*, 1991). The reason for this high female biased ratio was its relatively high longevity. Sex ratio of ladybeetles, *viz. Scymnodes lividigaster* and *L. galbula* in the laboratory was close to 50:50, but in the field it was biased towards female (Anderson, 1981), *Cleothera notata, C. onerata,*

Cleothera sp., *Hyperaspis* sp.1 and *Hyperaspis* sp.3 possessed a sex ratio (male:female) of 1:1, but in *Hyperaspis* sp. 2, sex ratio was 1:3 (Sullivan *et al.*, 1991). In *C. septempunctata* the sex ratio (male:female) was 1:1.4 (Singh and Singh, 1993), 0.60:1 and 0.72:1 (Zhou *et al.*, 1995).

Table 1.4: The emergence percent in the ladybeetles

Name of the ladybeetle	Temperature (°C)	R.H. in %	% Emergence	Reference
Adonia variegata	25±2	60±5	61.5	El-Hag and Zaitoon (1996)
Brumoides suturalis	27±1.5	60±5	100.0	Gautam (1990c)
Cheilomenes sexmactulata	25±2	65±5	71.8	Bind (1998)
Coccinella transversalis	25±2	65±5	83.3	Veeravel and Baskaran (1995)
Micraspis discolor	26	60-80	74.32	Agarwala *et al.* (1988)

Table 1.5: The longevity of male and female ladybeetles in days

Name of the ladybeetles	Longevity (Male)	Longevity (Female)	Reference
Adalia bipunctata	55.0	73.0	Kalushkov (1994)
Brumoides suturalis	33.25	40.50	Gautam (1990c)
Cheilomenes sexmaculata	80.0	108.0	Saha (1987)
Coccinella septempunctata	40.1	53.6	Kumar and Verma (1996)
Hyperaspis jacunda	100.0	101.5	Nsiama She *et al.* (1984)
Micraspis discolor	39.4	46.8	Prodhan *et al.* (1995)
Nephaspis oculatus	56.1	67.5	Liu-Tong-Xian *et al.* (1997)
Scymnus coccivora	61.7	66.8	Padmaja *et al.* (1995)
Scymnus nubilus	49.9	57.23	Kumar and Chakraborty (1997)
Stethorus gilvitrons	167.0	124.8	Ahmad and Ahmad (1989)

Female biased ratio in certain coccinellids results from maternally inherited male-killers with a diverse array of bacteria (Majerus *et al.*, 2000). Thus, usually, females outnumbered the males. Conversely, there are certain instances of male biased ratio, e.g. *Scymnus coccivora* (Padmaja *et al.*, 1995), *Hippodamia koebelei* and *H. convergens* (Vallejo *et al.*, 1990). The sex ratio (male:female) in *Scymnus nubilus* (Kumar and Chakraborty, 1997), *H. convergens, H. parenthesis* and *Hippodamia tredecimpunctata* (Linn.) (Elliott *et al.*, 1998) was estimated to be 1:1.

Factors effecting development and reproduction

(a) Prey Density

Coccinellids exhibit Type II functional response (Holling, 1959), in which, initially there is an increase in the number of prey consumed, but the rate of prey consumption decreased with increased prey density. The prey consumption of *C. sexmaculata* (Agarwala and Bardhanroy, 1997) and *Coelophora inaequalis* (Mora *et al.*, 1995) increased up to a certain point of increase in prey density, thereafter it became constant. Functional response of the fourth instar of *C. septempunctata* on *M. persicae, A. craccivora, A. gossypii, U. compositae* and *A. nerii* confirmed a curvilinear relationship between prey consumption and prey density (Omkar and Srivastava, 2001d; Omkar and Srivastava, 2002a). In *C. transversalis* and *C. sexmaculata*, almost 100% of prey consumption took place at initial prey densities of 20 to 50 and the number of eggs produced increases with increase in prey density. Handling time of *C. transversalis* and *C. sexmaculata* was found to be 8.4 min and 11.74 min, respectively and the attack rate was 1.10 and 1.12/ min respectively (Veeraval and Baskaran, 1997). Handling times ranged from 34.8 to 81.0 in *M. discolor* depending upon the densities of prey and were higher at lower prey densities. The interference constant, quest constant and area of discovery of *M. discolor, C. sexmaculata* and *P. quatuordecimpunctata* ranged from -0.1233 to -0.0333, 0.5191 to 0.6931 and 0.55 to 0.76, respectively (Akhtaruzzaman and Ahmad, 1998). In addition, the rate of prey consumption also depends on the distribution of the prey (Zebitz, 1998).

Coccinellids increased their number with increase in prey density (Ofuya and Akingbohungbe, 1988; Omkar and Pervez, unpubl.). Moreover, the increased predator density also increases the total prey consumption, though the rate of prey consumption by individual predator was highest with one predator (Aggregative numerical response) (Pandey *et al.*, 1984). Searching efficiency of predator also decreased with increase in prey as well as predator densities (Pandey *et al.*, 1984). Mean number of eggs laid per female per day increased curvilinearly with increased prey density (Agarwala and Bardhanroy, 1997; Veeravel and Baskaran, 1997). Feeding rate of *C. luntata* is related positively to the prey density. Number of prey consumed by each predator increase more steeply at lower prey density than at higher prey density. Fecundity of *C. lunata* also initially increased and then reached a plateau with increasing prey density (Ofuya and Akingbohungbe, 1988). Searching efficiency of *C. septempunctata* decreased linearly with increase in prey density (Tripathi *et al.*, 2000).

Females of *A. bipunctata* display a strong increase in food intake with aphid densities up to 10 aphids per day. The male showed a weaker response to increasing aphid availability. When 1 or 2 aphids per day were provided, the individuals of both the sexes ate the same quantity. But, at higher aphid densities, the males were relatively less voracious. The slope of regression of food consumption at different aphid densities for the male is significantly less than the slope for the female (Hemptinne *et al.*, 1996).

(b) Temperature

Studies on coccinellid behaviour in relation to temperature have explained their relative competitiveness and adaptability to the change in climatic conditions. Fluctuation in temperature strongly affects growth, development and reproduction in ladybeetles (Bind, 1998; Omkar and Pervez, 2002c). The daily mean temperature ranges of 20-27°C favours the activity of coccinellids (Freier and Triltsch, 1996). 27°C was found to be optimum for *M. discolor* owing to high and extremely viable reproductive rate and high survival ability of the immature stages (Omkar and Pervez, 2002; unpubl.).

High temperature, i.e., 28°C, as compared to low ones, i.e., 18°C and 23°C enables more food consumption (Ali, 1973; Ives, 1981). Feeding rates of *C. sexmaculata* increased linearly with the increase in temperature up to 32°C (Ali Khan and Yousuf, 1986). High temperature shortens the developmental period of coccinellids (Padmaja *et al.*, 1995; Jalali *et al.*, 1999; Jarosik *et al.*, 2000). Developmental rate has a linear relationship with the increasing temperature range from 15°C to 30°C (Baumgartner *et al.*, 1986; Ponsonby and Copland, 1996; Omkar and Pervez, unpubl.). The incubation period of C. *septempunctata* lasted for 2.3 and 13 days at temperatures 30°C and 15°C respectively (Butler, 1982). Their larval period lasted for 6.2 and 31.1 days at 32.5°C and 17°C, respectively. The pupal stages lasted for 2.8 and 13.0 days at 32.5°C and 17°C, respectively (Bulter, 1982). The total developmental period of *Hippodamia quniquesignata* was 66.5, 15.0, 11.9 and 10.6 days at 15.5°C, 27°C, 30°C and 32°C, respectively (Kaddou, 1960). The incubation period of C. *septempunctata* was 8.5 and 1.8 days at 15°C and 30°C (Singh and Malhotra, 1979). The average larval duration of C. *septempunctata* varied from 13-14 days to 10-11 days at 18.35°C to 23.60°C (Singh *et al.*, 1994). Total developmental period of *Coccinella trifasciata* varied from 44.5 to 11.1 days at 18°C to 34°C.

Increasing temperature shortens the longevity of ladybeetles probably due to increased biological and biochemical processes. The longevity of adult *Coccinella novemnotata* was 62.0, 47.6 and 21.0 days at 21.1°C, 26.7°C and 32.2°C, respectively (McMullen, 1967). Survival rate of female, C. *sexmaculata* varied from 65.7% to 6.1% with the increase of temperature from 30°C to 34°C (Anil Khan and Yousuf, 1986).

Temperature below 17°C is also not suitable for growth and immature survival of certain ladybeetles (Jean *et al.*, 1990). The mortality of *H. convergens* (83%) (Miller, 1992) and *H. parenthesis* (67%) (Orr and Obrycki, 1990) was maximum at the lowest temperature tested, i e. 17°C. Similarly, mortality was maximum (74%) at 15°C (Naranjo *et al.*, 1990). At all the tested temperatures, first instars suffered the highest mortality, while pupae had the lowest (Ponsonby and Copland, 1996, Omkar and Pervez, unpubl.). As far as tolerance is concerned, eggs of ladybeetle, *Chilocorus kuwanae* are tolerant to changes in temperature, whereas the fourth instars were the least (Yang *et al.*, 1998). In comparison to melanics, the typicals of C. *septempunctata* were more tolerant to high temperatures, i.e. 32° and 35°C probably due to their lower internal temperatures than melanics (Rhamhalinghan, 1987b). Newly emerged adults reared at 18-22°C weigh significantly more than the adults reared at 24°, 26° and 30°C (Orr and Obrycki, 1990). The size of the developing stages decreased with the increase in temperature from 18°C to 30°C (Kawauchi, 1979). The critical threshold temperature in *P. japonica* for development was found to be 8.44°C. The theoretical threshold for development of the primogenital stages of ladybeetles ranged from 9°C to 13°C (Obrycki and Tauber, 1978).

(c) Light

The intensity and duration of light also influences the growth, development and reproductive activity of ladybeetles. Long photoperiods enhance these biological processes. The metabolic processes are highest at 16L/8D and, therefore, it is considered as an optimum photoperiod (Takahashi, 1993). Adults were active only in daytime when the luminosity was more than 0.2 MJ/m2/0.5 h, and wind velocity was less than 10 m/s. Predation rate, oviposition rate and mating incidence were more at daytime. After sunset, the ladybeetles descend to the base of the plants with reduced activity (Takahashi, 1993).

In *C. septempunctata*, the ovarian development was increased at long day conditions, while at shorter days it was inhibited (Sakurai *et al.*, 1990). Occasionally, reverse photoperiodic responses were observed in the closely related species. For instance, *C. septempunctata* has a positive response to long day photoperiod for its growth and development, whereas *C. septempunctata brucki* prefers short day photoperiods (Sakurai *et al.*, 1991).

Conclusion

The published information on the ladybeetles revealed that: (1) The records of their occurrence form various biogeographical regions accentuate their global recognition, with some eurytopic species, such *C. septempunctata, H. axyridis, C. nigritus, R. cardinalis, etc.* capable of easily establishing themselves in a new environment, owing to their ecological plasticity. Evidence in light of their ecology suggests that these species are suitable as major bio-control agents. The occasional failure of these bioagents in the pest management programmes is an area that needs further attention. (2) The studies on dynamics of the prey and the coccinellid populations may provide basic information about the correlation and the influence of various abiotic and biotic factors upon them. (3) Coccinellids possess an extensive and intensive foraging pattern and use various olfactory-visual cues to locate prey-habitats. (4) Prey specificity and suitability studies revealed that the coccinellids have certain prey choice preferences. (5) The immense reproductive efficiency of coccinellids affirms their assertiveness in responding quickly to pest population by increasing their own reproductive output. (6) The analytical studies of functional and numerical responses of coccinellids show that they increase their feeding efficacy and number of individuals against increasing prey densities, (7) The controlled and optimum abiotic factors improve the survival, reproductive, foraging and predation efficacy of the predaceous coccinellids. The ecological evidence thus suggests that predaceous coccinellids are one of the important bio-control tools, hence are an essential component in integrated pest management programmes of those pests that fall within their prey range.

Reproduction in coccinellids needs further intensive investigations to reveal the influence of adult age, same prey species form different host plants, and other biotic factors on their reproductive attributes. More studies are needed on the competitive ability amongst the predatory ladybeetles and their heterospecific interactions in the predatory guild. The principal aim behind these studies should be whether a slight manipulation in the guild composition could provide an additive or synergistic effect on the predation of the preys by the heterospecific predators.

Acknowledgements

Authors are thankful to Prof. M.E.N. Majerus, Department of Genetics, University of Cambridge, Cambridge, U.K. for critically examining the manuscript and providing invaluable suggestions and OM and A.P to Council of Science and Technology, U.P. and Indian Council of Agricultural Research, New Delhi for financial assistance.

REFERENCES

Afroze, S. 1999. Preferential predation by *Anegleis cardoni* (Weise) (Coccinellidae: Coleoptera) on some aphid species, *J. Aphidol.,* 13: 95-97.

Afroze, S. and Shujauddin 1998. Bioecology of *Synia melanaria* Mulsant (Coleoptera: Coccinellidae), predating on *Coptosoma ostensum* Distant. *J. Ent. Res.,* 22(4): 329-336.

Agarwala, B.K. 1991. Why do ladybrids (Coleoptera: Coccinellidae) cannibalize? *J. Biosci.,* 16(3): 103-109.

Agarwala, B.K. and Bardhanroy, P. 1997. Oviposition behaviour and reproduction efficiency in ladybeetles (Coccinellide: Coleoptera). A case study of *Menochilus sexmaculatus* (Fabr.) *J. Aphidol.,* 11(1): 49-59.

Agarwala, B.K., Bhattacharya, S. 1995. Adaptive features of the predatory ladybird beetles in uncertain and competitive habitats. *Current Sct.,* 69: 587-591.

Agarwala, B.K., Bhattacharya, S. and Bardhanroy, P. 1997. Cluster laying of eggs does not increase the risk of cannibalism by first instar larvae in ladybird beetles. *Proc. Ind. Natn. Sci. Acad.,* 63(1): 1-10.

Agarwala, B.K., Bhattacharya, S. and Bardhanroy, P. 1998. Who eats eggs? Intra-versus iner-specific interactions in starving ladybird beetles predaceous on aphids. *Ethol, Ecol. Evol.,* 10(4): 361-368.

Agarwala, B.K., Das, S. and Chaudhuri, M.S. 1988. Biology and food relations of *Micraspis discoulor* (F.): An aphidophagous coccinellid of India. *J. Aphidol.,* 2(1&2): 7-17.

Agarwala, B.K. and Dixon, A.F.G. 1991. Cannibalism and interspecific predation in ladybirds. In: Polgar L., Chambers R. J., Dixon A.F.G., Hodek I. (eds.): *Behviour and impact of Aphidophaga. SPB Acad. Publ. The Hague,* 95-102.

Agarwala, B.K. and Dixon, A.F.G. 1993a. Kin recognition: egg and larval cannibalism in *Adalia bipunctata* (Coleoptera: Coccinellidae). *Eur. J. Entomol.,* 90: 45-50

Agarwala, B.K. and Dixon, A. F. G. 1993b. Why do ladybirds lay eggs in cluster? *Func. Ecol.,* 7: 541-548.

Agarwala, B.K. and Ghosh, A.K. 1988. Prey records of aphidophagous Coccinelliadae in India. A review and bibliography. *Tropical Pest Management,* 34: 1-14.

Agarwala, B.K. and Yasuda, H. 2000. Competitive ability of ladybird predators of aphids: A review of *Cheilomenes sexmaculata* (Fabr.) (Coleoptera: Coccinellidae) with a worldwide checklist of preys. *J. Aphidol.,* 14: 1-20.

Ahmad, Z.I. and Ahmad, R.F. 1988. Biological studies of predator, *Stethorus gilvifrons* Mulsant (Coleoptera: Coccinellidae) on the strawberry mite, *Teranychus turkestani* Ugarov and Nikoski (Acariformes: Tetranychidae). *J. Biol. Sci. Res.,* 20(1): 33-44.

Akhtaruzzaman, M. and Ahmad, M. (1998). Handling time, interference constant, quest constant and area of discovery of coccinellid predators *Thai J. Agric. Sci.,* 31(3): 342-351.

Ali, M. 1973. Influence of photoperiod and temperature on the food consumption of alfalfa beetle, *Subcoccinella vigintiquatorpunctata* (Coleoptera: Coccinellidae). *Acta Phytophthil. Acad. Hung.*, 8(1/2): 207-215.

Ali Khan, M.A. and Yousuf, M. 1986. Temperature and food requirements of *Cheilomenes sexmaculata* (Coleoptera: Coccinellidae). *Environ. Entomol.*, 15(4): 800-802.

Allawi, T.F. 1989. A list of predaceous Coccinellids collected in Jordan (record of the different coccinellid species found in Jordan: immature stage, emergence in the adult sage). *Dirasat (Jordan):* 16(7): 23-26.

Anderson, J.M.E. 1981. Sex ratio of adult *Scymnodes lividigaster and Leptothea galbula* (Coleoptera: Coccinellidae). *Entomophaga.* 26(1): 39-45.

Anonymous 1995. Technology for production of natural enemies. *Project Directorate of Biological Control,* Bangalore (India). pp. 221.

Arakaki, N. 1988. Egg protection with faeces in the ladybeetle, *Pseudoscymnus kurohime* (Miyatake) (Coleoptera: Coccinellidae). *Appl. Entomol. Zool.,* 23: 495-497.

Arioli, M.C.S. 1985. Coccinellids of Rio Grando do Sul, Brzil (Coleoptera: Coccinellidae). Part of dissertation. *Revista do Centro de Ciencias Rurais,* 15(1): 5-35.

Atwal, A.S. and Sethi, S.L. 1963. Biochemical basis for food preference of a predator beetle. *Sci.,* 32: 511-512.

Babu, T.R. and Azam, K.M. 1988. Effect of low holding temperature during pupal instar on adult emergence, pre-oviposition and fecundity of *Cryptolaemus montrouzieri* Mulsant (Coccinellidae: Coleoptera). *Insect Sci. Appl.* 9(2): 175-178.

Baumgaertner, J.U., Guitierrez, A.P. and Summers, C.G. 1981. The influence of aphid prey consumption on searching behaviour, weight increase, development time, and mortality of *Chrysopa carnea* (Neuroptera: Chrysopidae) and *Hippodmia convergens* (Coleoptera: Coccinellidae) larvae. *Can. Entomol.* 114: 1007-1014.

Bielawski, R. 1980. The ladybeetles (Coleoptera: Coccinellidae) of North Korea. *Fragm Faun (Warsaw).* 25(13): 221-254.

Bind, R.B. 1988. *Bioecology and behaviour of a ladybird beetle, Cheilomenes (=Menochilus) sexmaculata (Fabricius) (Coleoptera: Coccinellidae).* Ph. D. Thesis. University of Lucknow, Lucknow, pp. 164.

Bind, R.B. and Omkar 2000. Sexual dimorphism and sex-ratio in a ladybeetle, *Cheilomenes sexmaculata* (Fabricius). *Seventh National Symposium on Aphidology, 11-13 March, 2000 at DDU Gorakhpur University, Gorakhpur,* 65-66.

Bode, H. 1980. Aphids in winter wheat: abundance and limiting factors from 1976 to 1979. *Bull. WPRS* 3: 49-57.

Boraei, H.A., Metwally, S.M., Shenishen, Z. and Meshah, A.H. 1993. Seasonal abundance of coccinellid predators at Kafr El Sheibh Govenorate Egypt. *J. Agri. Res. Touta Univ.* 19(4): 833-840.

Brown, H.D. 1972. The behaviour of newly hatched coccinellid larvae (Coleoptera: Coccinellidae). *J. Entomol. Soc. South Africa* 35: 149-157.

Brown, M.W. and Miller, S.S. 1998. Coccinellidae (Coleoptera) in apple orchards of eastern West Virginia and the impact of inversion by *Harmonia axyridis. Ent. News.* 109(2): 143-151.

Butler, G.D. Jr. 1982. Development time of *Coccinella septempunctata* in relation to constant temperatures (Coleoptera: Coccinellidae). *Entomophaga.* 27(3): 349-353.

Camargo, F.C. 1937. Notas taxonomicas e biologicas sobra algunos Coccinellideos do genero *Neocalvia* Crotch, predatores de las larvas do genero *Psyllobora* Chevrolat (Col. Coccinellidae). *Revista Entomologia Rio de Janeiro,* 7: 362-377.

Canepari, C. 1997. Coccinellidae from the Nepal Himalayas. *Stuttgarter Beitr. Naturk. A* 565: 65.

Carer, M.C. and Dixon, A.F.G. 1984a. Honeydew an arrestant stimulus for coccinellids. *Ecol. Entomol.* 9: 383-387.

Carter, M.C. and Dixon, A.F.G. 1984b. Foraging behaviour of coccinellid larvae: duration of intensive search. *Entomal. Exp. Appl.* 36: 133-136.

Chakrabarti, S., Debnath, N. and Ghosh, D. 1995. Bioecology of *Harmonia eucharis* (Mulsant) (Coleoptera: Coccinellidae). An aphidophagous predator in Western Himalayas. *Entomon* 20(3&4): 191-196.

Chazeau, J. 1989. Complement to Coccinellini from New Guinea (Coleoptera: Coccinellidae). *Revue Francaise d'Entomologie (France).* 11(4): 157-168.

Chen, Z.H., Qin, J. D. and Shen, C.L. 1989. Effect of altering composition of artificial diets on the larval growth and development of *Coccinella septempunctata. Acta Entomol. Sinica,* 32(4): 385-392.

Chitradevi, L., Singh, T.K. and Varatharajan, R. 1997. Impact of predators on the field population of *Lipaphis erysimi* (Kalt.) infesting mustard. *J. Aphidol.* 11(1): 1-146.

Choudhury, D., David, B.A. and Singh, D.R. (1983). Studies on the host preference of *Cheilomenes sexmaculata* (Fabr.) (Coleoptera: Coccinellidae). *Comp. Physiol. Ecol.,* 8(4): 289-290.

Chunram, S. and Sasaji, H. 1980. A contribution to the Coccinellidae (Coleoptera) of Thailand. *Oriental Insects,* 14(4): 473-491.

Coderre, D., Provencher, L. and Tourneur, J.D. 1987. Oviposition and niche partitioning in aphidophagous insects on maize. *Can. Entomol.* 119: 195-203.

Debraj, Y., Singh, T.K. and Singh, H.J. 1997. Seasonal Occurrence, Life history and feeding efficiency of certain predaors of the cabbage aphids, *Brevicoryne brassicae* (Linn.) in north east India. *J. Aphidol.* 11(1): 143-146.

Devi, N., Raj, D. and Verma, S.C. 1996. Biology and feeding potential of *Coccinella septempunctata* Linn. (Coccinelllidae: Coleoptera) on cabbage aphid; *Brevicoryne brassicae Linn. J. Ent. Res.,* 20(1):23-25.

Dimetry, N.Z. 1974. The consequences of egg cannibalism in *Adalia bipunctata* (Coleoptera: Coccinellidae). *Entomophaga.* 19: 445-451.

Dixon, A.F.G. 1988. *Aphid Ecology.* Chapman & Hall. pp. 300.

Dixon, A.F.G. 2000. *Insect-prey dynamics, ladybird beetles and biological control.* Cambridge University press. p. 257.

Doumbia, M., Hemptinne, J.L. and Dixon, A.F.G. 1998. Assessment of patch quality by ladybird: role of larval tacks. *Oecologia.* 113(2): 197-202.

Easterbrook, M.A. 1998. The beneficial fauna of strawberry fields in southeast England. *J. Horticultural Sci. Biotechnol.* 73(1): 137-144.

El Hag, E.T.A. and Zatioon, A.A. 1996. Biological parameters for four coccinellied species in Central Saudi Arabia, *Biolgical Control, Theory and applications in pest management (USA).* 7(3): 316-319.

Elliott, N.C., and Kieckhefor, R.W. 1990. Dynamics of aphidophagous coccinellid assemblages in small grain fields in eastern South Dakota. *Environ. Entomol.* 19: 1320-1329.

Elliot, N.C., Kieckhefer, R.W., French, B.W. and Lee, J.H. 1998. A comparison of size, sex-ratio, and dispersal in three coccinellids. *Southwestern Entomologist.* 23(2): 137-145.

Erichsen, C., Samways, M.J. and Hattingh, V. 1991. Avoidance by the ladybird *Chilocorus nigriuts* (F.) (Coleoptera: Coccinellidae) of a doomed scale host resource. *Journal of Applied Entomology* 112: 493-498.

Espinha, I.G. and Torres, L.M. 1995. Observations on the control value of the *Stethorus punctillum* (Weise) as a predator of *Panonychus ulmi* (Koch) in Vila Real region (northeastern Portugal). *Boletin de Sanidad Vegetal Plagas (Espana).* 21(3): 337-347.

Evans, E.W., Stevenson, A.T. and Richards, D.R. 1999. Essential versus alternative foods of insect predators: benefits of a mixed diet, *Oecologia* 121(1): 107-112.

Ferran, A. and Deconchat, M. 1992. Exploration of wheat leaves by *Coccinella septempunctata* L. (Coleoptera: Coccinellidae) larvae. *J. Insect Behav.* 5: 147-159.

Ferran, A., Iperti, G. Lapchin, L. and Rabasse, J. M. 1991. Localization, behaviour and 'prey-predator' relationships for *Coccinella septempunctata* L. in a wheat field. *Entomophaga.* 36(2): 213-225.

Ferran, A., Ettifouri, M., Clement, P. and Bell, W.J. 1994. Sources of variability in the transition from extensive to intensive search in coccinellid predators (Coleoptera: Coccinellidae). *J. Insect Behav.* 7: 633-647.

Frazer, B.D. Gilbert, N., Ives, P.M. and Raworth, D.A. (1981). Predation of aphids by coccinellid larvae. *Can. Entomol.* 113: 1043-1046.

Freier, B. and Gruel, H.J. 1993. Occurrence and importance of coccinellids as beneficial arthropods in agrosystems. *Gesunde Pflanzen.* 45(8): 300-307.

Freier, B. and Triltsh, H. 1996. Climate chamber experiments and computer simulations on the influence of increasing temperature on wheat-aphid-predator interactions. *Implications of 'Global environmental change' for crops in Europe, 1-3 April 1996, Churchill College, Cambridge, UK. Aspects-of-Applied-Biology,* 45: 293-298.

Fursch, H. 2001. Contribution to the Coccinellide fauna of central and eastern African forests, with description of new species. (in deutscher Sprache). *Entomologische Zeitschrift* 111(3&4): 90-94.

Gautam, R.D. 1990a. Effect of substrata and age of *Menochilus sexmaculatus* (Fabr.) (Coleoptera: Coccinellidae) on its oviposition. *J. Biol. Control* 4(1): 8-10.

Gautam, R.D. 1990b. Influence of substrata and age of *Coccinella septempunctata* on its oviposition. *Indian J. Agric. Sci.* 60(4): 393-396.

Gautam, R.D. 1990c. Mass-multiplication technique of coccinellid predator, ladybird beetle *(Brumoides suturalis). Indian J. Agric. Sci.* 60: 747-750.

Gautam, R.D. 1994. *Biological pest suppression,* Westvill Publishing House, New Delhi. p. 219.

Geoghegan, I.E., Thomas, W.P. and Majerus, M.E.N. 1997. Notes on the coccinellid parasitoid *Dinocampus coccinellae* (Schrank) (Hymenoptera: Braconidae) in Scotland. *Entomologist* 116(3-4): 179-184.

Ghorpade, K.D. 1979. *Ballia eucharis* (Coleoptera: Coccinellidae) breeding on Cicadellidae (Homoptera) at Shillong. *Current Res.* 8: 113.

Gordon, R.D. 1978. West Indian Coccinellidae II. (Coleoptera): Some scale predators with keys to genera and species (Includes *Coccidophilus cariba, Viridigloba imitator, Nephaspis nigra* and *Colloeneis bennetti,* New Taxa, Antigua, Trinidad and Tobago). *Coleopterists' Bull.* 32(3): 205-218.

Gordon, R.D. 1985. The Coccinellidae (Coleoptera) of America north of Mexico. *J. New York Entomol, Soc.* 93, 1-912.

Gordon, R.D. 1996. South American Coccinellidae Part III: Correction and addition. *Ent. News* 107(4): 232.

Gordon, R.D. 1999. South American Coccinellidae Part IV: A systematic revision of the South American Diomini, new tribe. *Annales Zoologici (Warszawa)* 49 (Supplement 1): 1-219.

Gordon, R.D. and Hilburn, D.J. (1990). The Coccinellidae (Coleoptera) of Bermuda. *J. New York Entomol. Soc.* 98(3): 265-309.

Grunbawn, D. 1998. Using spatially explicit models to characterize foraging performance in heterogeneous landscapes. *Amer. Naturalist.* 151(2): 97-115.

Gruppe, A. and Roemer, B. 1988. The lupin aphid (*Macrosiphum albifrons* Essing, 1911) (Homoptera: Aphididae) in West Germany: Its occurrence, host plants and natural enemies, *Z. Angew. Entomol.* 106: 135-143.

Harmon, J.P., Losey, J.E. and Ives, A.R. 1998. The role of vision and colour in the close proximity foraging behaviour of four coccinellid species. *Oecologia* 115: 287-292.

Haseeb, M. and Murad, H. 1998. Pathogenecity of entomophagous fungus, *Beauveria bassiana* (Bals.) Vuill. To insect predators. *International Pest Cont.* 40(2): 50-51.

Hattingh, V. and Samways, M.J. 1995. Visual and olfactory location of biotopes, prey patches, and individual prey by the ladybeetle *Chilocorus nigritus. Ent. Exp. et Applic.* 75: 87-98.

Heniz, K.M. and Zalom, F.G. 1996. Performance of the predator, *Delphastus pusillus on Bemisia* resistant and susceptible tomato lines. *Ent. Exp. et Applic.* 81: 345-352.

Hemptinne, J.L., Dixon, A.F.G. and Coffin, J. 1992. Attack strategy of ladybird beetles (Coccinellidae): factors shaping their numerical response. *Oecologia* 90: 238-245.

Hemptinne, J.L., Dixon, A.F.G. Doucet, J.L. and Petersen, J.E. 1993. Optimal foraging by hoverfiles (Diptera: Syrphidae) and ladybirds (Coleoptera: Coccinellidae): mechanisms. *Eur, J. Ent.* 90(4): 451-455.

Hemptinne, J.L., Dixon, A.F.G. and Lognay, G. 1996. Searching behaviour and mate recognition by males of the two-spot ladybird beetle, *Adalia bipunctata. Ecol. Entomol.* 21: 165-170.

Hemptinne, J.L., Lognay, G., Gauthier, C. and Dixon, A.F.G. 2000. Role of surface chemical signals in egg cannibalism and intraguild predation in ladybirds (Coleoptera: Coccinellidae). *Chemology* 10: 123-128.

Hodek, I., Holman, J., Novak, K., Skuhravy V., Stary, P., Weismann, L. and Zeleny, J. 1996. The present possibilities and prospects of integrated control of *Aphis fabae* Scop. In: Hodek I. (ed.): *Ecology of aphidophagous insects.* Academia. Prague *and Dr. W. Junk.* pp. 331-335.

Hodek, I. and Honek, A. 1996. (ed.) *Ecology of Coccinellidae.* Kluwer Academic Publisher Dordrecht Boston London, pp. 464.

Holling, C.S. 1959. Some characteristics of simple types of predation and parasitism. *Can. Entomol.* 91: 385-398.

Holloway, G.J., Brakefield, P.M., Jong, P.W. de, Ottenheim, M.M., Vos, H. de, Kesbeke, F. and Peneburg, L. 1995. A quantitative genetic analysis of an aposeamtic colour pattern and its ecological implications. *Philosophical Transactions of the Royal Society of London. Series B-Biological Science.* 348: 373-379.

Honek, A. 1985. Activity and predation of *Coccinella septempunctata* adults in the field (Coleoptera: Coccinellidae). *Z. Angew, Entomol.* 100: 399-409.

Honek, A. 1989. Overwintering and annual changes of abundance of *Coccinella septempunctata* in Czechoslovakia (Coleoptera: Coccinellidae). *Acta Entamol. Bohemoslov.* 86: 179-192.

Hukusima, S. and Kamei, M. 1970. Effects of various species of aphids as food on development, fecundity and longevity of *Harmonia axyridis* Pallas (Coleoptera: Coccinellidae). *Res. Bull. Fac. Agric. Gifu. Univ.* 29: 53-66.

Hukusima, S. and Kouyama, S. 1974. Life histories and food habits of *Menochilus sexmaculatus* (Fabricius) (Coccinellidae: Coleoptera). *Res. Bull. Fac. Agric. Gifu. Univ.* 33: 75-82.

Hurst, G.D.D., Bandi, C., Sacchi, L., Cochrane, A.G., Bertrand, D. and Karaca, I. 1999a. *Adonia variegata* (Coleoptera: Coccinellidae) bears maternally inherited flavobacter that kill males only. *Parasitol.* 118(2): 125-134.

Hurst, G.D.D., McMeechan, F.K. and Majerus, M.E.N. 1998. Phoridae (Diptera) parasitizing *Coccinella septempunctata* (Coleoptera: Coccinellidae) select older prepupal hosts. *Eur. J. Entomol.* 95(1): 179-181.

Hurst, G.D.D., Vonder Schulenburg, J.H.G., Majerus, T.M.O., Bertrand, D., Zakharov, I. A., Baungaard, J., Volki, W., Stouthamer, R. and Majerus, M.E.N. 1999b. Invasion of one insect species, *Adalia bipunctata* by two different male-killing bacteria. *Insect Molecular Biol.* 8(1): 133-139.

Iperti, G. 1966. The choice of oviposition sites in aphidophagous Coccinellidae. In. Hodek I. (ed.), *Ecology of Aphidophagous Insects. Academia, Prague and Dr. W. Junk, The Hague*, pp. 121-122.

Iperti, G. 1986. Coccinellidae in France. *Phytoma (France).* 377: 14-22.

Iperti, G. and Paoletti, M.G. 1999. Biodiversity of predaceous Coccinellidae in relation to bioindication and economic importance. Special issue. Invertebrate biodiversity as bioindicators of sustainable landscapes. Practical use of invertebrates to assess sustainable land use. *Agriculture, Ecosystem and Environment.* 74(1-3): 323-342.

Iperti, G. and Prudent, P. 1986. Effect of the substrate properties on the choice of oviposition sites by *Adalia bipunctata* In: Hodek I. (ed.): *Ecology of Aphidophaga. Academia, Prague and Dr. W. Junk, Dordrecht.* pp. 143-149.

Ives, P.M. 1981. Estimation of coccinellid numbers nd movements in the field. *Can. Entomol.* 113: 981-997.

Jalali, S.K., Singh, S.P. and Biswas, S.R. 1999. Effect of temperature and female age on the development and progeny production of *Cryptolaemus montrouzieri* Mulsant (Coleoptera: Coccinellidae). *Entomon* 24(3): 293-296.

James, B.E. 2000. Contribution on certain aspects of Bioecology and behaviour of a ladybeetle, *Coccinella transversalis* Fabricius (Coccinellidae: Coleoptera). University of Lucknow, Lucknow, p. 190.

Jarosik, V., Honek, A. and Dixon, A.F.G. (2000). Rate isomorphy in insects and mites. In: A.F.G. Dixon (ed.). Insect Predator- Prey Dynamics, ladybird Beetles and Biological Control. *Cambridge Unviersity Press.*

Jean, C., Coderre, D. and Tourneur, J.C. 1990. Effect of temperature and substrate on survival and lipid consumption of hibernating *Coleomegilla maculata lengi* (Coleoptera: Coccinellidae). *Environ. Entomol.* 19(6): 1657-1662.

Jiao, Y., Chen, Y. Ye, S.D., Wang, Z.L., Wang, S.Y. and Mao, Y.F. 1997. A study on *Chilocorus bijugus* Mulsant (Coccinellidae: Coleoptera). *Forest Res.* 10(3): 328-331.

Jordanova, V.N. 1979. Ladybirds in the Vitosha, Bulgaria mountains (Coleoptera: Coccinellidae). *Acta. Zool. Bulg.* 10(1): 76-80.

Kaddou, I. 1960. The feeding behaviour of *Hippodamia quinquesignata* (kirby). *Univ. Calif. Publ. Ent.* 16: 181-232.

Kadona, O.K., Weyda, F. and Okuda, T. 1998. *Dinocampus (=Perilitus) coccinellae* teratocyte-specific polypeptide: its accumulative property, localization and characterization. *J. Insect Physiol.* 44(11): 1073-1080.

Kajita, Y., Takano, F., Yasuda, H. and Agarwala, B.K. 2000. Effects of indigenous ladybird species (Coleoptera: Coccinellidae) on the survival of an exotic species in relation to prey abundance. *Appl. Entomol. Zool.* 35(4): 473-479.

Kalushkov, P.K. 1994. Longevity, fecundity and development of *Adalia bipunctata* (Linn.) (Coleoptera: Coccinellidae) when reared on three aphid diet. *Anzeirger fur Schaedlingskunde Pfluanzenschultz Umweltschutz* 67(1): 6-7.

Kawauchi, S. 1979. Effects of temperature on the aphidophagous coccinellids. *Kurume. Univ. J.* 28(1): 47-52.

Kawauchi, S. 1985. Comparative studies on the fecundity of three aphidophagous coccinellids (Coleoptera: Coccinellidae). *Japan. J Appl. Entomol. Zool.* 29: 203-209.

Kearns, P.W.E., Tomlinson, I.P.M., Veltman, C.J. and O'Donald, P. 1992. Non-random mating in *Adalia bipunctata* (the two spot ladybird). II. Further tests for female mating preference. *Heredity* 68: 385-389.

Kindlmann, P. and Dixon, F.G. 1993. Optimal foraging in ladybird beetles (Coleoptera: Coccinellidae) and its consequences for their use in biological control. *Eur. J. Entomol.* 90(4): 443-450.

Koide, T. 1962. Observations on the feeding habit of the larva of *Coccinella septempunctata brucki* Mulsant. The feeding behaviour and number of prey fed under different temperatures. *Kontyu.* 30: 236-241.

KuliKova, L.S. 1984. Changes of insect fauna on cereals on newly established arable soils in Primorskii krai. In: Fauna I Ekologia Dalnego Vostoka (Vreditieli I Entomofagi) *Dalnevostonchnii Nauchnii Centre Akademii Nauk SSSR, Vladivostok.* pp. 37-42. (In Russian).

Kumar, A., Kumar, N., Siddiqui, A. and Tripathi, C.P.M. 1999. Prey-predator relationship between *Lipaphis erysimi* Kalt. (Hom.,

Aphididae) and *Coccinella septempunctata* L. (Col., Coccinellidae). II. Effect of host plants on the functional response of the predator. *J. Appl. Ent.* 123(10): 591-596.

Kumar, D. and Verma, G.C. 1996. Studies of the biology of various predators of mustard aphid, *Lipaphis erysimi* (Kalt.). *J. Insect. Sci.* 9(2): 175-177.

Kumar, S.M.V. and Chakraborty, N. 1997. Biology of *Scymnus nubilus* Mulsant (Coleoptera: Coccinellidae), a native predator of pink mealy bug, *Maconellicoccus hirsutus* (Green) (Hemiptera: Pseudococcidae). *J. Ent. Res.,* 21(4): 329-334.

Kuznetsov V.N. 1981. Fauna and ecology of Coccinellidae (Coleoptera) of the Kamchatka Peninsula. *In: Fauna Ekologia Nasekomykh Primorskogo Kraya i Kamchatki Vladivostok.* pp. 46-54.

Leary, R.F. and Allendorf, F.W. 1989. Fluctuating asymmetry as an indicator of stress: implications for conservation biology. *Trends Ecol. Evolution,* 4: 214-217.

Liu-Tong-Xian, Stanley, P.A., Hoelmer, K.A. and Osborne, L.S. 1997. Life history of *Nephaspis osculatus* (Coleoptera: Coccinellidae), a predator of *Bemisia argentifolii* (Homoptera: Aleyrodiae). *Ann. Ent. Soc. Amer.* 90(6): 776-782.

Lucas, E., Coderre, D. and Vincent, C. 1997. Voracity and feeding preferences of two aphidophagous coccinellids on *Aphis citricola* and *Tetranychus urticae. Ent. Exp. et Applic.* 85(2): 151-159.

Lusis Ya Ya 1961. On the biological meaning of colour polymorphism of *Adalia bipunctata L. Latv. Entomol.* 4: 2-29.

Lusis Ya Ya 1973. The choice of oviposition sites in female of three species of *Adalia* (Coleoptera: Coccinellidae) by breeding in laboratory. *Zool. Muz. Rakasti* 11: 55-63.

Majerus, M.E.N. 1994. Female promiscuity maintains high fertility in ladybirds (Coleoptera: Coccinellidae). *Entomologist's-Monthly Magazine.* 130 (1564-67): 205-209.

Majerus, M.E.N., Henrich, J., Schulenburg, G.V.D. and Zakharov, I.A. 2000. Multiple causes of male-killing in a single of the two-spot ladybird, *Adalia bipunctata* (Coleoptera: Coccinellidae) from Moscow. *Heredity* 84: 605-609.

Majerus, M.E.N. and Hurst, G.D.D. 1997. Ladybirds as a model system for the study of male killing symbionts. *Entomophaga,* 42(1-2): 13-20.

Majerus, M. and Kearns, P. 1989. *Ladybirds.* Richmond Publishing Co. Ltd. England. p. 103.

Malcolm, S.B. 1992. Prey defence and predator foraging. In *Natrual Enemies: The population biology of predators, parasites and diseases,* M.J. Crawley, pp. 458-475. Oxford: Blackwell Scientific Publication.

Manzar, A., Lal, M.N. and Singh, S.S. 1998. Population dynamics of mustard aphid, *Lipaphis erysimi* (Kalt.) on *Brassica* crop in relation to abiotic and biotic factors. Shaspa 5(1): 21-24.

McConnel, J.A. and Kring, T.J. 1990. Predation and dislodgment of *Schizaphis graminum* (Homoptera: Aphididae), by adult *Coccinella septempunctata* (Coleoptera: Coccinellidae). *Environ. Entomol.* 19: 1798-1802.

McMullen, R.D. 1967. The effects of photoperiod, temperature and food supply on rate of development and diapause in *Coccinella novemnotata. Can. Entomol.* 99: 578-59-86.

Michel, B. 1992. Information of some Coccinellidae (Coleoptera) in Paraguay, *Cotton Fibres Tropicales* 47(4): 301-304.

Milevoj, L. 1997. Effects of food on the adult coccinellids, *Coccinella septempunctata* L. *Zhornik Biotechniske Fakultele Univerze Ljubljani, Kmetijstvo,* 69: 137-140.

Miller, J.C. 1992. Temperature-dependent development of the convergent lady beetle (Coleoptera: Coccinelllidae). *Environ. Entomol.* 21: 1139-1142.

Mills, N.J. 1981. Essential and alternative foods for some British Coccinellidae (Coleoptera). *Entomol. Gaz.,* 32: 197-202.

Mills, N.J. 1982. Voracity, cannibalism and coccinellid predation. *Ann. Appl. Biol.* 101: 144-148.

Ming, L.S. 1971. *The Biology and Ecology of some aphidophagous Coccinellidae.* Ph.D. thesis, University of Malaya, Kuala Lumpur.

Mondal, N.U. and Kar, K.R. 1983. Studies on searching behaviour of some predaceous coccinellids (Coleoptera: Coccinellidae) larvae on plants. *Bangladesh Assoc. Adv. Sci (Dhaka). Proc. 8th Bangladesh Sci. Confr. (Dhaka), Bangladesh, BASS.* 155.

Mora, J.G., Gapud, V.P. and Velasco, L.R.I. 1995. Life history and voracity of *Coelophora inaequalis* (Fabricius) (Coleoptera: Coccinellidae) on *Aphis craccivora* Koch. (Hemiptera: Aphididae) *Philippine Entomologist,* 9(5): 523-553.

Nakamuta, K. 1987. Diel rhythmicity of prey search activity and its predominance over starvation in the ladybeetle, *Coccinella septempunctata brucki. Physiol. Entomol.* 12: 91-98.

Naranjo, S.E., Gibson, R.L. and David, D.W. 1990. Development, survival and reproduction of *Scymnus frontalis* (Coleoptera.

Coccinellidae), an important predator of Russian wheat aphid at four fluctuating temperature. *Ann. Entomol. Soc. Amer.* 83(3): 527-531.

Nsiama she, H.D., Odebiyi, J.A. and Herren, H.R. 1984. The biology of *Hyperaspis jacunda* (Coleoptera: Coccinellidae), an exotic predator of cassava mealy-bug, *Phenacoccus manihoti* (Homoptera: Pseudococcidal) in Southern Nigeria. *Entomophaga* 29(1): 87-93.

O'Donald, P. and Majerus, M.E.N. 1992. Non-random mating in *Adalia bipunctata* (The two spot ladybird). III. New evidence of genetic preference. *Heredity* 69: 521-526.

O'Donald, P. and Muggleton, J. 1979. Melanic polymorphism in ladybirds maintained by sexual selection. *Heredity,* 43: 143-148.

Obata, S. 1986. Mechanism of prey finding in the aphidophagous ladybeetle, *Harmonia axyridis,* Pallas (Coleoptera: Coccinellidae). *Entomophaga.* 31: 303-311.

Obata, S. 1987. Mating behaviour and sperm transfer in ladybeetle, *Harmonia axyridis* Pallas (Coleoptera: Coccinellidae). *Appl. Ent. Zool.* 22(4): 434-442.

Obata, S. and Hidaka T. 1987. Ejection and ingestion of the spermatophore by the female beetle, *Harmonia axyridis* Pallas (Coleoptera: Coccinellidae). *Can. Entomol.* 119: 603-604.

Obata, S. and Johki, Y. 1991. Comparative study on copulatory behaviour in four species of aphidophagous ladybirds. *Proceedings of the 4th meeting of IOBC W.G. Ecology of Aphidophaga* 207-212.

Obrycki, J.J., Giles, K.L. and Ormord, A.M. 1998. Interactions between an introduced and indigenous coccinellid species at different prey densities. *Oecologia* 117: 279-285.

Obrycki, J.J. and Kring, T.T. 1998. Predaceous Coccinellidae in biological control. *Ann. Rev. Entomol.* 58: 839-845.

Obrycki, J.J., Ormord, A. M. and Giles, K.L. 1997. Partial life table analysis for larval *Coleomegilla maculata* (De geer) and *Coccinella septempunctata* L. (Coleoptera: Coccinellidae) in Alfalfa. *J. Kansas Entomol. Soc.* 70(4): 339-346.

Obrycki, J.J. and Orr, C.J. 1990. Suitability of three prey species for Nearcytic populations of *Coccinella septempunctata, Hippodamia variegata* and *Propylea quatuordecimpunctata* (Coleoptera: Coccinellidae). *J. Econ. Entomol.* 83: 1292-1297.

Obrycki, J.J. and Tauber, M.J. 1978. Thermal requirements for development of *Coleomegilla maculata* (Coleoptera: Coccinellidae) and its parasites *Perilitus coccinellae* (Hymenoptera: Braconidae). *Can. Entomol.* 110(4): 407-412.

Ofuya, T.I. and Akingbohungbe A.E. 1988. Functional and numerical responses of *Cheilomenes lunata* (Fabricius) (Coleoptera: Coccinellidae) feeding on Cowpea aphid, *Aphis craccivora* Koch (Homoptera: Aphididae). *Insects Sci. Appl.* 9: 543-546.

Olszak, R.W. and Niemczyk, E. 1987. Predaceous Coccinellidae associated with aphids in apple orchards. *Ekol. Pol. (Poland). Polish. J. Entomol.* 34(4): 711-721.

Omkar and Bind, R.B. 1998. Prey preference of a ladybird beetle, *Cheilomenes (=Menochilus) sexmaculata* (Fabr.). *J. Aphidol.* 12: 63-66.

Omkar, James, B.E. and Srivastava, S. 1999. Prey preference of a ladybeetle, *Coccinella transversalis* Fabricius (Coleoptera: Coccinellidae). *Adv. Bios.* 18(2): 117-122.

Omkar, Mishra, G. and Pervez, A. 2002. Intraguild predation by ladybeetles: an ultimate survival strategy or an aid for advanced aphid bio-control? *Prof. S.B. Singh Comm. Vol. Zool. Soc. India.* 77-90.

Omkar and Pervez, A. 2000a. Biodiversity of predaceous coccinellids (Coleoptera: Coccinellidae) in India: a review. *J. Aphidol.* 14(1&2): 41-67.

Omkar and Pervez, A. 2000b. Sexual dimorphism in *Propylea dissecta* (Mulsant) (Coccinellidae: Coleoptera). *J. Aphidol.* 14(1&2): 139-140.

Omkar and Pervez, A. 2001. Prey preference of a ladybeetle, *Micraspis discolour* (Fabricius). *Entomon.* 26(2): 195-197.

Omkar and Pervez, A. 2002a. A Review on the ecology of an aphidophagous ladybird beetle, *Coccinella septempunctata* Linnaeus (Coleoptera: Coccinellidae). *J. Aphidol.* (In Press).

Omkar and Pervez, A. 2002b. Reproduction in predatory ladybeetles: A review. In *Perspectives in Animal Ecology and Reproduction Vol I,* Daya Publishing House Delhi, pp. 181-205.

Omkar and Pervez, A. 2002c. Influence of temperature on age specific fecundity of the ladybeetle, *Micraspis discolour* (Fabricius). *Insect Sci. Appl.* 22(1): 61-65.

Omkar and Srivastava, S. 2001a. Comparative predatory potential of a ladybird beetle, *Coccinella septempunctata* Linn. on six prey species. *Biological Memoirs* 59-63.

Omkar and Srivastava, S. 2001b. Ovipositional preference of *coccinella septempunctata* Linnaeus (Coccinellidae: Coleoptera). *J. Aphidol.* 15(1&2): 5-8.

Omkar and Srivastava, S. 2001c. First Record of Egg and Pupal Parasitoids of *Coccinella Transversalis* Fabricius with a Note on the Parasitization in Ladybeetles. *J. Adv. Zool.* 22(2): 129-130.

Omkar and Srivastava, S. 2001d. Functional response of a ladybeetle, *Coccinella septempunctata* Linnaeus to three aphid species. *J. Adv. Zool.*, 22(1): 40-45.

Omkar and Srivastava, S. 2002a. Functional response of the ladybeetle *Coccinella septempunctata on Uroleucon compositae* (Theobald) and *Aphis nerii* Boyer de Fonscolombe. *Prof. S.B. Singh Comm. Vol. Zool. Soc. India* 115-128.

Omkar and Srivastava, S. 2002b. The reproductive behaviour of an aphidophagous ladybeetle, *Coccinella septempunctata* Linnaeus. *Eur. J. Entomol.* 99(4): 465-470.

Omkar and Srivastava, S. 2002c. Influence of prey species on certain biological attributes of a ladybird beetle, *Coccinella septempunctata* Linnaeus Bio-control (Accepted).

Omkar, Srivastava, S. and James, B.E. 1997. Prey preference of a ladybeetle, *Coccinella septempunctata* Linnaeus (Coleoptera: Coccinellidae). *J. Adv. Zool.* 18(2): 96-97.

Orr, C.J. and Obrycki, J.J. 1990. Thermal and dietary requirements for development of *Hippodamia parenthesis* (Coleoptera: Coccinellidae). *Environ. Entomol.* 19(5): 1523-1527.

Osawa, N. 1989. Sibling and non-sibling cannibalism by larvae of a ladybeetle *Harmonia axyridis* Pallas (Coleoptera: Coccinellidae) in the field. *Res. Pop. Ecol.* 31: 153-160.

Osawa, N. and Nishida, T. 1992. Seasonal variation in elytral colour polymorphism in *Harmonia axyridis* (the ladybird beetle): the role of non-random mating. *Heredity* 69: 297-307.

Padmaja, Ch., Ramesh Babu, T., Reddy, D.D.R. and Sriramulu, M. 1995. Biology and predation of *Scymnus coccivora* Aiyar (Coleoptera: Coccinellidae) on mealy bugs. *J. Ent. Res.* 19(1): 79-81.

Pandey, K.P., Kumar, A., Singh, R., Shanker, S. and Tripathi, C.P.M. 1984. Numerical response and area of discovery of a predator, *Coccinella septempunctata* L. *Z. Angew. Entomol.* 97: 418-423.

Phoofolo, M.W. and Obrycki, J.J. 1998. Potential for intraguild predation and competition among predatory Coccinellidae and Chrysopidae. *Ent. Exp. Applic.* 89: 47-55.

Polis, G.A. and Holt, R.D. 1992. Intraguild perdition: the dynamics of complex trophic interactions. *Trends in Ecology Evolution* 7: 151-154.

Ponsonby D.J. and Copland M.J.W. 1995. Olfactory responses by the scale insect predator *Chilocorus nigritus* (F.) (Coleoptera: Coccinellidae). *Bio-control Sci. Tech.* 5: 83-93.

Ponsonby, D.J. and Copland, M.J.W. 1996. Effect of temperature on development and immature survival in the scale insect predator, *Chilocorus nigritus* (F.) (Coleoptera: Coccinellidae). *Bio-control Sci. Tech.* 6(1): 101-109.

Pope, R.D. 1973. *Colesptera Coccinellidae and Sphindidae.* Royal Entomological Society of London Handbooks for the identification of British insects, vol. V, part 7.

Poprawski, T.J., Legaspi, J.C. and Parker, P.E. (1998). Influence of entomopathogenic fungi on *Serangium parcesetosum* (Coleoptera: Coccinellidae): an important predator of whiteflies (Homoptera: Aleyrodidae). *Environ. Entomol.* 27(3): 785-795.

Parsana, G.J., Kapadia, M.N. and Butani, P.G. 1997. Association of Groundnut aphid, *Aphis craccivora* Koch with coccinellid predators and their relation to abiotic factors. *J. Aphidol.* 11(2): 199-200.

Prodhan, M.Z.H., Haque, M.A., Khan, A.B. and Rahman, A.K.M.M. 1995. Biology of *Micraspis discolour* (F.) (Coleoptera: Coccinellidae) and its susceptibility to two insecticides, *Bangladesh, J. Entomol.* 5(1-2): 11-17.

Raimundo, A.A.C. and Harten van, A. 2000. An annotated checklist of the Coccinellidae (Insecta: Coleoptera) of Yemen. *Fauna of Arabia* 18: 211-243.

Rhamhalinghan, M. 1986. Seasonal variation in ovariole numbers/ ovary in *Coccinella septempunctata* L. (Coleoptera: Coccinellidae). *Proc. Indian Natn. Sci. Acad.* 52: 619-623.

Rhamhalinghan, M. 1987a. The behaviour of *Coccinella septempunctata* L. Associated with the exit of *Coccinellimermis Rubtzov* (Nematoda: Mermithidae): 1. Prior to the emergence of the parasite. *Indian Zoologist* 10: 95-97.

Rhamhalinghan, M. 1987b. Feeding behaviour of *Coccinella septempunctata* L. *var. confusa* Wiedemann (Coleoptera: Coccinellidae) in relation to temperature-I. Pre-oviposition period. *J. Ent. Res.* 11(2): 178-183.

Rhamhalinghan, M. 1998. The reproductive behaviour of *Coccinella septempunctata* L. var. *confusa* Wiedemann (Coleoptera: Coccinellidae). *J. Adv. Zool.* 19(1): 17-21.

Richards, A.M. 1980. Sexual selection, guarding and sexual conflict in a species of Coccinellidae (Coleoptera). *J. Aust. Entomol. Soc.* 19: 26.

Rodriguez, S.C. and Miller, J.C. 1995. Life-history traits in *Hippodamia convergens* (Coleoptera: Coccinellidae) after selection for fast development. *Biological control,* 5(3): 389-396.

Rowlands, M.L.J., Chapin, J.W. 1978. Prey searching behaviour in adults of *Hippodamia convergens* (Coleoptera: Coccinellidae). *J. Georgia Entomol. Soc.* 13: 309-315.

Saha, J.L. 1987. Studies on the fecundity, hatchability, and mortality of *Menochilus sexmaculatus* Fabr. (Coleoptera: Coccinellidae). *J. Aphidol.* 1: 47-50.

Sakurai, H., Nakanishi, H. and Takeda, S. 1990. Studies on the ovarian development of the ladybeetle, *Coccinella septempunctata:* effect of environmental factors on oogenesis. *Res. Bull. Fac. Agric. Gifu. Univ.* 55: 83-90.

Sakurai, H., Yoshida, N., Kobayashi, C. and Takeda, S. 1991. Effect of temperature and day length on oviposition and growth of ladybeetle *Coccinella septempunctata bruckii. Res. Bull. Fac. Agric. Gifu. Univ.* 56: 45-50.

Samways, M.J. and Tate, B.A. 1984. Sexing of *Chilocorus nigritus* (F.) (Coccinellidae). *Citrus and Subtropical Fruit Journal* 607: 4-5.

Savoiskaya, G.I. 1965. Biology and prospective of utilization of coccinellids in the control of aphids in south eastern Kazakhistan orchard. *Trudy Inst. Zashch. Rast., Alma-Ata* 9: 128-156.

Savoiskaya, G.I. 1970. Introduction and acclimatisation of some coccinellids in the Alma-Ata reserve. *Trudy Alma-Atin. Gos. Zapov.* 9: 138-162.

Shah, M.A. 1982. The Influence of plant surfaces on the searching behaviour of coccinellid larvae. *Ent. Exp. Appl.* 31: 377-380.

Shah, M.A. 1983. A stimulant in *Berberis vulgaris* inducing oviposition in coccinellids. *Ent. Exp. Appl.* 33: 119-120.

Shantibala, K. and Singh, T.K. 1991. Studies on the vertical distribution of aphidophagous coccinellids (Coleoptera: Coccinellidae) of Manipur and Nagaland. *J. Aphidol.* 5(1&2): 39-43.

Sharma, D.K., Varma, G.C. and Kishore, L. 1997. Feeding capacity of predators of mustard aphid, *Lipaphis erysimi. J. Aphidol.* 11(2): 171-174.

Sharma, R.P. and Yadav, R.P. 1994. Population dynamics of bean aphid (*Aphis craccivora* Koch.) and its predatory coccinellid complex in relation to crop type (lentil, *Lathyrus* and faba bean) and weather conditions. *J. Ent. Res.* 18(1): 25-36.

Singh, D. and Singh, H. 1993. Biology of ladybird beetle, *Coccinella septempunctata Linn. An. Biol. Ludhiana.* 9(2): 250-253.

Singh, M.P. and Devi, C.S. 1996. Food preference of C. *septempunctata* (Linn.) on aphids. *Insect Environ.* 2(1): 5-6.

Singh, P. and Singh, H. 1994. Predatory potentiality of coccinellids, *Coccinella septempunctata* Linn. and *Hippodamia variegata* (Goeze) over mustard aphid, *Lipaphis erysimi* (Kalt.). *Crop Research Hissar* 7(1): 120-124.

Singh, R. and Malhotra, R.K. 1979. Bionomics of *Coccinella septempunctata* Linn. *Indian J. Entomol.* 41(3): 244-249.

Singh, S., Singh, L.K.C. and Singh, T.K. 1993. Biology and feeding potential of *Oenopia quadripunctata* Kapur, a coccinellid predator of an oak aphid, *Tuberculatus nervatus* Chakraborthy and Roy Chowdhury, *J. Adv. Zool.* 14(1): 7-11.

Singh, S.B. and Singh, H.M. 1998. Natural enemies associated with insect pests of lemon, Citrus lemon Burm. *Shashpa.* 5(1): 109-110.

Singh, V.S., Yadav, R.R. and Singh, R. 1994. Post embryonic development, survival rate and predation potential of *Coccinella septempunctata* Linn. in relation to the mustard aphid *(Lipaphis erysimi* Kalt.) *J. Entomol. Res.* 18(1): 5-10.

Smith, B.C. 1971. Effects of various factors on the local distribution and density of coccinellid adults on corn (Coleoptera: Coccinellidae). *Can. Entomol.* 103: 1115-1120.

Smith, B.C. and Williams, R.R. 1976. Temperature relations of adult *Coleomegilla maculata lengi and Coleomegilla maculata medialis* (Coleoptera: Coccinellidae) and responses to ovipositional stimulants. *Can. Entomol.* 108(9): 925-930.

Srivastava, S. 2000. Certain aspects of Bioecology and ethology of a ladybeetle, *Coccinella septempunctata* Linnaeus (Coccinellidae: Coleoptera). University of Lucknow, Lucknow. p. 160.

Stewart, L.A., Hemptinne, J.L. and Dixon, A.F.G. 1991. Reproductive tactics of lady beetles: relationships between egg size, ovariole number and developmental time. *Funct, Ecol,* 5: 380-385.

Sugiura, K. and Takada, H. 1998. Suitability of seven aphid species as prey of *Cheilomenes sexmaculata* (Fabricius) (Coleoptera: Coccinellidae). *Japan J. Appl. Ent. and Zool.* 42(1): 7-14.

Sullivan, D.J., Castillo, J.A. and Billotti, A.C. 1991. Comparative biology of six species of coccinellid beetle (Coleoptera: Coccinellidae) predaceous on the mealybug, *Phenacoccus herreni* (Homoptera: Pseudococcidae), a pest of cassava in Colombia, South America. *Environ. Entomol.* 20: 685-689.

Takahashi K. 1987a. Differences in oviposition initiation and sites of ladybeetles *coccinella septempuncata brucki* Mulsant and *Harmonia axyridis* (Pallas) (Coleoptera: Coccinelllidae) in the fields. *J. Eco. Ent.* 90(5): 1102-1105.

Takahashi K. 1987b. Cannibalism by the larvae of *Coccinella septempunctata bruckii* Mulsant (Coleoptera : Coccinellidae) in mass rearing experiments. *Japan. J. Appl. Entomol. Zool.* 31: 201-205.

Takahashi, K. 1993. Daily behaviour of adult *Coccinella septempunctata brucki* (Coleoptera: Coccinellidae). *Japanese J. Entomol.* 61(4): 839-845.

Takahashi K. and Natio A. 1984. seasonal occurrence of the aphids and their predators (Coleoptera: Coccinellidae) in alfalfa fields. *Bull. Natn. Grassland Res. Inst.* 29: 62-66.

Tao, C.C. and Chui, S.C. 1971. Biological control of citrus, vegetables and tobacco aphids. *Spec. Publ. Taiwan Agric. Res. Inst.* 10: 1-110.

Thalji, R.A. 1994. Occurrence and distribution of aphidophagous coccinellids (Coleoptera: Coccinellidae) on agricultural crops and wild plants in Vojodina (Yugoslavia). *Zastita Bilja.* 45(210): 279-291.

Thalji, R.A. 1997. Notes on parasitism of ladybirds (Coleoptera: Coccinellidae) in Vojvodina (Serbia, Yugoslavia). *Zhornik Matice Srpske Za Prirodne Nauke.* 93: 67-75.

Tomlinson, T.P.M. 1989. *Theoretical and Experimental Studies of the Evolution of Mate Choice.* Ph.D. Thesis. University of Cambridge. Cambridge.

Tomlilnson, T.P.M. 1996. A study of non-random mating in a British population of the two-spot ladybird with a high frequency of the melanic morph. *J. Insect Behav.* 9(4): 549-556.

Triltsch, H. 1996. On the parasitization of the ladybird *Coccinella septempunctata* L. (Coccinellidae: Coleoptera). *J. Appl. Entomol.* 120: 375-378.

Tripathi, C.P.M., Siddiqui, A., Dubey, R.J., Sinha, G., Kumar, N. and Rai, S. 2000. Effect of food plant on the searching efficiency, mutual interference and killing power of the ladybird beetle, *Coccinella septempunctata* Linn., a predator of *Lipaphis erysimi* (Kalt.) *Seventh National Symposium on Aphidology, 11-13 March, 2000 at DDU Gorakhpur University, Gorakhpur,* 69.

Ueno, H. 1994. Fluctuating asymmetry in relation to two fitness components, adult longevity and male mating success in a ladybird beetle, *Harmonia axyridis* (Coleoptera: Coccinellidae). *Ecol. Entomol.* 19(1): 87-88.

Vallejo, M.C.T., Jose-Luis, C.S., Jesus, R.N. and Hiram, B.M. 1990. Comparative study of the biology and development of *Hippodamia convergens* and *Hippodamia koebelei* (Coleoptera: Coccinellidae). *Agrociencia (Mexico). Serie Protection Vegetal.,* 1(3): 23-34.

Veeravel, R. and Baskaran, P. 1995. Cannibalistic behaviour and survival rate of two coccinellid predators, *Coccinella transversalis* Fabricius *and Menochilus sexmaculatus* (Fabricius). *Indian J. Ecol.* 22(2): 113-117.

Veeravel, R. and Baskaran, P. 1997. Functional and numerical responses of *Coccinella transversalis* and *Cheilomenes sexmaculata* Fabr. Feeding on the melon aphid, *Aphis gossypii* Glov. *Insect Sci Appl.* 17(3/4): 335-339.

Vohland, K. 1996. The influence of plant structure on searching behaviour in the ladybird, *Scymnus nigrinus* (Coleoptera: Coccinellidae). *Eur. J. Entomol.* 93: 151-160.

Werren, H.H., Hurst, G.D.D., Zhang, W., Breeuwer, J.A.J., Stouthamer, R. and Majerus, M.E.N. (1994). Rickettsial related male-killing in the ladybird beetle *(Adalia bipunctata). J. Becteriol.* 176: 388-394.

Wheeler, A.G. Jr. and Hoebeke, E.R. 1995. *Coccinella novemnotata* in Northeastern North America: Historical occurrence and current status (Coleoptera: Coccinellidae). *Proc. Entomol. Soc. Wash.* 97(3): 701-716.

Won, H.J., Cho, S.Y., Hong. O.K., Park, H.C., Kim, H.J. and Park, Y.C. 1996. Parasitoids of the aphidophagous ladybeetle, *Coccinella septempunctata. Korean J. Entomol.* 26(3): 243-247.

Yang, X. MioQing, S. ZhenZhong, S. and Jiwen, X. 1998. Effects of temperature on experimental population of *Chilocorus kuwanae* Silvestri. *Zool. Res.* 19(1): 39-44.

Yasuda, H. and katsuhiro, S. 1997. Cannibalism and interspecific predation in two predatory ladybirds in relation to prey abundance in the field. *Entomophaga* 42(1/2): 153-163.

Yasuda, H. and Ohnuma, N. 1999. Effect of cannibalism and predation on the larval performance of two ladybird beetles *Ent. Exp. Applic.* 93(1): 63-67.

Yu, G.Y. and Pang, H. 1997. A catalogue of Coccinellidae of Taiwan (Coleoptera). *J. Taiwan Museum.* 50(1): 1-19.

Zebitz, C.P.W. 1988. Efficiency of coccinellid larvae in dependence from prey distribution. *Mitteilungen der Deutschen Gesellschaft feur Allgemeine und Angewardte Entomologie.* 6(4-6): 454-457.

Zhang, Z. and Saboori, A. 1996. A new host record of *Allothrombin triticium* larvae (Acari: Trombidiidae) ectoparasitic on *Coccinella septempunctata* (Coleoptera: Coccinellidae). *Systematic Appl. Acarol.* 1: 207-208.

Zhou, X. and Carter, N. 1992. The ecology of coccinellids on farmland *Aspects Appl. Biol.* 31: 133-138.

Zhou X, Carter, N. and Powell W. 1994. Seasonal distribution and aerial movement of adult Coccinellids on farmland. *Biocontrol Sci. Tech.* 4(2): 167-175.

Zhou X, Honek, A., Powell W. and Carter, N. 1995. Variations in body length, weight, fat content and survival in *Coccinella septempunctata* at different hibernation sites. *Ent. Exp. Applic.* 75: 99-107.

2

Status of the Asian Giant Honey Bee *Apis dorsata* F. and its Conservation in Southern Part of the Deccan Peninsula, Karnataka, India

S. Basavarajappa

Department of Applied Zoology, Kuvempu University, B.R. Project- 577115, Karnataka, India

ABSTRACT

Evaluation of Asian giant honeybee *Apis dorsata F.* population in different habitats in southern part of the Deccan Peninsula revealed significant variation with respect to the nest site, nesting archi-tecture, comb size and population density. The bee flora consisted of diversified tree, shrub and herb vegetation which have played a key role in shaping the size of colonies of *A. dorsata* in the wild at a well defined ecological boundary. However, in spite of being the giant bodied open nester and efficient traveller in large area to forage on innumerable number of plant species, the maladies viz., high degree of human interference, nest site destruction, hive hunting, pests and predators problem, pesticides poisoning and other unidentified human associated disturbances continue to be the bottlenecks in the wild for the survival of *A. dorsata*. The availability of the preferred flora to allocate nutritional resources (viz., pollen, nectar, water and etc.) for the growth and development of the colony, congenial nest site and climatic fluctuations in different habitats have played a key role in accelerating the pollination service and hive products productivity in southern part of the Deccan Peninsula, Karnataka. Further, the conventional hive hunting practices employed so far have achieved little success in harnessing the full potential of the rock bee colonies as revealed by the many instances witnessed in the field.

Recent developments in giant honeybee hive hunting methods have come out with novel strategies to overcome the problems of the conventional hunting approaches to improve the efficiency, reduce the loss and maximize hive products yield. The proven source of hive products (viz., honey and wax) in different habitats with respect to season is being made to know the potential of the rock bee colonies. Moreover, improved biological and eco-friendly methods have come out with good results to minimize the pests and predators problems in the colonies of *A. dorsata.* Therefore, the situation demands the need to try these new techniques more vigorously with other countries in the world. Hence, ground work is being laid for exploring the possibilities of using under utilized resources in the waste land, forest area, cultivated and uncultivated land and human dwelling places which are probably of unknown limitation by initiating studies on feral insect *A. dorsata,* its forage plants and eco-friendly service in agri-horticultural-forest areas and at the human dwelling places for the conservation of *A. dorsata* in their preferred habitats with the economic and ecological interest in the years to come.

Introduction

The honey bees (Genus *Apis* Linnaeus) are poikilothermic social insects, well known for their excellent and efficient pollination service to varied floristic taxa in nature. The rock bee is the most widely known among the honey bee *Apis* species in India. It is indigenous to Indian sub-continent, commonly referred to as the Asian giant honey bee owing to its large body size (20mm in length) (Engel, 1999). It belongs to genus Apinae, subgenus Megapis. The subgenus Megapis is monotypic and represented by *Apis dorsata* Fabricus (Engel, 2002). *A dorsata* represents four sub-species viz., *Apis d. binghami* Cockerel; *A.d. breviligula* Maa; *A.d. laboriosa* Smith and *A.d. dorsata* Fabricus are well know in different parts of the world. Of all, only the sub species *A. dorsata* is found in India (Engel, 2002).

A.d. dorsata builds large colonies in the open and most often high in trees usually affixed to the sub-surface to the various objects. Its hives yield more honey than any other *Apis* species. More than 60-70% of the honey and virtually all of the wax harvested in India is produced taken from the huge colonies of *A. dorsata* (Thakar and Tonapi, 1961; Singh, 1980). In Karnataka, 50-90% of all honey is collected from wild colonies of the rock bee, *A. dorsata* (Nicola Bradbear and Reddy, 1998).

However, in recent years the importance of *Apis dorsata* is declining gradually due to the introduction of the exotic bee *Apis mellifera* (Rathore and Kapur, 2002). Furhter, people of some agro-climatic zones are unaware about its importance and have not tried to benefit fully from its presence. Because of its well recognized defensive behaviour, it has been relatively poorly studied (Otis, 1991). This review presents results of observations made on *A. dorsata* some parts of southern part of Deccan Peninsula, Karnataka.

Study Area

Karnataka state is situated between 11° 30′ and 9° N latitude and between 74° and 78° E. longitude. The geographical area is 19.2 million hectares (Kamath, 1991), and from the apicultural point of view the following distinct regions viz., Coastal area, western ghats and Deccan plateau were made (Anonymous, 1998). In the Deccan plateau, the present study was made.

Physiographically, the study area (Channagiri and Bhadravathi - the part of Davangere and Shimoga districts respectively) occupies roughly a central position in the state of Karnataka, and partly constitutes the southern part of the Deccan Peninsula (Fig. 2.1). It lies in between 130° 27′-14° 39′ N latitudes and 74° 38′-76°40[1] E longitudes with an altitude of 671 to 750 meters above msl. The climate is semi-arid and arid type with an annual rainfall of 871.8 to 1100 mm (Kamath, 1991). Water from Bhadra Reservoir has become life line for people of this region (Basavarajappa, 1998a and b).

Materials and Methods

Totally 10 different habitats were selected in the proposed study area (Fig. 2.1). The selected habitats are as follows: (1) Cultivable land with limited water source. (2) Dry deciduous forest. (3) Residential area with huge tall trees. (4) Non-residential area with huge buildings. (5). Bhadra Reservoir (B.R. Dam). (6) Irrigated cultivable land with scattered coconut gardens. (7) Rainfed land with road side trees. (8). Canal bank trees. (9) Scrubby vegetation. (10) Partially inhabited human dwelling sites nearby rock cliff and natural vegetation selected so as to represent the existing conditions of this region. All these habitats

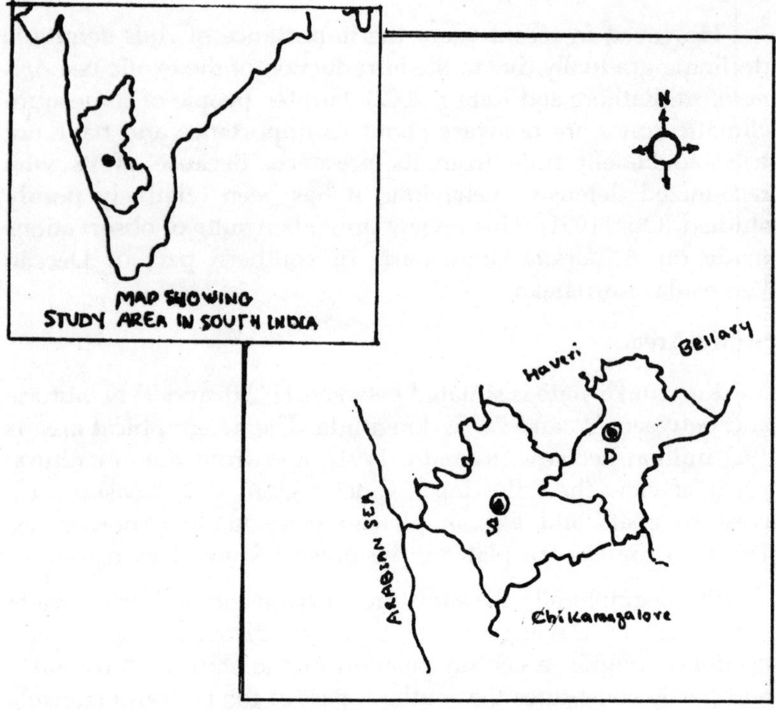

Fig. 2.1: Map showing study area

were surveyed for the available flora and the existing natural colonies of *Apis dorsata*. In each habitat, three permanent observation sites (1 km^2 each) were selected and in each site regular observations were made at fortnightly intervals during 1999-2001. As the vegetation distribution was not even in all these habitats (Basavarajappa, 2002), an all out search method was followed to record the available source of bee forage and density of honey bee colonies in each and every habitat. The available flora was identified by following standard methods, and classified as per Koteswara Rao (1998), nesting sites, site preference, status of flora and bee colonies viz., comb elevation, size, population density, honey, wax production, honey hunted combs and their destruction's were recorded and calculated by following methods of Saha, (1992). The pests were identified by following standard methods. The Friedman two-way analysis of

Table 2.1: Status of *A. dorsata* flora in southern part of the Deccan Peninsula, Karnatak.

Sl. No.	Common name	Botanical name	Family	Plant	Flora type	Source (P/N)	Apicultural value	Season	Occur in Habitats
1	2	3	4	5	6	7	8	9	10
1.	Mango	*Mangifera indica*	Anacardiaceae	T	C	Both	++	Sp, Sm	1,3,6,7,8
2.	Niligeri	*Eucalyptus citriodora*	Myrtaceae	T	C, UC	Both	+++	W	2,3,4,8,9
3.	Tamarind	*Tamarindus indica*	Ceaslpinaceae	T	C,UC	Both	++	R	4,6,9
4.	Coconut	*Cocos nucifera*	Plamaceae	T	C	Both	+++	Ty	1,3,4,6,7,8
5.	Srigandha	*Santalum album*	Santalaceae	T	UC	P	+	SP	2,9
6.	Papaya	*Carica papaya*	Caricaceae	T	C	Both	++	Ty	1,2,6,9,10
7.	Adake	*Areca catecha*	Araceaceae	T	C	Both	+++	Ty	1,6,7,9
8.	Banana	*Musa paradisiaca*	Musaceae	T	C	Both	+++	Ty	1,6,7,9
9.	Cashew	*Anacardium occidentalis*	Anacardiaceae	T	UC	P	++	W, Sp	1,2,6,7,9
10.	China Rose	*Hibiscus rosa-sinensis*	Malvaceae	S	O	Both	+++	Ty	1,3,4,10
11.	Cotton	*Gossypium arboreum*	Malvaceae	S	C	Both	++	R,W	7
12.	Lady's finger	*Abelmsoschus esculentus*	Malvaceae	S	C	Both	++	Ty	1,3,6,7,10
13.	Lemon	*Citrus aurantifolia*	Rutaceae	S	C	Both	+	Ty	1,3,6,7,10
14.	Oleander	*Nerium indicum*	Apocyanaceae	S	O	Both	+	Ty	1,3,4,10
15.	Waterbund weed	*Inpomoea reptans*	Convolvalaceae	S	Wd	Both	++	Ty	1,3,4,10
16.	Jasmine	*Michelia champaka*	Oleaceae	S	O	Both	+	Ty	1,3,4,6,7,9
17.	Sensitive plant	*Mimosa pudica*	Mimosaceae	H	Wd	P	+++	Ty	1,2,3,6,7,8,10
18.	Water lily	*Nymphaea lotus*	Nymphaeaceae	H	Wd	Both	+	Ty	2,6

(contd.)

Table 2.1: contd. . . .

1	2	3	4	5	6	7	8	9	10
19.	Thorn apple	*Datura fastuosa*	Solanaceae	H	Wd	Both	+	Ty	1,6,7
20.	Silk cotton	*Salmalia malabarica*	Bombacaceae	T	UC	Both	+++	Sp	2
21.	Dwarf gold mohur	*Caesalpina pulcherrima*	Caesalpinaceae	T	UC	Both	++	Sp, Sm	3,4,7,10
22.	Madar	*Calotropis gigantea*	Asclepidaceae	S	Wd	Both	++	Ty	1,3,6,10
23.	Rice	*Oryza sativa*	Poaceae	H	C	Both	++	R, Sm	1,6
24.	Thatchigrass	*Saccharum spontaneum*	Poaceae	H	Wd	P	++	Ty	1,6,8
25.	Congress weed	*Parthenum hysterophorus*	Poaceae	H	Wd	P	++	Ty	1,3,6,7,8,10
26.	Castor	*Ricinus communis*	Euphorbiaceae	S	C	Both	++	Ty	1,3,7
27.	Heechala mara	*Pheonx vulgaris*	Arecaceae	T	Wd	Both	++	Ty	1,3,7
28.	Kolavalika	*Hygrophila longifolia*	Acanthaceae	H	Wd	Both	+	Ty	1,6,7,9
29.	Maize	*Zeamays* sp.	Poaceae	H	C	P	+	R, Sm	1,6,7
30.	Soapnut	*Sapindus emerginatus*	Sapindaceae	T	UC	Both	+	Sp	2,3,4,9
31.	Rose wood	*Dalbergia sisso*	Papilionaceae	T	UC	Both	+	Sp, Sm	2,3,4,9
32.	Rose wood	*Dalbergia latefolia*	Papillionaceae	T	UC	Both	+	Sp, Sm	2,3,4,9
33.	Drum stick	*Moringa oleifera*	Moringaceae	T	C	Both	++	W, Sp	1,3,4,6,7,1
34.	Bamboo	*Bambusa bambos*	Poaceae	T	UC	Both	+++	R,W	9,2
35.	Guava	*Psidium guava*	Myrtaceae	T	UC	P	++	W	2
36.	Jack fruit	*Artocarpus heterophyllum*	Moraceae	T	C	Both	+	W, Sp	1,4,6,7,10
37.	Jawar	*Sorghum vulgare*	Poaceae	H	C	Both	++	R, Sm	1,6,7

38.	Amla	*Emblica officnalis*	Euphorbiaceae	T	C	Both	+	R, Sm	2
39.	Rain tree	*Somanea saman*	Mimosace	T	UC	N	++	R, w	2,8
40.	Nerale	*Syzgium cumini*	Myrtaceae	T	C,UC	Both	++	Sm	1,2,3,7
41.	Camela foot tree	Bauhinia vahlii	Papilionaceae	T	UC	Both	+	R, w	2,7
42.	Pomgranate	*Punica granatum*	Punicaceae	SA	C	P	+	Ty	3,6,7
43.	Indian beech	*Pongamia pinnata*	Leguminaceae	T	UC	N	++	R	1,3,4,6,7
44.	Inidan beech	*Pongamia glabra*	Leguminaceae	T	UC	N	++	R	1,2,4,6,7
45.	Citrus	*Citrus reticulata*	Ruteceae	S	C	Both	++	Ty	1,3,7,10
46.	Arjuna	*Terminalia arjuna*	Combretaceae	T	UC	Both	++	Sp, Sm	2,6,7
47.	Nalla maddi	*T. tomentosa*	Combretaceae	T	UC	Both	++	Sp, Sm	5,6,10
48.	Yellow myrobalan	*T. chebula*	Combretaceae	T	UC	Both	++	Sp, Sm	2,6,7
49.	Thorn apple	*Datura stramonium*	Solanaceae	H	Wd	P	++	Ty	1,6,7
50.	Night Jasmine	*Nyctanthes arbor-tristis*	Oleaceae	T	O	Both	++	Sp, Sm	5,6,10
51.	Teak	*Tectona grandis*	Verbenaceae	T	C, UC	N	++	Sp	2,7
52.	Rose	*Rosa* sp.	Rosaceae	S	O	P	++	Ty	4,10
53.	Tulasi	*Ocimum sanctum*	Labiate	H	O	N	++	Ty	1,4,10
54.	Acacia	*Acacia arabica*	Mimoseaceae	T	UC	P	++	Sp, Sm	1,2,6,7
55.	Silk cotton	*Ceiba pentandra*	Bombacaceae	T	UC	N	++	W, Sp	2
56.	Cassia	*Cassia auriculata*	Leguminosae	S	Wd	Both	++	Ty	1,2,6,7
57.	Gajjaga	*Caesalpinia bonduc*	Leguminosae	S	Wd	Both	++	Ty	1,2,6,7
58.	Citrus	*Citrus acida*	Rutaceae	S	C	Both	++	Ty	1,2,6,7
59.	Ficus	*Ficus religiosa*	Moraceae	T	UC	Both	++	Sp	1,2,9
60.	Kasthuri Jali	*Acacia farmesiana*	Mimosaceae	T	UC	P	++	Sp, Sm	1,2,9,10
61.	Muthuga	*Butea monosperma*	Fabaceae	T	UC	P	++	Sp, Sm	2
62.	Neem	*Azadirachta indica*	Meliaceae	T	C, UC	Both	+	Sp	1,2,6,7,10

(contd.)

Table 2.1 : concld.

1	2	3	4	5	6	7	8	9	10
63.	Lantana	*Lantana camara*	Verbenaceae	S	Wd	N	+	Ty	1,2,6,7,8,9
64.	Sonchus	*Tridax procumbens*	Compositae	H	Wd	P	++	Ty	1,2,6,7,8,9
65.	Sunflower	*Helianths* sp.	Compositae	S	C	Both	+++	R, Sm	1,6,7
66.	Black night shade	*Solanum nigrum*	Solanaceae	H	Wd	N	+	Ty	1,2,6,7
68.	Cucumis	*Cucumis* sp.	Compositae	H	Wd	N	+	Ty	1,2,6,7
69.	Corriander	*Corriandrum sativum*	Apiaceae	H	C	P	++	Sp, W	1,2,6,7
70.	Opuntia	*Opuntia* sp.	Cataceae	S	Wd	Both	+	Ty	2,7
71.	Jotropa	*Jotropa* sp.	Euphorbiaceae	H	Wd	Both	+	Ty	2,7
73.	Musturd	*Brassica nigrum*	Cruciferae	H	C	Both	++	Sp	1,6,7
74.	Fry wood tree	*Albizia lebbek*	Mirnosaceae	T	UC	Both	++	Sp	2
75.	Ber	*Ziziphus mauritiana*	Rhamnaceae	T	UC	Both	++	Sp	2
76.	Eucalyptus sp.	*Eucalyptus tereticorms*	Myrtaceae	T	C,UC	Both	+++	Sp	2
77.	Bilijali	*Acacia leucophloea*	Mimoseaceae	T	UC	Both	++	R,W	1,2,3,6,7,8,9
78.	Akash moni	*A. aluriculiformis*	Minoseaceae	T	UC	Both	++	Sp, Sm	1,2
79.	Istrachi-babool	*A. tortitis*	Mimoseaceae	T	UC	Both	++	Sp, Sm	1,2
80.	Cutuch	*A. catechu*	Mimoseaceae	T	UC	Both	++	Sp, Sm	1,2
81.	Basuti	*Adhatoda zeylarica*	Mimoseaceae	T	UC	Both	++	Sp, Sm	1,2
82.	Milk weed	*Asclepias curassacaica*	Asclepidaceae	S	UC	Both	+	Sp, Sm	1,2,6,7
83.	Silk cotton	*Bombax ceiba*	Bombacaceae	T	UC	N	+++	Sp	2
84.	Gulrnohr	*Pelonix regia*	Caesalpinaceae	T	O	N	+++	Sp, Sm	1,2,4,9

85.	Deccan Hemp	*Cannabis sativa*	Cannaba-ceaceae	T	C	Both	++	Sp	6,7
86.	Rain way creeper	*Ipomea pulchella*	Convolvul-laceae	H	UC	Both	++	Ty	2
87.	Onion	*Allium cepa*	Amaryllidaceae	H	C	Both	+++	Sm	6,7
88.	Garlic	*A. sativum*	Amaryllidaceae	H	C	Both	+++	Sm	6,7
89.	Chrysan-thamum	*Chrysanthemum* sp.	Asterceae	H	O	Both	++	R, W	6,7
90.	Radish	*Raphanus sativus*	Brassicaceae	H	C	Both	+	Ty	6,7
91.	Tumip	*Brassica rapa*	Brassicaceae	H	C	Both	++	Ty	7
92.	Sweet potato	*Ipomoea batatus*	Convolvulaceae	H	C	Both	+	R, Sm	7
93.	Pumpkin	*Citrulus vulgaris*	Cucurbetaceae	H	C	Both	+++	Ty	6,7
94.	Water melon	*Cucurbita maxima*	Cucurb etaceae	H	C	Both	++	W	7
95.	Gram	*Cicer arietinum*	Papilionaceae	H	C	Both	++	R, Sm	7
96.	Soya bean	*Glyciene max*	Papilionaceae	H	C	Both	++	R, Sm	7
97.	Garden pea	*Pisum sativum*	Papilionaceae	H	C	N	++	W	1,6,7
98.	Sesrnum	*Sesamum* sp.	Pedaliaceae	S	C	Both	++	W	1,6,7
99.	Wild rose	*Rosa macrophylla*	Rosaceae	S	C	Both	++	R,W	1,7
100.	Tomato	*Lycopersicum esculenturn*	Solanceae	H	C	N	++	Ty	1,7
101.	Pulse	*Cajanus cajana*	Papilionaceae	H	C	Both	++	Ty	1,6,7
102.	Bamboo	*Bombuse arundinaceae*	Graminae	Bush	UC	Both	+++	W, R	2,9
103.	Bamboo	*Dendrocalamus stricts*	Graminae	Bush	UC	Both	+++	W, R	2,9
104.	Ficus	*Ficus bengalensis*	Moraceae	T	UC	Both	++	Sp, Sm	2,9
105.	Terminalia	*Terminalia bellarica*	Verbinaceae	T	UC	Both	++	Sp, Sm	2,9

Note: Plant-T: Tree; S: Shrub; H: Herb. Flora; C: Cultivable; UC: Uncultivable; O: Ornamental; Wd: Weed.
Source: P: Pollen; N: Nectar; Both: Pollen & nectar. Apicultural value: Potential: +++; Major: ++; Minor: +.
Season: Sp: Spring (Feb. & march); Sm: Summer (April to June); R: Rainy (June to Sept.); W: Winter (Oct. to Jan.). Ty: Throughout the year.

variance (X^2 test) was used to test the statistical significant difference in the status of *A. dorsata* population between different habitats. The results were compared by following standard methods (Saha, 1992).

Results and Discussion

Floral status

Of Karnataka's total land area of 19.2 million hectares, a considerable amount of productive land is used for cropping and which includes both irrigated and rainfed area. Of all, nearly half of the crop field is rain dependent and multifarious. Most of the crop plants yield good quantity of pollen and nectar from the month of August to until the onset of summer. In case of irrigated areas the blooming period sometimes overlap with the rainfed crop plants too (Basavarajappa, 2002). The yield of pollen and nectar is mainly depended on the type of crop cultivated in different habitats or at the vicinity of respective habitats 1 to 10. More than eighty flowering plants (Table 2.1) which belongs to 46 families were found potential to yield pollen and nectar to *A. dorsata*. Detailed aspects on status of flora viz., apicultural value, pollen and nectar source, flora type, their occurrence during various seasons are shown in Table 2.1. About 34.4% cultivated flora (i.e., food crops, fruit yielding plants), 32.3% uncultivated flora (i.e., naturally grown tree, shrub and herb species on and nearby the cultivable land), 17.1% weed flora, 8.6% ornamental flora and remaining 7.6% both cultivated and uncultivated flora identified as rock bee forage in this part of the Deccan peninsula (Table 2.2). It is evident that, *A. dorsata* offer large flowers/ inflorescences secreting plenty of nectar and pollen producing plants. Such flowers are handled effectively by giant bee and collect large amount ($X=35.5\pm10.3$ml) of nectar loads to have energy requirements (Sihag, 2000).

Pollen calendar

The floral calendar or pollen calendar is defined as the availability of pollen plants with respect to their blooming season in the area (Verma, 1990). The floral calendar of some major plant species shown the continuous flow of pollen and nectar around the year (Table 2.3; Fig. 2.2 and 2.3). Pollen plants in the sense that, whichever the plant would benefit for the yielding of pollen and nectar, and if they are orderly placed in a year so represents the pollen calendar of that area (Basavarajappa and Krishnamurthy, 2002a). Thus, it gives an idea about the available plant resources, their flowering season and the duration of flowering. Therefore, bee flora and its pollen calendar has got much importance in apiculture. It may become a ready reckoner to the peoples of this region who are very much concerned with beekeeping and bees conservation (Table 2.1) Further, the pollen calendar differs from place to place and region to region and it is essential to have a pollen calendar of every area/region from the bee's conservation point of view. Various workers (Siddapppa Setty and Bawa, 2002; Sihag, 2002; Sihag and Koteswara Rao, 1998) have studied the pollen calendar of different geographical regions in India. Further, Basavarajappa and Krishnamurthy (2002a) have prepared a pollen calendar of B.R. Project area.

Table 2.2: Status of Gaint bee (*A. dorsata*) flora in southern part of Deccan Peninsula of Karnatak

Sl. No.	Type of Flora	Nature Flora	Species Belongs to			
			Tree	Shrub	Herb	Total
1.	Cultivated flora	Pollen	0.0	0.95	2.90	3.90
		Nectar	0.0	0.00	1.90	1.90
		Both	7.6	8.60	12.4	28.6
		Total	**7.6**	**9.55**	**17.2**	**34.4**
2.	Uncultivated flora	Pollen	5.70	0.0	0.0	5.70
		Nectar	5.70	0.0	0.0	5.70
		Both	17.1	0.0	3.8	20.9
		Total	**28.5**	**0.0**	**3.8**	**32.3**
3.	Ornamental flora	Pollen	0.00	0.95	0.00	0.95
		Nectar	0.95	0.00	1.90	2.90
		Both	0.95	2.90	0.95	4.80
		Total	**1.90**	**3.85**	**2.85**	**8.60**
4.	Weed flora	Pollen	0.00	0.00	4.80	4.80
		Nectar	0.00	0.95	1.95	1.90
		Both	0.95	5.71	3.80	10.4
		Total	**0.95**	**6.66**	**9.55**	**17.1**
5.	Cultivated and Uncultivated flora	Pollen	0.95	0.00	0.0	0.95
		Nectar	0.95	0.0	0.0	0.0
		Both	5.70	0.0	0.0	5.70
		Total	**7.6**	**0.0**	**0.0**	**7.6**
	Grand Total		46.6	20.0	33.4	100.0

Note: Values are on percent basis

Table 2.3: Flora calendar of the major plant species

Plant	Flowering period	Source
Coconut (*Cocos nucifera*)	Jan.-Dec.	PN
Adake (*Areca catecha*)	Jan.-Dec.	PN
Banana (*Musa paradisiaca*)	Jan.-Dec.	PN
Sensitive plant (*Mimosa pudica*)	Jan.-Dec	PN
Papaya (*Carica papaya*)	Jan.-Dec.	P
China rose (*Hibiscus rosa-sinensis*)	Jan.-Dec.	PN
Oleander (*Nerium indicum*)	Jan.-Dec.	PN
Waterbund weed (*Ipomea reptans*)	Jan.-Dec.	N
Heechalamara (*Pheonix vulgaris*)	Jan.-Dec.	PN
Congress weed (*Parthenium hysterophorus*)	Jan.-Dec.	PN
Drum stick (*Moringa olefera*)	Dec.-April, Aug.-Nov.	PN
Niligiri (*Eucalyptus* sp.)	Aug.-Dec., Feb.-March	PN
Mango (*Mangifera indica*)	Jan.-April	PN
Tamarind (*Tamarindus indica*)	Jan.-April	N
Paddy (*Oryza sativa*)	March-May	PN
Sunflower (*Helianthus annus*)	Oct.-Dec.	PN
Tomato (*Lycopersicum esculentum*)	Nov.-March	P
Kolavalika (*Hygrophila longifolia*)	Jan.-Dec.	PN
Nerale (*Syzygium cumini*)	March-May	P
Jawar (*Sorghum vulgare*)	March-April, Aug.-Sept.	P
Cotton (*Gossypium arboreum*)	Sept.-Dec.	PN
Maize (*Zeamays*)	March-April, Aug.-Sept.	P
Indian beech (*Pongamia* sp.)	June-Sept.	N
Jali (*Acacia* sp.)	Feb.-April	P
Silk cotton (*Ceiba* sp.)	Feb.-April	N

Note: N: Nectar; P: Pollen; PN: Pollen and Nectar.

Interestingly, the contribution of flower plants which belongs to different families are varied considerably. The percent contribution of flora from different families are presented in the Table 2.4. The Mimoseaceae family members contributed maximum (9.5%) followed by Papilionaceae (6.7%), Leguminaceae (5.6%) and the Myrtaceae, Solanaceae, Poaceae and Moraceae members contributed 3.7% energy source viz., pollen and nectar in the form of food to the rock bees in this region (Table 2.4). In general, tree species (46.6%) followed by herbs (33.4%) and shrubs (20.0%) constituted the main floral component of *A. dorsata* in this region. About 16.0% and 13.0% of the major pollen and nectar producing species respectively have been identified in this region. However, very high (71%) number of plants were of both pollen and nectar producing species (Fig. 2.3). Further, bee forage during winter

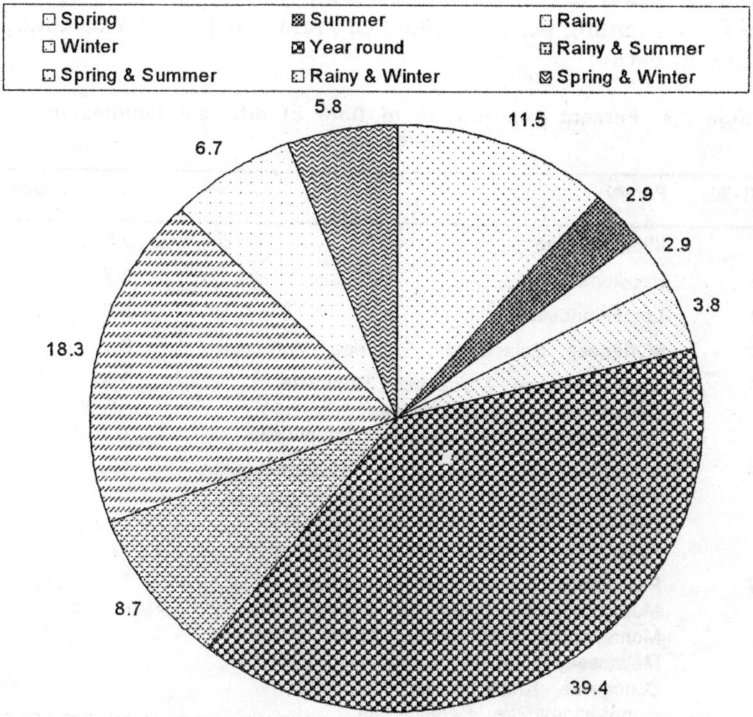

Spring
Summer
Rainy
Winter
Year round
Rainy & Summer
Spring & Summer
Rainy & Winter
Spring & Winter
5.8
11.5
6.7
2.9
2.9
3.8
18.3
8.7
39.4

Fig. 2.2: Floral availability during different seasons.

Pollen
Nectar
Both
16%
13%
71%

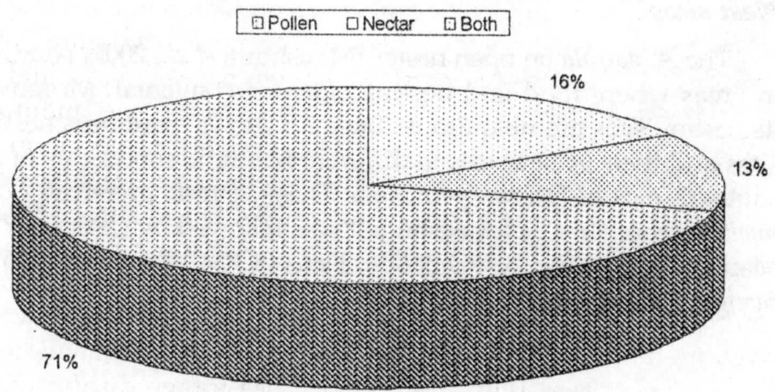

Fig. 2.3: Source of Pollen and Nectar.

(i.e., December to February) is considerably less (Fig. 2.2) compared to the remaining period of the year, and this period was called "dearth period".

Table 2.4. Percent contribution of flora of different families in the study area

Sl. No.	Family	%	Grade
1.	Mimoseaceae	9.5	1
2.	Papilionaceae	6.7	2
3.	Leguminaceae	5.6	3
4.	Myrtaceae, Solanaceae, Poaceae, Moraceae	3.7	4
5.	Caeaslpinaceae, Malvaceae, Rutaceae, Convolvulaceae Euphorbiaceae, Combretaceae, Verbenaceae, Compositae	2.8	5
6.	Amaryilidaceae, Brassicaceae, Cucurbetaceae, Anacardiceae, Araceaceae, Apocyanaceae, Oleaceae, Bombacaceae, Asclepidaceae, Asteraceae, Aacanthaceae, Rosaceae	1.9	6
7.	Palmaceae, Santalaceae, Caricaceae, Musaceae, Nymphaeaceae, Sapindaceae, Moringaceae, Punicaceae, Labiate, Fabaceae, Meliaceae, Apiaceae, Cactaceae, Graminae, Cruciferae, Rhamnaceae, Asteraceae, Cannabaceaceae, Pedaliaceae	0.9	7

Status of Giant Honeybee colonies

Nest sites

The *A. dorsata* an open nester (Moushumi *et al.*, 2002) common in areas where food and nesting sites are congenial. Moreover, its nesting sites are ubiquitous. Table 2.5 shows giant bees nesting sites and their preferences in different habitatic conditions of the southern part of Deccan peninsula. The *A. dorsata* preferred *Cocos nucifera* more (16.4%) for nesting. It was followed by *Cebia* (13.7%), Mango (12.3%), *Eucalyptus* (9.6%), Ficus (8.3%), Tamarind (5.6%), *Syzigium* (4.1%) trees (Table 2.5).

Table 2.5: Nesting site diversity of *A. dorsata* in Southern part of Deccan Peninsula

Sl. No.	Nesting site	% Preference
I.	**Tree species:**	
1.	*Cocos nucifera*	16.4
2.	*Ceiba* sp.	13.7
3.	*Mangifera indica*	12.3
4.	*Eucalyptus* sp.	9.6
5.	*Ficus* sp.	8.3
6.	*Tamarind* sp.	5.6
7.	*Syzizum* sp.	4.1
8.	*Pongamia* sp.	2.7
9.	*Acacia* sp.	2.7
10.	*Mimosa* sp.	2.7
11.	*Terminalia* sp.	2.7
12.	*Artocarpus* sp.	1.4
II.	**Natural hillock:**	
13.	Rock cliff	1.4
III.	**Man made establishments:**	
14.	Educational Institutes, Bridges, Iron pillars, Dam etc.	12.3
15.	Water tanks	2.7
16.	House	1.4

The *Pongamia, Acacia, Mimosa* tree species and the rock cliff as well as concrete surface such as the eaves of a huge buildings, water tanks, water canal bridges and houses were also used for nesting. Thus, the nesting choice of *A. dorsata* was highly diversified in this part of Deccan peninsula. The natural (active and abandoned) colonies hanging vertically with different orientations (Table 2.6). The combs fixed underneath a broad support such as a thick limb of a tree or the eaves of a man made establishments. The combs were observed at different elevations (Table 2.6) and it was varied from 19 to 70.1 meters (X=35.69±4.74 meters), fixed completely in an unconcealed sites underneath a broad support viz., walls of huge buildings, Power house at Bhadra Reservoir (B.R. Dam), educational institutes, close to human dwelling places, the limbs of big trees, at the edge of canal banks, on trees in gardens, forests and rock cliffs (Table-2.5), Fig. 2.4 indicates the status of *A. dorsata* combs with relation to elevation in different habitats. Majority of the cases the combs were at lower elevations in Maidan (Channagiri area) region compared to hill (some parts of forest areas of Bhadravathi taluk) region. Hence, the distribution and altitudinal location varied in the plains and hilly areas. Similar observations were made by Basavarajappa (1998a and b), Engel (2002). Preference of nesting sites on coconut tree (Muthuraman *et. al.*, 2002), and as high as 10 metres above the ground level is generally found as part of the defence strategy of *A. dorsata*, (Naresh Kumar and Reddy, 2002); and to support this, majority of the colonies were found above 19 metres from the ground in this region (Table 2.7 and Fig. 2.4).

Table 2.6: Orientation of combs of *A. dorsata*

Sl. No.	Direction	(%) Combs
1.	East-West	66.1
2.	South-North	17.6
3.	Northeast-Southwest	10.3
4.	Northwest-Southeast	6.0

Note: Each value is mean of 75 observations.

Morphological features of comb

The shape of the comb is more or less semicircular or cuneiform and hanging vertically from the surface of different objects. The width of the comb (which corresponds to the line of attachment) is generally greater than the length. The details on morphological features of combs are predicted in Tables 2.7 and 2.8. The comb size of 75 measured colonies in this region varied from 282.7 to 1188.8 inch (X=710.01 ± 122.9 inch) (Table 2.7). Each comb has several thousands of hexagonal thin walled fragile cells (made up of wax secreted by the wax glands of worker's abdomen) arranged in two opposite rows on a common base. The hive size varied considerably both in width and height. The average width of the comb at the point of attachment to the support was more and in the middle region the length was also more compared to lateral sides of the comb (Tables 2.7 and 2.8). The cells are mainly intended for the storage of honey and pollen specially in the upper portion of the comb while those in lower part are of brood rearing. However, there are no special cells for lodging the adults which generally keep clustering or moving

Table 2.7: Status of *A. dorstata* colonies in southern part of Deccan peninsula

Habitat	Comb Elevation (in Mtrs.)	Comb Size (in Inch)	Bee Population (X 00000)	Live combs (in Nos.)	Honey prodn. (in kg.)	Wax prodn. (in gm.)	Abondened Colonies (in Nos.)
01.	31.9	463.26	2.0846	3.4	4.21	418.3	1.1
02.	39.6	282.70	1.2722	7.5	7.3	875.8	2.4
03.	34.9	296.68	1.3350	1.3	0.81	811.16	0.6
04.	29.4	799.43	3.5974	4.2	4.81	811.16	1.5
05.	51.4	558.2	2.7118	33.3	24.4*	2667.8	7.4
06.	19.0	1008.07	4.5363	2.3	5.2	458.97	1.2
07.	29.1	397.38	1.7445	5.1	8.17	321.34	1.9
08.	24.4	1188.8	4.8636	15.4	50.46*	2207.2	4.7
09.	27.1	1414.5	6.0576	3.8	8.27	822.9	2.0
10.	70.1	691.14	3.1101	6.7	5.9	903.05	2.4
Mean	35.69	710.01	3.1313	8.3	11.95	965.92	2.52
	±	±	±	±	±	±	±
SE	4.74	122.9	0.005	3.04	4.72	260.42	0.63
CV	42.0	54.7	0.51	116.0	124.85	85.19	79.36
X^2	25.74S	23.72S	17.38S	24.08S	13.89S	23.5S	4.29NS

Note: S: Values are Significant (P>0.001; NS; Not significant.

* Average yield/l aggregate (i.e., 10 colonies/100 sq. metre area)

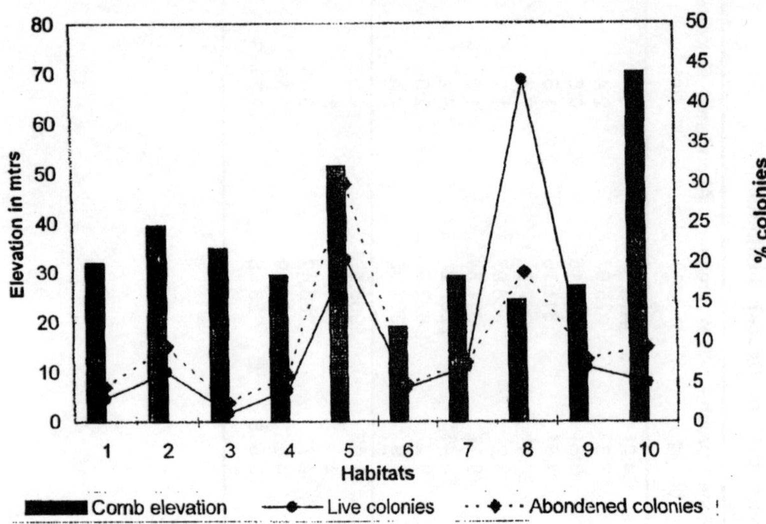

Fig. 2.4: *A. dorsata* **colonies in relation to elevation in different habitats.**

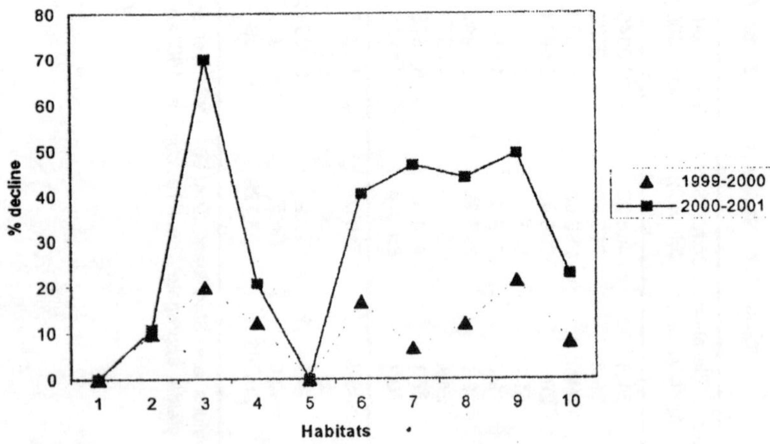

Fig. 2.5: Per cent decline of *A. dorsata* **in different habitats.**

about on the surface of the comb with an average number of 25 cells per inch^2. Thus, the comb size varied significantly (X^2=23.72, df=9, P>0.001) among different habitats. Further, the population size also varied and it was directly correlated with the size of comb (Table 2.7) among different habitats (Basavarajappa, 2002).

Table 2.8. Some Morphometric details of *A. dorsata* comb

Characters	Measurement
Total comb size (inch)	710.01 ± 122.9
Average width of the comb (cm)	55-74.12
Average length of the comb (cm)	45-51.75
Average weight of the comb (g)	323-515
Cell depth at storage part (cm)	4.5-7
Cell depth at brood part (cm)	1.8-2
Capacity of the honey storage/cell (ml)	0.8-1.5
Cells/inch^2	25

Source: Basavarajappa, 1998; Basavarajappa and Krishnamurthy, 2002.

Colonies of *A. dorsata*

Further, among the different habitats, the density of *A. dorsata* colonies were highest per km^2 area in the habitats 5 and 8 and lowest in habitat 3. However, in habitats like 1, 2, 4, 6, 7, 9 and 10 where the colonies density was normal (Table 2.6). It was observed that, there was a significant correlation between elevation of the colony and the density (r=0.35, df=9, P>0.05). Further, the colonies density was significantly (t=2.306, df=9, P>0.05) varied among the habitats and it was directly correlated with the height of the nesting place (Fig. 2.4).

Hive products of *A. dorsata*

The amount of honey stored in cells were highly varied. It was varied from 0.81 to 1.5 ml (Table 2.8). It was directly correlated with the size of the hive. Accordingly the honey production per comb varied from 0.81 to 50.46 kg. (X=11.95 ± 4.72 kg) per colony except in habitats 5 and 8 where, per one aggregate on an average 5 colonies per 100^2 metre area recorded (Table 2.7). The honey production was high (11.95 ± 4.72 kg) compared to wax (965.92 ± 260.40 gm) production (Table 2.7). Thus, the proven source of honey and wax production varied significantly in different habitats. Basavarajappa and Krishnamurthy (2002b) have recorded the honey and wax production potential in B.R. Project area of Shimoga .district (Table 2.8). The wax and honey production varied considerably in different habitats. The amount of honey production per colony and per unit area was directly correlated with the number of live colonies inhabited in the respective habitats. Moreover, similar observations was also made by Basavarajappa (1988a and b) in Maidan region of Karnataka. It is noteworthy to mention that, the honey production per colony was considerably low in Maidan area compared to hilly area of Deccan peninsula of Karnataka (Basavarajappa, 1998b) and it was below the standard norms compared to other parts of the world (Ruttner, 1998). Thus, there is a need to educate the farmers about importance of honey bees in nature and their role in maintaining the ecological balance. Further, more emphasis should be given to enrichment of bee flora of this region so as to encourage the movement and conservation of existing bee flora and bee colonies in this area.

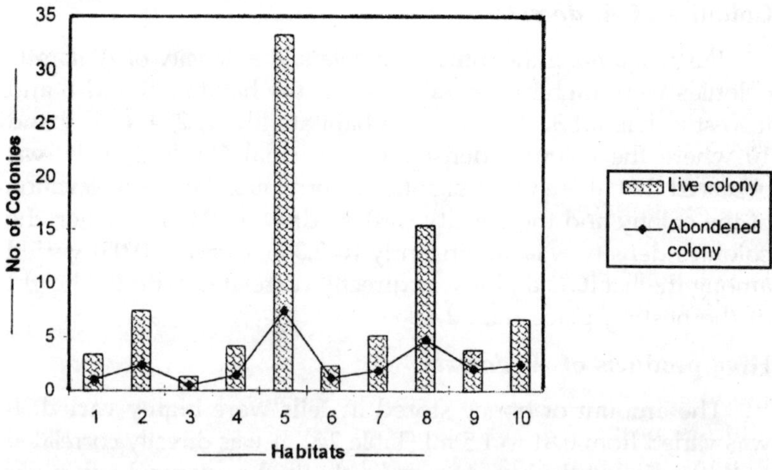

Fig. 2.6 Status of live and abondened colonies in southern part of Deccan Peninsula.

Colonies decline

More than twenty nesting places were observed as potential for the hunting in the study area, and were classified as partially hunted and severely hunted sites. In hunted sites, the colonies were burnt, fire torched frequently for honey harvest and other reasons by local farmers/villagers. Honey hunting is being practised by forest dwellers, tribal (Anonymous, 1998), honey hunters, villagers (Basavarajappa, 2001), to gather honey by following different indigenous methods (Basavarajappa, 1998a and b). Further, the honey gatherers who use smoke, fire torching to drive and burn some of the bees, cut the combs, squeeze the honey and melt the combs for bees wax (Anonymous, 1998). However, the nest site destruction, hive hunting, fire torches to drive away and burn bees on the combs, intensive agricultural operations include cultural practices and other human activities have contributed the severed depletion of natural colonies in habitats 2, 3, 4, 6, 7, 8, 9 and 10 (Fig. 2.5). The honey hunters harvest not only storage areas but the entire comb including brood. This is devastating for colony reproduction and survival (Thapa *et al.*, 2000). Reasons for this was not understood properly. However, it was suspected that, the prevailed conditions might have not disturb the bees during nesting and provided congenial conditions to stay for longer period. In habitats 2, 3, 4, 6, 7, 8, 9 and 10, where the activities viz., intensive agriculture operations in agricultural areas includes cultural practices, harvesting of coconut and areca nut, removal of old leaves led to the destruction of combs on tree species. Basavarajappa (1998a) have recorded the destruction of *A. dorsata* combs on coconut (56.8%), Mango (23.2%), Tamarind (5%) and natural vegetation viz., *Eucalyptus* (4.5%) and other tree species such as *Acacia, Syzigium, Bombax,* Neem, *Dalbergia* and other man made establishments where the comb destruction was about 10.5%. Thus, harvesting of honey from the rock bee hives in a destructive fashion poses the appearance of abandoned colonies (Shetty and Murali, 1998) in this region (Table 2.7 and Fig. 2.6). Hence, there is a need to educate farmers, honey gatherers about the scientific honey hunting techniques as devised by Sattagi *et al.* (2002) and to create awareness on the rock bees in agri-horticultural ecosystems (Basavarajappa, 1998a and b; Sihag and Chaudhary, 2002; Neupane, 2002; Muthuraman *et al.*, 2002) and to know their pollination potential (Solomon raju, 2002). This kind of activities would help us to conserve the rock bee population in this new millennium (Sihag, 2002).

Table 2.9. Wax and Honey production potential in B.R. Project area

Habitats	Wax/comb (g.)	Wax/acre (kg.)	Honey/comb (ml.)	Honey/acre (lt.)
I	308.5	6.480	328.5	6.900
II	308.3	7.400	306.2	7.350
III	345.0	4.140	375.0	4.500
IV	314.0	14.130	333.3	15.000
V	330.8	15.220	356.5	16.400
Mean	321.32	9.474	339.9	10.030
±	±	±	±	±
SE	7.20	2.19	11.86	2.37

Note: Each value is a mean of 16 observations.
Source: Krishnamurthy, 2002.

Pests and Predators on comb

Besides human interference, the pests and predators also caused havoc to the honey bee population in this region. Krishnamurthy (2002), Basavarajappa and Krishnamurthy (2002b) have recorded more than seven species as serious enemies to rock bee *A. dorsata* in some parts of Deccan Peninsula. The wasp *(Vespa* sp.), Greater wax moth *(Galleria mellonella)* and, the Bee eater *(Merops orientalis)*, Purple Sun Bird *(Nectarinia asiatica)*, Blue Rock Pigeon *(Columba livia)*, House Sparrow *(Passer domesticus)*, Fly catcher and Wood pecker were found to cause considerable damage to Giant bee colonies in habitats 1, 3, 6 and 9. Of all the infestation occurred due to the greater wax moth *G. mellonella* was very high (35.2 ± 3.14) in this area (Table 2.10). Hanumantha Swamy *et al.* (2002) have studied the comparative biology of greater wax moth on combs of different species of honey bees in southern part of Karnataka. However, record of the infestation of wax moth on giant bee colonies provided an insight about the population decline of *A. dorsata*. As a result a considerable number of colonies became victimized and deserted due to the absconding of bees. These activities affected the population density and production of hive products in this region. Hence, the predation as well as uncontrolled human activities have interfered and played a pervasive role in reducing the colonies density and aggregates of giant honeybee on certain trees, cliff and buildings in habitats 1, 3, 6 and 9.

Table 2.10. Greater wax moth *G. mellonella* infestation on *A. dorsata* comb

Habitat	Infestation (%)
I (Human dwelling places)	30.0
II (Man-made establishments)	58.3
III (Agri-horticultural area)	25.0
IV (Forest area)	44.4
V (Rainfed area)	18.7
Mean	35.28
+	+
SE	7.14

Note: Each value is a mean of 16 observations.

Bee Poisoning

Furthermore, the honey bee population is seriously harmed by pesticides (Jagadish *et al.*, 2002). Nicola Bradbear and Reddy (1998) have reported the extensive use of pesticides for the control of various pests in different agro-ecosystems of Karnataka. The commonly used chemical insecticides are viz., Aldrin, Carbofuran, Dimethoute, Endosulfan, Fenitrothion, Fenvalerate, Parathion, Phorate, Methyl Bromide, Phosphamidon, Monocrotophos etc., in the form of granules, dustings, wetable-powder, and liquid sprays in various agro-ecosystems of Karnataka (Basavarajappa, 2001). Large-scale use of rainfed crops leads to killing of large number of bees (Anitha and Shivaram, 2002; Venkatareddy and Reddy, 2002). In view of these observations, there is a need to create an awareness among farmers and bees importance in agriculture (Subbalakshmi and Lokanadhan, 1998; Shivakumar and Gowda, 2002). Thus, farmers need to know ways of avoiding pesticide-killing of wild bees in this part of Deccan peninsula. Some suggestions are made to reduce the poisoning effect on *A. dorsata*.

1. Selective use of pesticides in agriculture fields which are less harmful to bees.
2. Use of pesticides before the on set of wild bees intensive foraging activity.
3. Avoiding more use of chemical pesticides in the field.
4. Encouraging use of natural products.
5. Following eco-friendly way of controlling pest population.

6. Discouraging use of chemical pesticides near by honey bee colonies.

Conclusion

As *A. dorsata* provides non-consumptive benefits of conserving botanical resources (Nikam, 2002) by dealing with primary functions of the ecosystem, it is necessary to know its status so as to enact measures to restore its population at rural areas (Shinkarankutty, 2002), and in other habitats. In the recent past, due to the fast destruction of the bee habital and floral resources and poisoning of the bee environment, a large number of bee species are under the threat of extinction (Sihag, 2002). This loss of bee diversity would affect the reproductive success of several plant species in the cultivated fields (Fabunmi, 2002). In the present century, therefore, there is a dire need to conserve the bee pollinators. The present study emphasized the need to rely on conservation of bee forage and bee environment and in-turn to conserve the bees.

Acknowledgements

Author express his sincere gratitude to Dr. K. Chidhananda Gowda, Vice-Chancellor, and Sri. Praveen Chand Pandy Registrar, Kuvempu University, Shankaraghatta for their encouragement. Thanks are also due to the Chairman, Dept. of Applied Zoology, K.U.B.R. Project for providing necessary facilities and support.

REFERENCES

Anitha, M. and Shivaram, V. 2002. Studies on the chronic intoxication of selected insecticides to the Indian honeybee *Apis cerena indica* F. Proc. 6th AAA Inter. Con. and World APIEXPO-2002, Bangalore. p. 195.

Anonymous, 1998. Existing Apicultural Practices within *Karnataka. Mission Report by AIBA for FAO of UN*, Pune. pp. 6-7.

Basavarajappa, S. 1998a. Studies on the depletion rate of natural hives of rock bee *Apis dorsata* Fabr. in Maidan area of Karnataka. FAO-Workshop. Sustn. Beekeeping Develop. AIHF (Apiexi' 98), Dharwad. p. 40.

Basavarajappa, S. 1998b. Status of natural colonies of *Apis dorsata* in Maidan region of Karnataka. *Indian Bee J.* 60(3): 143-146.

Basavarajappa, S. 2001. Chemical insecticidal pollution in agro-ecosystems of Belagere area, Karnataka. (Unpublished data).

Basavarajappa, S. 2002. Floristics of *Apis dorsata* Fabr. in an agri-horticultural ecosystem of Belalagere, Karnataka. Proc. 89th ISC., Lucknow, p. 24.

Basavarajappa, S. and Krishnamurthy, D.V. 2002a. Status of flora of wild bee *Apis dorsata* F. in B.R. Project area, Karnataka, India. 90th ISC., Bangalore. (Communicated).

Basavarajappa, S. and Krishnamurthy, D.V. 2002b. Status of wild bee *Apis dorsata* F. in B.R. Project area, Karnataka, India. 90th ISC., Bangalore. (Communicated).

Engel, M.S. 1990. The taxonomy of recent and fossil honey bees (Hymenopter: Apidae; *Apis). J. Hym. Res.* 8: 165-169.

Engel, M.S. 2002. The honey bees of India, Hymenoptera: Apidae. *JBNHS*. 99(1): 3-7.

Fabunmi, T.Z. 2002. Tackling food crisis and rural poverty with honey bee resource. Proc. 6th AAA Inter. Con. and World APIEXPO-2002, Bangalore. pp. 199-200.

Gambel, J.S. 1967. *Flora of the Presidency of Madras*. (2nd ed.) vol. I and II. Botanical Survey of India. pp. 1-500.

Hanumantha Swamy, B.C., Rajagopal, D. and Gowda, B.L.V. 2002. Comparative biology of greater wax moth, *Galleria mellonella* on combs of different species of honeybees. Proc. 6th AAA Inter. Con. and World APIEXPO-2002, Bangalore, p. 125.

Jagadish, K.S., Gowda, A.N.S., Govindan, R. and Shantala, J. 2002. Beekeeping and insecticidal pollution-An appraisal of the current scenario. Proc. 6th AAA Inter. Con. and World APIEXPO-2002, Bangalore. p. 181.

Kamath, U.S.. 1991. *Karnataka State Gazetteer*, Govt. Press, Bangalore.

Koteswara Rao, S.R. 1998. Bee flora of Chinanpalli, Andhra Pradesh. Indian Bee J. 60(3): 162-164.

Krishnamurthy, D.V. 2002. Status of giant honeybee *Apis dorsata* F. colonies in B.R. Project area. M.Sc., Project Report. Kuvempu Univ. B.R. Project. pp. 31-36.

Moushumi, S.S., Juergen, T., Stefan, F., Christian, W. and Gadagkar, R. 2002. The open nesting honeybee: A very special bee. *Proc. 6th AAA Inter. Con. and World APIEXPO-2002*, Bangalore. p. 38.

Muthuraman, M., Suresh, D.D., Thangavelu, G.K. and Srinivasan, G. 2002. Rock bees in coconut gardens. Proc. 6th AAA Inter. Con. and World APIEXPO-2002, Bangalore. p. 44.

Naresh Kumar, G.M. and Chandrasekhara Reddy, C. 2002. Nest density, dimension and distribution of *Apis dorsata* F. in relation to height and direction. Proc. 6th AAA Iner. Con. and World APIEXPO-2002, Bangalore. p. 52.

Neupane, K.R. 2002. Foraging preference of honeybee species to selected horticultural crops. Proc. 6th AAA Iner. Con. and World APIEXPO-2002, Bangalore, p. 69.

Nicola Bradbear and Reddy, M.S. 1998. Existing Apicultural Practices within Karnataka. Mission Report. FAO & UN, Pune. pp. 1-20.

Nikam, T.B. 2002. A need to conserve honeybees for sustainable agro-horticultural and forestry development. Proc. 6th AAA Iner. Con. and World APIEXPO-2002, Bangalore. p. 211.

Otis, G.W. 1991. A Review of the diversity of species within *Apis* Diversity in the Genus. Westview Press, Boulder. pp. 29-49.

Rathore, R.R.S. and Kapur, S. 2002. Impact of Beekeeping activities and bee diversity. Proc. 6th AAA Iner. Con. and World APIEXPO-2002, Bangalore. p. 25.

Ruttner, F. 1998. *Biogeography and Taxonomy of* honey bees. Springer-Verlag, Berlin. pp. 75-80.

Saha, T.K. 1992. Biostatistics in Theory and Practice. Emkay Publications, Delhi. pp. 5-104.

Sattagi, H.N., Kulkarni, K.S., Rajasekhar, D.W. and Kambrekar, N. 2002. Techniques to remove honey comb from rock bees *Apis dorsata* (F.) colonies for honey extraction. Proc. 6th AAA Inter. Con. and World APIEXPO-2002, Bangalore. p. 52.

Shankarankutty, 2002. Beekeeping for Rural. Proc. 6th AAA Inter. Con. and World APIEXPO-2002, Bangalore. p. 216.

Shivakumar, H.R. and Gowda, A.N.S. 2002. Eco-friendly approaches to overcome the problem of bee poisoning by pesticides. Proc. 6th AAA inter. Con. and World APIEXPO-2002, Bangalore. p. 194.

Shetty, S.R. and Murali, K.S. 1998. Extractions of honey from rock bee *(Apis dorsata)* by Soliga tribes in B.R. Hills. III Cong. of IUSSI Indian Chap. Nat. Symp. Div Soc. Insects and Arthropods and Func. of Eco. Mudageri, Karnataka. p. 34.

Siddappa Setty, R. and Bawa, K.S. 2002. Characteristics of honey resources in a tropical forest: productivity and extractions of *Apis dorsata* honey in Biligiri Rangayan hills wildlife sanctuary, India. Proc. 6th AAA Inter. Con. and World APIEXPO-2002, Bangalore. p. 54A.

Sihag, R.C. 2000. Ecobiology of the giant honey bee (*Apis dorsata)* in semi-arid sub tropical climates of India. Proc. 4th AAA Inter. Con. Kathmandu, Nepal. pp. 50-52.

Sihag, R.C. 2002. Conservation of bee diversity–challenges in the new Millennium. Proc. 6th AAA Inter. Con. and World APIEXPO-2002, Bangalore, p. 190.

Sihag, R.C. and Chaudhary, N. 2002. Diversity, Abundance, Foraging behaviour, activity duration, number of loose pollen grains and foraging efficiency of different pollinators visiting Onion *(Allium cepa* L.) blossoms. Proc. 6th AAA Inter. Con. and World APIEXPO-2002, Bangalore. p. 55.

Singh, Y. 1980. Beekeeping in Uttar Pradesh - a Review. Proc. Int. Con. Apic. Trop. Climate. 2:211-226.

Solomon raju, A.J. 2002. The rock honeybee *Apis dorsata* and its pollination potential. Proc. 6th AAA Inter. Con. and World APIEXPO-2002, Bangalore. p. 37

Subbalakshmi and Lokanadhan, K. 1998. Strategies to increase crop and honey production through crop management. III Cong. of IUSSI Indian Chap. Nat. Symp. Div. Soc. Insects and Arthropods and Func. of Eco. Mudageri, Karnataka. p. 26.

Thakar, L.V. and Tonapi, K. V. 1961. Nesting behaviour of Indian honey bees. I. Differentiation of worker, queen and drone cells of the comb of *Apis dorsata* Fabr. Bee World. 42: 61-67 and 71.

Thapa, R., Wongsiri, S., Oldroyd, B.P. and Prawan, S. 2000. Migration of *Apis dorsata* in northern Thailand. Proc. 4 the AAA Inter. Con. Kathmandu, Nepal. pp. 39-43.

Venkatareddy, E. and Reddy, C.C. 2002. Mortality of honeybee pollinators visiting insecticide treated crops. Proc. 6th AAA Inter. Con. and World APIEXPO-2002, Bangalore. p. 184.

Verma, L.R. 1990. Beekeeping - an integrated mountain development. Oxford and IBH Publishing Co. Ltd. New Delhi. pp. 57-80.

3

Relative Food Preference for Various Mulberry Varieties in *Spodoptera litura* (F) and *Diacrisia obliqua* Walk

J.S. Tara and Baldev Sharma

P.G. Department of Zoology, University of Jammu, Jammu-Tawi-180006

ABSTRACT

Experiments conducted on six mulberry varieties viz. Sujanpur, Berhampore, local, Chack-majra, KNG and Goshorami for *Spodoptera litura* suggests that on the basis of percentage of larvae survival, larval period (12-20 days), percentage of larvae pupated (53.33%) and their growth index values (3.73), the Sujanpur variety appears to be the most preferred and KNG the least preferred varieties by the larvae, while in case of *Diacrisia obliqua,* on the basis of total fresh leaves consumed by a single larva during the entire period of experiment (170.94 mg.) again suggest that the Sujanpur variety is the most and Chack-majra (138.64 mg.) the least preferred varieties. Thus, on the basis of their preference the authors recommends the cultivations of pest resistant i.e., the Chack-majra variety for cultivation in the various mulberry nurseries in Jammu region of J & K State.

Key Words: *Spodoptera litura, Diacrisia obliqua,* Mulberry varieties.

Introduction

Various methods have been used to ascertain the comparative preference for a particular host plant or its variety by an insect pest attacking it. Katiyar *et al.* (1975), Lal and Mukharji (1978, 1979), Bassi (1972) Pandey *et al.* (1968), Bhardwaj and Kushwaha (1976) and Gupta *et al.* (1979). Perumal *et al.* (1972) has used the overall life cycle development including larval period, percentage of larval survival and pupation, growth index values and percentage of emergence etc. as indices to pin-point the most preferred host. Katiyar *et al.* (1975), Lal and Mukharji (1975) studied the relative preference of larva of *D. obliqua* by using disc method, while Bassi (1972) have studied the food preference in both *S. litura* and *D. obliqua* using cauliflower and raddish leaves respectively.

So far no attempt has been made by any worker to study the food preference in both *S. litura* and *D. obliqua* which have been recorded as serious pests of mulberry in Jammu region. Sharma and Tara (1985), and (1988) have compared the consumption and utilization of mulberry leaves by the larvae of both these pests and have studied the effect of different mulberry varieties on the larval survival, growth and pupation in *S. litura* in the area.

However, Chitra and Rao (1996) have made some observations by using plant extracts on the consumption and utilization of food by *Spodoptera litura* whereas Mishra and Srivastava (1998) have made a comparison of food consumption and utilization by *S. litura* between mulberry and castor and found the castor to be preferred food than mulberry.

Materials and Methods

For *Spodoptera litura,* experiments were conducted on six mulberry varieties viz., Sujanpur, Berhampore, Goshorami, Chackmajra, KNG and local. Thirty newly hatched larvae were reared in batches of ten forming three replication on each variety. Each set of larvae was reared separately. Fresh mulberry leaves were given to the larvae daily in the morning. Data regarding the growth and development were calculated by recording length and weight of 5 randomly selected larvae from each replication at different intervals viz. 4th, 8th, 12th and 16th days after hatching. Relative percentage of larval survival between 1-4 days, 5-8 days, 9-12 days 13-16 days, 17-20 days and 21-22 days after hatching was also recorded. Relative duration of larval period, relative duration of pupal period, average pupal period and weight of pupae were also recorded. Relative growth index value was calculated according to the formula adopted by Srivastava

(1959), Pandey *et al.* (1968). Bhardwaj and Kushwaha (1976) and Lal and Mukharji (1978) i.e. by dividing the percentage of larvae pupated (n) with the average larval period (AV).

For *Diacrisia obliqua,* in a rearing tray known number of larvae were released in the centre at evening. At equal distance from the centre known amount of mulberry leaves (by weight) of Sujanpur, Chack-majra and local varieties were kept. Next morning (i.e. after a gap of 11-12 hours) all the larvae from rearing tray were collected back and kept on fast till evening. The left over leaves of mulberry varieties were weighed separately to find out the amount of leaves eaten during the night.

Simultaneously in the another tray under similar conditions control experiment without larvae was also carried and the leaves of the mulberry varieties kept in this tray were also weighed in the morning to estimate the amount of water loss. This experiment was continued till 9 days and from this data the weight of fresh leaves of each variety consumed daily by one larva was calculated. Preference of larvae for a particular variety was assessed on the basis of total amount of fresh leaves consumed by one larva during the entire period of the experiment.

Results and Discussions

After conducting the experiments as discussed, the data so obtained have been discussed in Tables 3.1-3 (Figure 3.1 and 3.2). A look at the Tables 3.1 and 3.2 reveals that the varietal preference by *S. litura* to various mulberry varieties planted in Jammu division suggests that on the basis of percentage of larvae survival, larval period, percentage of larvae pupated and their growth index values the Sujanpur variety shows shorter larval period (12-20 days), highest percentage of survival of larvae, highest percentage of pupae formation (53.33%) and highest growth index value (3.73) compared to all other varieties under comparison. KNG variety on the other hand is the least preferred variety as its feeding larvae shows combination of factors, such as longer larvae period (14-21 days), lowest percentage of larval survival (Fig. 3.1) lowest percentage of pupae formation 26.30% (Fig. 3.2) and lowest growth index value (1.47).

Similarly, Perumal *et al.* (1972) has also considered the overall life cycle and development including larval period, percentage of

Table 3.1: Larval survival of *Spodoptera litura* on different days after hatching when fed on different varieties of mulberry

Varieties of mulberry	Percentage of survival at various age intervals (in days)					
	1-4	5-8	9-12	13-16	17-20	21-22
Sujanpur	93.33	86.66	73.33	40.00 (12th day)	20.00	Pup.
Berhampore	90.00	80.00	70.00	56.66 (15th day)	43.33	Pup.
Goshorami	80.00	70.00	63.33	53.33 (15th day)	30.00	3.33
KNG	63.33	60.00	46.66	30.00 (14th day)	20.00	3.33
Chack-majra	73.33	63.33	60.00	50.00 (14th day)	30.00	Pup.
Local	80.00	63.33	63.33	36.66 (14th day)	16.00	Pup.

Note: Figures in brackets indicate the starting day of pupation.

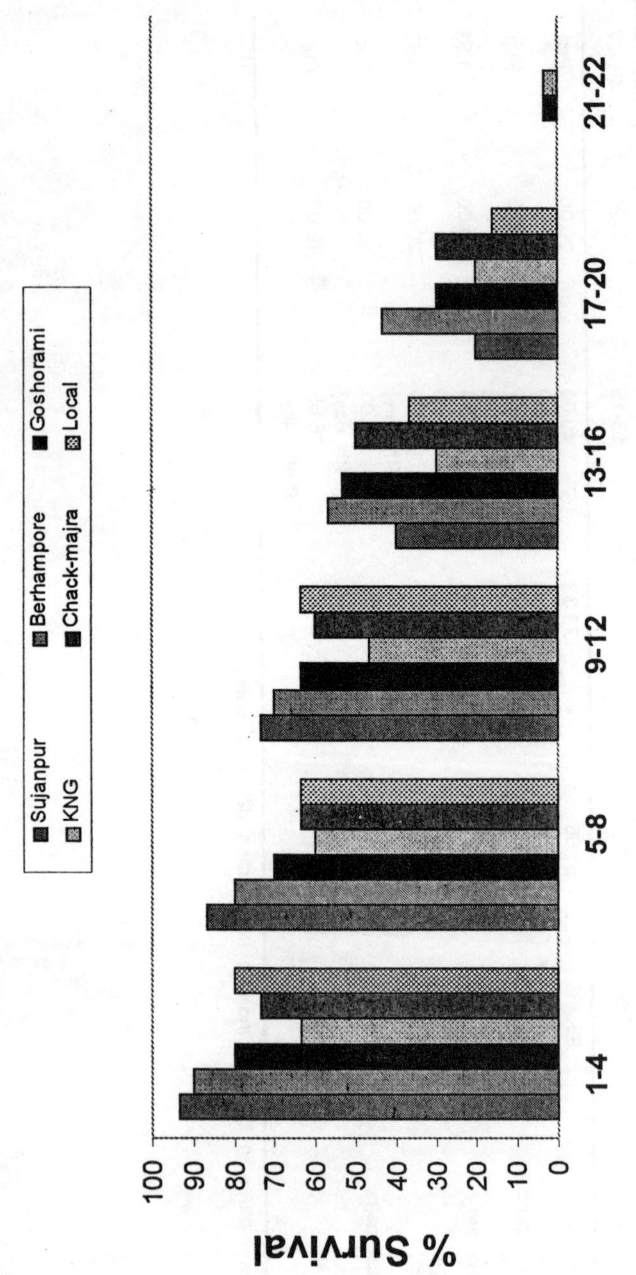

Fig: 3.1: Day After Hatching

larval survival and pupation, growth index values and percentage of emergence etc. as indices to pin-point the most preferred host.

So, on the basis of these parameters, various varieties of mulberry can be arranged in descending order of preference by *S. litura* as Sujanpur, Berhampore, local, Chack-majra, Goshorami and KNG.

Table 3.2: Average total larval period, percentage of pupation and growth index in *Spodoptera litura* feeding on varieties on mulberry

Mulberry varieties	Range of larval period (in days)	Average larval period (in days) Av.	Percentage of larval Pupation (n)	growth index (n/Av)
Sujanpur	12-20	15.81	53.33	3.73
Berhampore	15-20	17.81	46.66	2.61
Chack-majra	14-19	17.08	40.00	2.34
Local	14-20	16.75	40.00	2.38
Goshorami	15-21	18.00	40.00	2.22
KNG	14-21	17.87	26.30	1.47

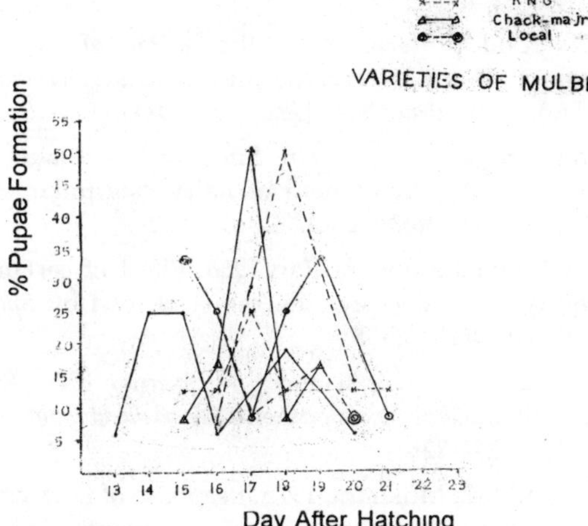

Fig. 3.2

For *D. obliqua,* the authors observed that on the basis of the total fresh leaves consumed by a single larva during the entire period of experiment, the Sujanpur variety appears to be most preferred variety as the larva consumed maximum (170.94 mg.) of leaves of this variety and is followed by local variety, (leaves consumption being 145.66 mg.), while the Chack-majra variety is the least preferred variety showing minimum leaf consumption by one larva (138.64 mg.).

So, from the foregoing observations, the authors conclude that Chack-majra variety is the most resistant variety (as it is least preferred by *S. litura* and *D. obiqua* and hence, it is recommended for cultivation in the mulberry nurseries in Jammu Division under Sericulture Development Department of J & K State. The same variety i.e. Chack-majra has also been recommended by these authors (Sharma and Tara, 1988) elsewhere.

Acknowledgements

Authors are thankful to Prof of. Y.R. Malhotra, the then Head of the Department of Bio-Sciences, University of Jammu, Jammu, for his whole hearted interest and facilities provided for the work. Authors are also thankful to J & K State Sericulture Development Department for providing financial assistance.

REFERENCES

Bassi, Asha 1972. Food preferences in the larvae of two moths: *Spodoptera litura* F. (Fam. Noctuidae and *Diacrisia obliqua* Walk. (Fam. Arctiidae). *J. Bombay. Nat. Hist. Soc.*, 71(1): 161-163.

Bhardwaj, S.C. and K.S. Kushwaha 1976. Effect of host plants on the larval and post-larval development of hairy caterpillar. *Amscata lineola* Fab. *Ind. J. Ent.* 38(3): 230-232.

Chitra, K.C. and S. Ramakoteshwara Rao 1966. Effect of certain plant extracts on the consumption and utilization of food by *Spodoptera litura. J. Insect. Sci.* 9(1): 55-58.

Gupta, B.M., R.K. Dabi, H.C. Gupta and S.K. Sharma 1979. Effect of host plants on the larval development *Diacrisia obliqua* Walk. *J. Entomol. Res.* 3(2): 224-226.

Katiyar, OP., L., Lal and S.P., Mukharji, 1975. Response of newly hatched caterpillar of *Diacrisia obliqua* to certain host plants. *Ind. J. Ent.* 37(1): 57-59.

Lal, Lakshman and S.P., Mukharji 1978. Growth potential of *Diacrisia obliqua* Walker in relation to certain food plants. *Ind. J. Ent.* 40(2): 177-181.

Lal, Lakshman and S.P., Mukharji 1979. Studies on the relative food preference of *Diaccrisia obliqua* Walker (Arctiidae: Lepidoptera). *Zoologische Beitrage* 25: 19-21.

Mishra, S.K. and R.P., Srivastava 1998. Comparative studies on growth and development of Tobacco caterpillar, *Spodoptera litura* (Fab.) on mulberry and castor. *J. Insect. Sci.* 11(1): 69-70.

Pandey, N.D. Daya Ram Yadav and T.P.S. Teotia 1968. Effect of different food plants on the larval and post larval development of *Diacrisia obliqua* Walker. *Ind. J. Ent.* 30(3): 229-234.

Perumal, R.S., K., Asaf Ali and T.R., Subramaniam 1972. Effects of feeding different host plants on the larval and post larval development of *Pericallia ricini* F. *Madras Agric. J.* 59: 324-328.

Sharma, Baldev and J.S., Tara 1984. Leaf yield studies of three mulberry varieties infested with *Bautocera rufomaculata* De. Geer (Coleoptera: cerambycidae) in Jammu division, *J & K State. Zool. Orient.* 1(1): 5-8.

Sharma, Baldev and J.S. Tara 1985. Insect pests of mulberry plants (*Morus* sp.) in Jammu region of Jammu and Kashmir State. *Ind. J. Seri.* 24(1): 7-11.

Sharma, Baldev and J.S. Tara, 1988. Comparison of consumption and utilization of mulberry leaves in two moths, *Spodoptera litura* (F) and *Diacrisia obliqua* Walker. *Ind. J. Ent.* 50(3): 336-342.

Srivastava, B.K. 1959. Growth potential of *Laphygma exgua* in relation to certain food plants. *Madras Agric. J.* 46(7) : 255-259.

4

Micro-arthropod Abundance and its Effect on Leaf Litter Decomposition and Nitrogen Release Pattern in a Tropical Deciduous Forest

M.C. Gupta and A. Wanganeo*

Remote Sensing Application Centre (M.P.C.S.T), Bhopal - India
**Environmental Impact Assessment Unit, Department of Limnology, Barkatullah University, Bhopal - India*

Introduction

Decomposition of plant litter, in nature occurs with the active involvement of micro-organisms and invertebrates as mediated by various abiotic environmental factors. These factor play a significant role in mass loss of litter (Witkamp and Drift, 1961; Witkamp and Crossley, 1966) and dynamics of mineral nutrients (Seastedt, 1984, Reddy, 1989). Soil and litter fauna regulates the decomposition and nutrient release process by their active involvement. Besides this fungi and bacteria are also responsible for most of the organic matter breakdown, together with protozoans, nematodes, annelids and arthropods. The soil faunal elements as such, act as catalysts, enhancing energy and nutrient influxes. The feeding activity of micro-arthropods increase the surface area of the substrate by exposing more surface area to microbial attack. Micro-arthopods are known to control the distribution, abundance and activity of litter fungi and bacteria, thus affects the decomposition process. Micro-arthopods while grazing serves to disperse propagules of the grazed species and their excretions may control the rate and proportion of nutrient return to the grazed site (Lussenhop, 1992).

The purpose of this investigation was to evaluate the role of litter and soil micro-arthropods in regulating decomposition and nitrogen release process. The present study was conducted in a tropical deciduous Teak bearing forest in central India, within, Vindhyan ranges (22° 5′-22° 12′ N and 77° 17′-77° 23′E).

Material and Methods

The study area has two types of forest stands viz., mixed teak forest and teak forest. Freshly fallen leaf litter of *Tectona grandis, Terminalia tomentosa, Buchanania latifolia, Diospyros melanoxylon* were collected separately, each, besides mixed leaves of other species occurring in the study area, during March-April 1995 and air dried. The litter bag method of Crosseley and Hoglund (1962) was followed for sampling of litter bags at different stages of decomposition. The nylon net (2 mm pore size) bags each of 10 cm × 10 cm were filled with 5 g air dried leaf litter and placed randomly on the soil surface underneath the canopy of the non-degraded forest sites (teak and mixed teak forest) of the catchment area in the month of first week of June 1995. Sub samples of the air dried litter were oven dried at 80°C to calculate the initial oven dry weight.

A similar set of experiment was set for the exclusion of arthropods. Litter bags of the species were treated with Phorate - G (insecticide, repellent) to exclude the arthropods. Three bags of each leaf litter species were recovered from both treated and non-treated (control) sites, seasonally (Rainy, Winter and Summer). Total three recoveries were made. Last recovery was made in June 1996.

Modified method of Macfadyen (1961) using galvanised sheets was used, one for each sample (litter bag). The collected arthropods were kept separately in vials containing 70% alcohol. The left over material was thoroughly searched under the binocular to count the densities of each group of micro arthropods both in treated (with Phorate-G) and non-treated (Control) conditions. The results were expressed as number of arthropods/ m^2. After extraction was over, the dry weight (80°±5 for 48 hr.)

of the leaf litter samples after carefully removal of the soil particles, were taken for mass losses. The leaf litter samples were then ground to powder for chemical analysis. Nitrogen concentration was estimated by micro-kjeldhal method as given in Jackson (1967).

The mean Relative Decomposition Rate (RDR) was calculated after Singh and Shekhar (1989), using formula

$$RDR\ (mg/g/d) = In\ (W_1 - W_0)/(t_1 - t_0)$$

where

W_0 = Weight of leaf litter present at time t_0

W_1 = Weight of leaf litter present at time t_1

t_1-t_0 = Sampling intervals (days).

The above equation gives the values of relative weight change rates; the negative sign of the values, however, is ignored.

The daily instantaneous decay rate (k) of leaf material for the entire study period was calculated through the negative exponential decay model of Olson (1963);

$$W_1/W_0 = expt(^{-kt})$$

where

W_0 = initial weight and W_1 = weight remaining after time t. Further, as suggested by Olson (1963) the time required for 50, 95 and 99% weight loss was calculated as: t_{50} = 0.693/k, t_{95} = 3/k and t_{99} =5/k.

Result

Seasonal densities of litter arthropods are presented in Tables 4.1 and 4.2. The litter arthropods were grouped into three categories (1) mites (2) collembola and (3) other groups. The densities of the mites are very high compared to collembolans and other group at both the sites and in all the three seasons (Rainy, Winter and Summer). During rainy season maximum density of the arthropods population were recorded in all the four species and in the mixed leaf litter at both the sites. Maximum density of arthropods population (2573-2999 m^{-2}) was found in non-treated leaf litter bags of *Tectona grandis* at two sites, during rainy season followed by *Terminalia tomentosa* (1132-

2123m^{-2}) during rainy season at both the sites. The population density in rest of the species i.e. *Buchnania latifolia* and *Diospyros melanoxylon* and mixed leaf litter varied between 444-2243 m^{-2}. During this period population density of arthropods were higher at mixed teak forest site than in teak forest sites. The population density of mites (*acarina)* varied between 68% to 95% of the total micro-arthropods population in different leaf litter species during rainy season.

Table 4.1: Micro-arthropod Population in Treated and Non-Treated Decomposing Leaf Litters in Mixed Teak Forest

Arthropods (no./m^2)	Non-treated (Control)			Treated (Phorate-G)		
	Rainy	Winter	Summer	Rainy	Winter	Summer
Tectona grandis						
Mite	2200	1280	666	367	206	88
Collembola	133	24	44	467	44	22
Others	240	48	133	33	22	22
Total	2573	1352	843	887	272	132
Buchanania latifolia						
Mite	911	400	33	289	122	228
Collembola	222	105	88	133	66	0
Others	200	128	67	107	22	22
Total	1333	633	448	529	198	310
Terminalia tomentosa						
Mite	2000	980	400	633	222	110
Collembola	0	66	44	0	0	0
Others	133	22	22	167	66	0
Total	2123	1068	466	800	288	110
Diospyros melanoxylon						
Mite	1444	688	310	667	185	67
Collembola	267	180	200	0	22	22
Others	177	66	0	33	66	0
Total	1888	934	510	700	273	89
Mixed litter						
Mite	1889	1011	466	443	205	44
Collembola	111	122	44	0	0	22
Others	224	66	22	33	22	0
Total	2243	1199	532	466	227	66

Table 4.2: Microarthropod Population in Treated and Non-treated Decomposing Leaf Litters in Teak Forest

Arthropods (no./m²)	Non-treated (Control)			Treated (Phorate-G)		
	Rainy	Winter	Summer	Rainy	Winter	Summer
Tectona grandis						
Mite	2844	1188	510	133	166	88
Collembola	44	66	66	0	0	44
Others	111	44	22	0	0	0
Total	2999	1298	598	133	166	132
Buchanania latifolia						
Mite	355	380	110	267	105	22
Collembola	0	90	44	0	0	0
Others	89	22	0	33	0	0
Total	444	492	154	300	105	22
Terminalia tomentosa						
Mite	911	350	133	267	88	0
Collembola	44	44	0	0	0	22
Others	177	68	66	0	0	0
Total	1132	462	199	267	88	22
Diospyros melanoxylon						
Mite	644	310	88	333	66	0
Collembola	44	22	66	0	0	0
Others	227	122	22	67	0	0
Total	915	454	176	400	66	0

The collembolans accounted only 0-18% of the total litter micro arthropod population at two sites indifferent species. In the cool dry period the population density of the leaf litter arthropods decreased. During winter season the population density of the arthropods varied in between 934-1352 m^{-2} in mixed teak forest and in between 454-1298m^{-2} in teak forest in four different leaf litter species and mixed half litter. Mites are the most abundant arthropod group during this period also (63% to 95% of the total arthropod population in different leaf litter species at both the sites). Collembola population varied between 1.9% to 19.3% of the total population. During summer season the population density of arthropod decreased sharply. During this period arthropod population varied between 154-843 m^{-2} in different leaf litter species at two sites. During this season also, the mites were the most abundant group of arthropods constituting 61-87% of the total population. However, collembola contributed only 0-39% of the total population.

Effect of Repellent on Arthropod Population

A sharp decline in the population density of arthropods were seen in all the leaf litter species at both the sites in the treated bags. At mixed teak forest site during rainy season population density varied between 466-887 m^{-2} in treated bags. The mites again contributed the maximum number. A 60%-70% decrease in population as compared to control (non-treated) was recorded at mixed teak forest indifferent decomposing leaf litter during rainy season (Fig's 4.1-5). At teak forest the population density during rainy season varied between 133-400 m^{-2} in different treated leaf litter bags. Reduction of 56-95% population density, as compared to control (non-treated) was recorded indifferent decomposing leaf litter species during this season.

During winter season the reduction in population density was 69%-81% in different species as compared to control (Fig's. 4.1-5). Near similar results were obtained for Teak forest site also.

The decrease in population was 78%-87% as compared to control in different species in teak forest. During summer season in mixed teak forest of treated decomposing leaf litter, the population densities were between 66-132 m^{-2} in different decomposing leaf litter species with a reduction of 30-87% in population densities, compared to control. During summer in teak forest population density of 0-132 m^{-2}, with a decrease of 77%-100% population as compared to control were recorded in different treated decomposing leaf litter bags.

Effect of partial Exclusion of Arthropod Population on Decomposition of Leaf Litter

(1) Mass Loss

A significant decrease in mass loss were found in all the decomposing leaf litters, whose left litter has treated with Phorate-

Table 4.3: Percent mass remaining Nirtogen Content and Relative Decomposition Rate (RDR) of differenent species in two forest

	Mixed Teak Forest							Teak Forest					
	Non Tereated Control			Treated (Phorate-G)				Non Treated			Treated		
Season	% mass remaining	RDR mg/g/d	Nitrogen (mg) content	% mass remaining	RDR mg/g/d	Nitrogen (mg) content	Leaf litter species	% mass remaining	RDR mg/g/d	Nitrogen (mg) content	%mass remaining	RDR mg/g/d	Nitrogen (mg) content
Rainy	68.48	3.57	41.08	74.48	2.77	40.96	*Tectona grandis* 42.5	64.00	4.21	40.32	75.10	2.70	38.68
Winter	35.20	5.99	27.10	68.30	0.78	39.27	(initial nitrogen content)	32.50	6.16	21.45	70.12	0.62	36.46
Summer	18.60	4.52	14.78	60.20	0.89	39.13		20.90	2.13	14.00	65.00	0.54	36.40
Rainy	81.26	1.95	46.72	93.06	0.68	46.99	*Buchanania latifolia* 49.5	50.50	6.45	33.08	60.60	4.73	39.09
Winter	63.00	2.29	39.37	89.10	0.39	46.33	(initial nitrogen content)	35.00	3.33	24.15	58.50	3.21	38.03
Summer	43.50	2.63	28.49	86.60	0.20	44.16	*Terminalia tomentosa* 52.0	28.14	1.58	19.88	57.10	0.17	35.69
Rainy	36.32	9.55	25.42	53.64	5.88	38.86	(initial nitrogen content)	36.30	9.56	26.32	44.92	7.55	30.10
Winter	29.20	1.96	21.12	52.30	0.23	34.51	*Diospyros melanoxylon* 40.0	33.10	0.84	24.99	42.80	0.44	29.53

Summer	23.50	1.54	17.50	50.60	0.24	33.90		26.80	1.50	19.97	42.00	0.13	28.42
Rainy	53.70	5.87	32.75	71.30	3.19	39.92		49.32	6.67	33.29	54.68	5.70	34.98
Winter	34.10	4.09	26.25	66.40	0.64	39.17	(initial nitrogen content)	29.50	4.67	21.68	53.00	0.03	32.60
Summer	23.00	2.79	15.14	64.80	0.17	38.55	Mixed leaf 45.0	24.60	1.23	18.57	51.50	0.20	32.19
Rainy	64.48	4.14	38.68	75.08	2.70	37.54		–	–	–	–	–	–
Winter	38.00	4.76	24.51	71.70	0.41	37.28	(initial nitrogen content)	–	–	–	–	–	–
Summer	26.70	2.50	17.48	68.50	0.32	46.99		–	–	–	–	–	–

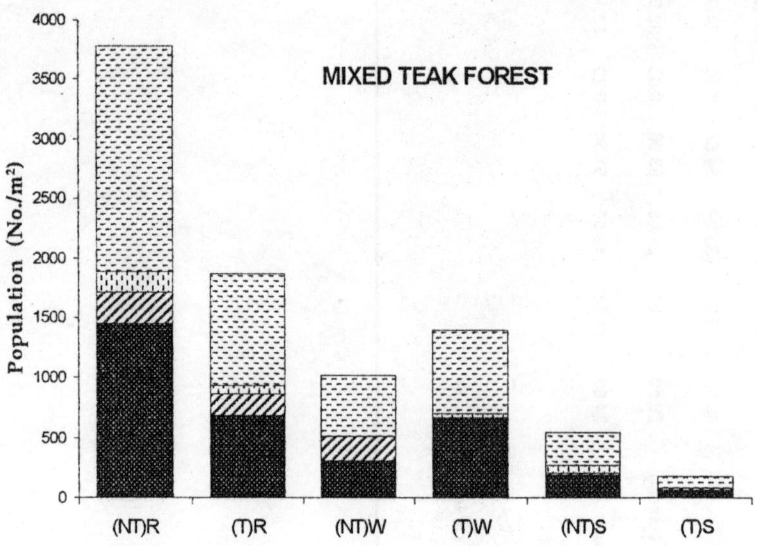

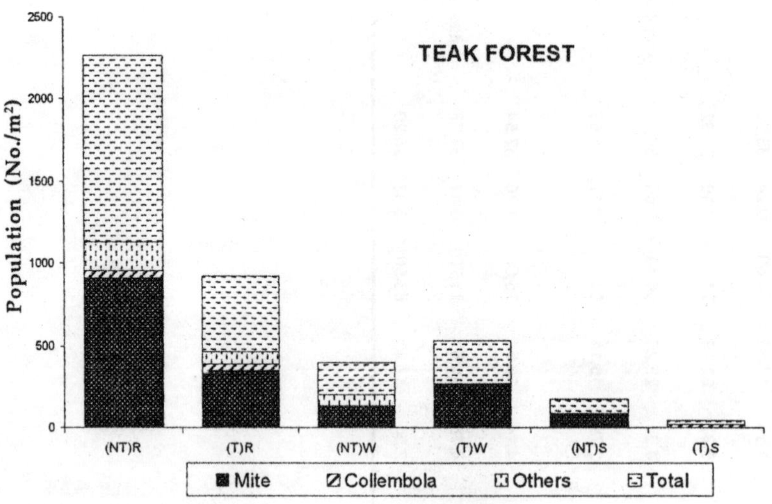

R = Rainy; W = Winter; S = summer; NT = Not treated; t = Treated (Phorate-G)

Fig. 4.1: Microarthropod Population in Treated and Non treated Decomposing Leaf litter of *Terminalia tomentosa* in Two Forests

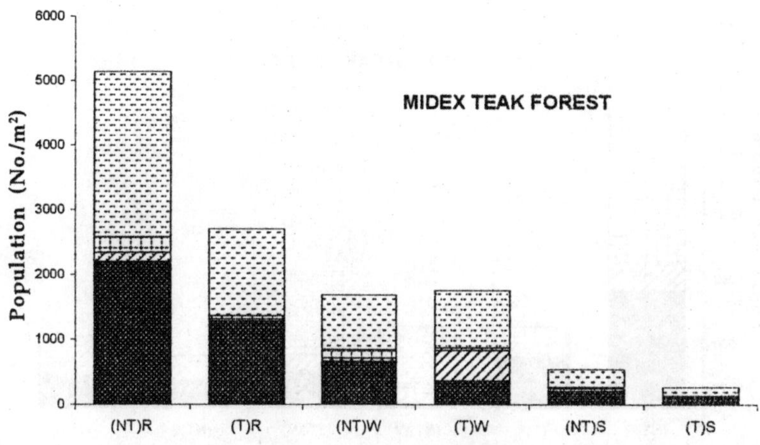

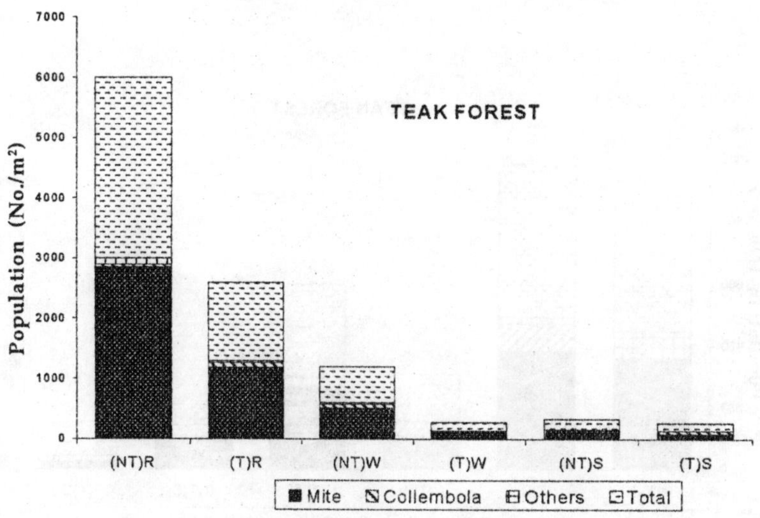

R = Rainy; W = Winter; S = summer; NT = Not treated; t = Treated (Phorate-G)

Fig.4. 2: Microarthropod Population in Treated and Non treated Decomposing Leaf litter of *Terminalia grandis* in Two Forests

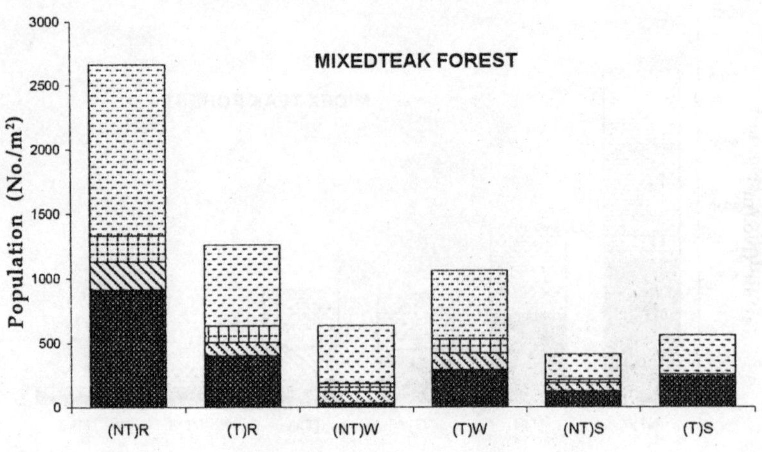

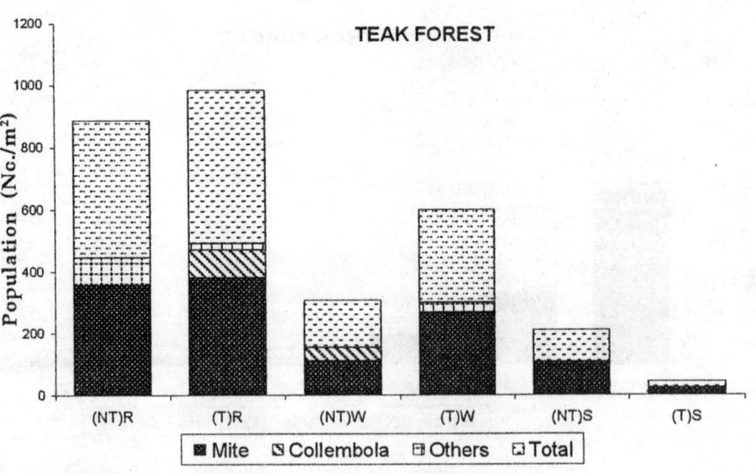

R = Rainy; W = Winter; S = summer; NT = Not treated; t = Treated (Phorate-G)

Fig. 4.3: Microarthropod Population in Treated and Non treated Decomposing Leaf litter of *Buchanania latifolia* in Two Forests

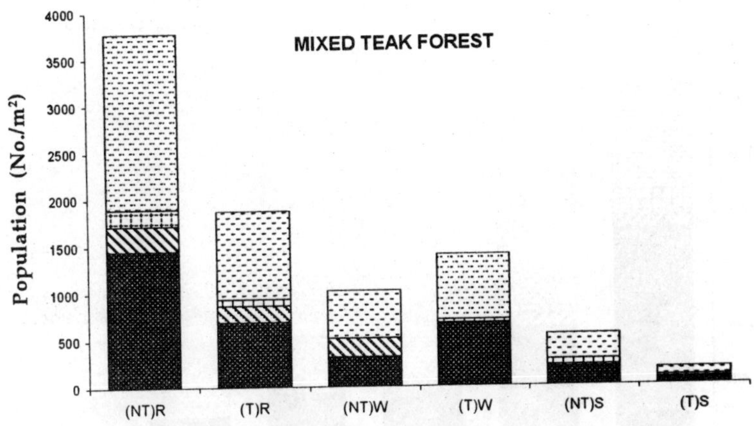

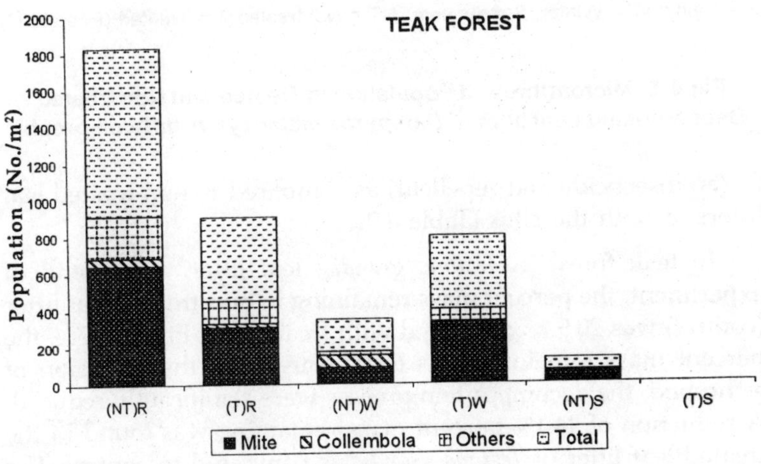

R = Rainy; W = Winter; S = summer; NT = Not treated; t = Treated (Phorate-G)

Fig. 4.4: Microarthropod Population in Treated and Non treated Decomposing Leaf litter of *Diospyros melanxylon* in Two Forests

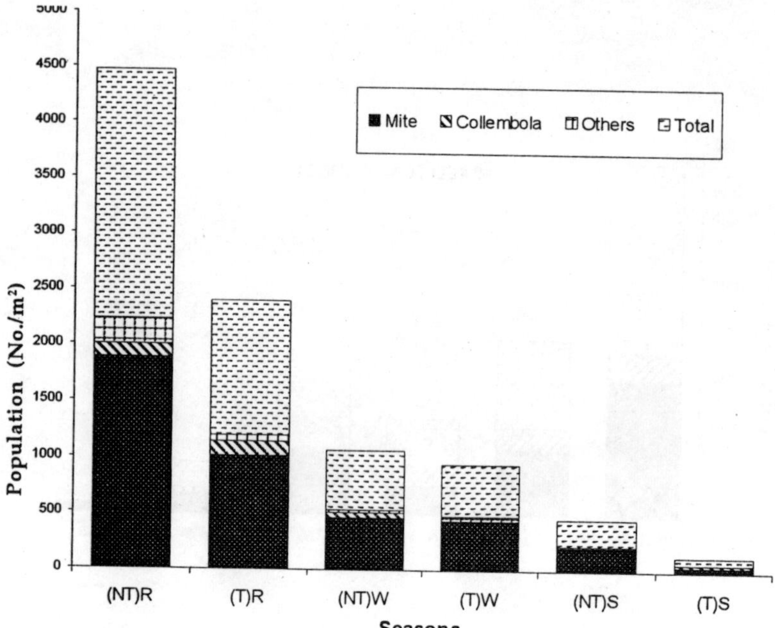

R = Rainy; W = Winter; S = summer; NT = Not treated; t = Treated (Phorate-G)

Fig.4. 5: Microarthropod Population in Treated and Non treated Decomposing Leaf litter of *Diospyros melanxylon* in Two Forests

G, (an insecticide and repellent) as compared to non-treated leaf litters at both the sites (Table 4.3).

In teak forest in *Tectona grandis,* leaf litter, at the end of experiment, the percent mass remaining in non-treated leaf litter (control) was 20.9 as compared with treated leaf litter where, the percent mass remaining was 65.0. Thus, with the exclusion of arthropod, the decomposition process were significantly reduced. A reduction of 44.1% percent mass remaining was found in the treated leaf litter of *Tectona grandis* as compared to control. The decrease in percent mass remaining were observed in all the three seasons (rainy, winter and summer) in all other decomposing leaf litter of *Buchnania latifolia, Terminalia tomentosa, Diospyros melanoxylon* and mixed leaf litter, as compared to non-treated leaf litter. The reduction of percent mass remaining were 15-28.96 in other four leaf litter compared to the control at the end of the experiment (Table 4.3). Effect of decomposition process on the exclusion of arthropods population reveals following trend for the reduction of mass losses as compared to control.

Tectona grandis > *Buchanania latifolia* > *Diospyros melanoxion* > *Terminalia tomentosa*.

At mixed teak forest more or less same effect were seen for the role of arthropod in the decomposition process. Maximum effect was found in *Buchanania lalitolia* where, percent mass remaining was 86.6 in treated, as compared to the non-treated (control) in which percent mass remaining were 43.5 percent at the end of the experiment. Decrease of percent mass remaining were found in all the three seasons in all the other litter bags. Reduction in percent mass remaining ranged between 27.1-43. in different litter bags during summer (Table 4.3). At this site mass loss due to arthopods exclusion followed the order:

Buchanania ltifolia >. *Diospyros melanoxylon* > *Mixed leaf litter Tectona grandis* ? *Terminalia latifolia*.

(2) On Mean Relative Decomposition Rate

At mixed teak forest reduction in mean relative decomposition rate (mg/g/d) with the treatment of Phorate-G were observed in all the cases (Table 4.3). During rainy season relative decomposition rate varied between 1.95 mg/g/d-9.55 mg/g/d, in control (non-treated) leaf litter as compared to treated leaf litter wherein, it was between 0.78 mg/g/d- 5.87 mg/g/d. During winter season relative decomposition rates were 1.96 mg/g/d-5.99 mg/g/d in control leaf litter as compared to treated leaf litter where mean relative decomposition rates were 0.22 mg/g/d- 0.78 mg/g/d in different species. At the end of experiment, during (Summer) the mean relative decomposition rates were 1.54 mg/g/d to 4.52 mg/g/d in different species in control (non-treated) as compared to treated leaf litter. RDR were 0.172 - 0.89 mg/g/d (Table 4.3). Similar decrease in relative decomposition rates were observed at teak forest also, in treated leaf litter with the phorate-G as compared to control (non-treated). At teak forest high RDR (9.55 mg/g/d in control leaf litter were found in *Terminalia tomentosa*, with decrease in RDR but still high RDR (7.54 mg/g/d) in treated leaf litter during the rainy season.

During the winter and summer season similar trend as in case of mixed teak forest were found in teak forest in different leaf litter bags (Table 4.3).

(3) On Decay Rate

A significant decrease of decay rate in treated leaf litter as compared to the non-treated leaf litter were observed at both the sites, in all decomposing leaf litters. At mixed teak forest the decay rate (k) were in between 0.00232-0.00469 in control (non-treated) as compared to treated leaf litter were it was 0.000401 - 0.001902 in different decomposing leaf litter (Table 4.4).

Similar decrease of decay rate in treated leaf litter as compared to non-treated (control) were found in teak forest also (Table 4.4). It was found that 95 percent decomposition (3/k) of the plant leaf material required much more time when the leaf litters were excluded from the arthropod populations by treating them with the Phorate-G.

In mixed teak forest 95% decomposition of non-treated (control) leaf material required between 639 days - 1293 days for different species, however, when arthropods where partially excluded, 95% decomposition of same plant leaf material required 1577 days -7481 days for different species (Table 4.4).

In teak forest control (non-treated) leaf litter, 95% decomposition will require 648 days - 841 days in different species. Whereas, when the same plant leaf material were excluded of arthropods by treating them with Phorate-G a longer time (1235 days - 2487 days) will be required to complete the decomposition process (95% decomposition). Decay rate and time required for decomposition is presented in (Table 4.4).

(4) On Nitrogen Release Pattern

The effect of arthropods exclusion on nitrogen release pattern is well marked. At mixed teak forest 42% to 65.88% nitrogen release occurred at the end of experiment in non-treated leaf litters as compared to treated leaf litters, where only 3.6 to 34.8 percent release of nitrogen has been noted in different species of leaf litter. At teak forest 59.84 to 67.05% nitrogen got released at the end of experiment in non-treated leaf litters of different species. In treated leaf litters only 14.35 to 45.34% release of nitrogen was

Table 4.4: Decay Rate (K) and Time (Days) Required for Decomposition of Different Leaf Litters in two Forests

Mixed Teak Forest									Take Forest							
Non Treted Control				Treated (Phorate-G)					Non Treated Control				Treated Phorate-G)			
	Days				Days					Days				Days		
Decay Rate (K)	50% Decom. (t_{50}) 0.693/K	95% Decom. (t_{95}) 3/K	99% Decom (t_{99}) 5/K	Decay Rate(K)	50% Decom. (t_{50}) 0.693/K	95% Decom. (t_{95}) 3/K	99% Decom. (t_{99}) 5/K	Species	Decay Rate (K)	50% Decom. (t_{50}) 0.693/K	95% Decom. (t_{95}) 3/K	99% Decom (t_{99}) 5/K	Decay Rate (K)	50% Decom. (t_{50}) 0.693/K	95% Decom. (t_{95}) 3/K	99% Decom. (t_{99}) 5/K
0.004690	148	640	1066	0.001410	491	2128	3546	*Tectona grandis*	0.004384	158	684	1141	0.001206	575	2488	4146
0.002320	299	1293	2155	0.000401	1728	7481	12469	*Buchananin latifolia*	0.003565	194	842	1403	0.001569	442	1912	3187
0.004040	172	743	1238	0.001902	364	1577	2629	*Terminalia tomenstosa*	0.003688	188	813	1356	0.002429	285	1235	2058
0.004105	169	731	1218	0.001211	572	2477	4129	*Diospyros melonoxylon*	0.003928	176	764	1273	0.001858	373	1615	2691
0.003688	188	813	1356	0.001056	656	2841	4735	Mixed leaf	–	–	–	–	–	–	–	–

recorded. The nitrogen concentration (% dry weight) and standing state of nitrogen (mg) during different seasons in different decomposing leaf litter species at two sites have been presented in Table 4.5 and 4.3 respectively.

Table 4.5: Mean Nitrogen Concentrations (% dry mass) During Decomposition of Different Leaf Litters in Two Forests

Season	Mixed Teak Forest		Teak Forest	
	Non-treated	Treated	Non-treated	Treated
Tectona grandis (Initial Conc. 0.85)				
Rainy	1.20	1.10	1.26	1.03
Winter	1.54	1.15	1.32	1.04
Summer	1.59	1.30	1.39	1.12
Buchanania latifolia (Initial Conc. 0.99)				
Rainy	1.15	1.01	1.31	1.29
Winter	1.25	1.04	1.38	1.30
Summer	1.31	1.02	1.39	1.35
Terminalia tomentosa (Initial Conc. 1.04)				
Rainy	1.40	1.30	1.45	1.34
Winter	1.45	1.32	1.51	1.38
Summer	1.49	1.34	1.49	1.45
Diospryos melanoxylon (Initial Conc. 0.8)				
Rainy	1.22	1.12	1.35	1.28
Winter	1.54	1.18	1.47	1.23
Summer	1.34	1.19	1.51	1.25
Mixed leaf (initial Conc. 0.9)				
Rainy	1.20	1.00	–	–
Winter	1.29	1.04	–	–
Summer	1.31	1.08	–	–

Discussion

The seasonal densities of arthropods population varies considerably due to moisture availability. The population density in different leaf litter species agree to that reported by Reddy (1981) and Reddy and Venkataiah (1989) in pine and eucalyptus litter. Reddy (1992) worked out the arthropod population in different months of the year and found that the densities were highly dependent on moisture availability and their number significantly reduced during hot summer. Our result is in agreement with the work of Reddy (1992). Higher population densities in comparison to Reddy were observed in different litter bags in the present study. Reddy (1992) also reported that in Neem leaf litter Acarina (mites) and Collembola were the major groups of micro-arthropods inhabiting the Neem leaf litter, representing < 82% of the total micro-arthropods. The higher values of densities in the present study might be to higher palatability of these leaf litter.

Reports on changes in the densities of edaphic micro-arthropods in relation to seasonal variation in rainfall and other microclimatic factors such as, litter moisture and temperature were made by Seastedt and Crossley (1980), Reddy (1989) and Reddy and Venkataiah (1989), and time of the year and litter conditions by Crossley and Hoglund (1962) and Reddy and Venkataiah (1989). As expected similar pattern were observed for different groups of arthropods across the seasons (Tables 1 and 2), which may indicate that the higher rainfall and litter moisture were favourable for increase in the densities of micro-arthropods.

Significant variation was reported on comparing our result with that of Reddy 91992), who used Neem leaf as an insect repellent, wherein he found no insect repellent properties in the Neem leaf and no decrease in population of arthropods. In the present study a significant reduction in the population (up to 87% and in some case 100%) of arthropods were noticed during different seasons in different decomposing leaf litter bags at both the sites.

Vats and Handa (1983) reported different groups of soil litter arthropods in a tropical deciduous forest stand of Kurukshetra University, dominated by *Dalbergia sisso*. They worked out different groups of arthropods population in different seasons. In their case study the relative humidity remains same and therefore, seems to have no influence on the seasonal variations in population density. However, they generally believed that population density on a whole is fairly high in rainy season.

Densities range from below 50,000 m^{-2} in the tropic and the other areas where organic matter was in low amount (Madge, 1969, Santos, *et al.* 1978) to numbers 300 m^{-2} or more in boreal forests (Mitchell 1977, Persson *et al.* 1980).

The organic matter ingested by the faunal elements is voided as faecal pellets which support rich microbial growth that again assist in the conversion of organic matter Comminution, channelling and mixing, excretion and grazing are the essential modes of activity of the micro-arthropods, that mediate physical and chemical break down (Crossley 1977). Litter is broken to nearly 100-200 mm sizes by the feeding activity of meso-arthopods and to even lesser dimensions of 20-50 mm by the micro-arthopods (Swift *et al.* 1979). In the present study higher micro-arthropod population and faster decomposition in different non-treated leaf litters as compared to treated leaf litter (Table 4) is also supported by the observation made by Swift *et al.* (1979).

Arthropod exclusion study show a significant reduction of mass losses in different species at two sites. At mixed teak forest the reduction of mass loss was 27% - 43% in different species. However, at teak forest the reduction of mass loss due to partial exclusion of arthropod was 15%-44% in different species. A number of studies have been done to work out the role of arthropods on decomposition process. Effect range from values not significantly different from zero to contribution of almost 70% of the decay rate (Seasted *et al.* 1983). Additional experiments of shorter duration have also indicated variable but often very significant effects of micro-arthropods on decomposition rates or mass loss of litter (Parkinson 1980, Santos and Whitford 1981 Strogan 1978), Vossbrink *et al.* (1979), Ward and Wilson (1973), Whitford *et al.* (1982), Zlotin (1971), Zlotin and Khodashova (1973). Effect of micro-arthropods vary when the same litter type is studied in different years (Seastedt and Crossley) (1983).

As the mass loss is related to the reactive decomposition rate, any change in decomposition (weight loss or mass remaining) is going to effect the relative decomposition rate. A significant slow down of relative decomposition rate in all the leaf litter type at both the sites in all the seasons, when treated with insecticide - Phorate-G, reflect, relatively high impact of arthropod population, on decomposition process.

A relatively higher value of decay rate in non-treated (control) leaf litter as compared to treated leaf litter is observed in all the leaf litter. Thus, much more time is required to decompose completely a particular substrate. In the present study, the different leaf litter *(Tectona grandis, Buchananina latifolia, Terminalia tomentosa, Diospyros melanoxylon* and mixed leaf) requires 639-1293 days at mixed teak forest in non-treated leaf litter, however, same substrate requires 1577-7481 days when arthropods were excluded from the decomposing leaf litter.

At teak forest, 648-841 days are required to complete the decomposition (95%) of the leaf litter without excluding the arthropod population. However, same substrate require longer period of time (1235-2487 days) to complete the decomposition (95%), when arthropods were partially excluded. A reasonable answer for longer period required for decomposing leaf litter is that, the dominant trophic function of micro-arthropods are dominant fungal feeder (mycophagous) predominant. Most oribatid mites, collembolans, and free living astigmatic mites have well developed mouth parts capable of fragmenting organic matter while feeding on the microflora adhering to this detritus. Fragmentation and comminution (i.e. reduction to small, fine particles) have very important consequences to decomposition and mineralization process particularly by creating new surface area for microbial activity and microbial colonization (Elkins and Whitford 1982). This group of arthropods contains species that are predacious and/or mycophagus. These groups also contain species that regulate densities of collembolans (Edwards *et al.* 1976) and nematodes (Santos *et al.* 1981), whose densities in turn may also effect decay rate and thus completion of decomposition processes. The micro-arthopods represented by diverse group of taxa inhabiting in litter play a significant role in mediating decomposition process (Crossley 1977, Swift *et al.*, 1979, Seastedt and Crossley 1980, Ineason *et al.* 1982, Anderson *et al* 1983, Seastedt 1984.

In the present study, at mixed teak forest 60.4% - 66.34% of nitrogen and at teak forest 53.57% to 67.05% nitrogen is released at the end of the experiment, in non-treated (control) litter bags, however, in treated (partially micro-arthropod excluded) leaf litter the percent release of nitrogen at the end of experiment was only 3.6 to 34.8 at mixed teak forest and a teak forest, 14.35 to 45.34 in different leaf litter bags. The result clearly indicate the strong influence of arthropod on the release pattern of nitrogen in decomposing leaf litter. The relatively higher concentrations of nitrogen (%) in decomposition leaf litter of different species during rainy, winter and summer season of non-treated (control) leaf litter as compared to that of treated leaf litter also suggest strong influence of arthropods in release pattern of nitrogen. Micro-arthopods such as Collembolans, Acarina, Symphylids and pauropods are abundant in different layers of litter and their abundance is influenced by several factors including the levels of moisture and nutrients and the prevalence of secondary plant substance in the chemical environment. The micro-arthopods cause fragmentation of litter and subsequently mediate nutrient cycling is evident in several instances (Swift *et al.* 1979, Seastedt 1984).

The relative balance of input minus output determines either the increase or decrease in the mineral nutrient of decomposed litter (Seastedt and Crossley 1980). The outputs are directly increased by consumption of litter by micro-arthropods and indirectly enhanced by fragmentation, stimulation of more microflora and other mechanism (Crossley 1977). The input may occur through the immobilization and recruitment of nutrients by microbial colonies. Nitrogen can potentially be mineralised by the grazing and excretory activities of protozoans, nematodes and perhaps, micro-arthropods. Removal of excretory products from the immediate environment requires leaching or volatilization, and in most instances, this nitrogen is probably rapidly re-immobilized by microbes on the litter. Thus, micro-arthropod feeding activities on microflora probably result in rapid recycling of most nitrogen within the system (i.e. within litter bags). These studies indicate either very high comminution effects, extensive grazing of microbial tissues by micro-arthropods or both (Reddy and Venkataiah 1989), Crossley 1977, Crossley and Hoglund 1962, Seastedt and Crossley 1980). Thus litter bag experiments have adequately proved that the abundance of micro-arthropods play a significant role in the seasonal dynamics of nutrients (nitrogen) during the process of decomposition.

REFERENCES

Anderson, J.M., P. Ineson and S.A. Huish. 1983. The effects of animal feeding activities on element release from deciduous forest litter and soil organic matter. In: Ph. Lebrun, H.M. Andren, A.De Medts, Gegoire-Wibo and G. Wauthy (ed.), New Trends in Soil Biology.

Proc. VIII Int. Coll. of Soil Zool. Imp. Dieu-Brichort, ottignies-Louvain-la-Neuve. pp. 87-100.

Crossley, D.A. Jr. and M.R. Hoglund. 1962. A litter bag methcd for the study of micro-arthropods inhabiting the leaf litter. *Ecology*. 43: 571-573.

Crossley, D.A. Jr. 1977. The role of terrestrial saprophagous arthropods in forest soils. Current status of concepts. In: W.J. Mattson (ed.), *The Role of Arthropods in Forest Ecosystems*. Springer-Verlag, New York. pp. 49-56.

Edwards, C.A., E.B. Dennis and D.W. Empson. 1967. Pesticides and the soil fauna; effects of aldrin and DDT in an arable field. *Ann. Appl. Boil*. 60: 11-22.

Elkins, N.Z. and W.G. Whitford. 1982. The role of micro-arthropods and nematodes in decomposition in semi arid ecosystem. *Oecologia*. 55: 303-310.

Inseson, P., Leonard, M.A. and Anderson, J.M. 1982. Effect of collembolans grazing upon nitrogen and cation leaching from decomposing leaf litter. *Soil boil. Biochem*. 14: 601-60

Lussenhop, J. 1976. Soil arthropod response to Praise burning soil. *Ecology*. 57: 88-98.

Lussenhop, J. 1992. Mechanisms of micro-arthropod - microbial interactions is soil. *Adv. Ecol*. 23: 1-33

Madge, D.S. 1965. Leaf-fall and litter disappearance in tropical forest. *Pedobiologia*. 5: 273-288.

Madge, D.S. 1969. Litter disappearance in forest and savanna. *Pedobiologia*. 9:288-299.

Mc Brayer, J.F., J.M. Ferns, L.J. Metz., C.S. Gist and B.W. Carnoby. *et al*. 1977. Decomposer invertebrate population in U.S. forest biomes. *Pedobiologia*. 17: 89-96.

Mitchell, M.J. 1977. Population dynamics of oribatid mites (Acari, Cryptostigmata) in an aspen woodland soil. *Pedobiologia*. 17: 305-319.

Olson, J.S. 1963. Energy storage and the balance of producers and decomposers in ecological systems. *Ecology*. 44: 322-328.

O'Neill, R.V. and D.L. Angelis. 1980. *Comparative productivity and biomass*

Parkinson, D. 1980, *Aspects of microbial ecology of forest ecosystems*. pp. 109-117.

Persson, T., E. Baath, M. Clarnolin, H. Lundkvist, B.E. Soderstrom and B. Sochlenius. 1980. Trophic structure, biomass dynamics and carbon metabolism of soil organisms in a Scots pine forest. In: *Structure and function of Northern Coniferous Forests. An Ecosystem Study* (ed.). T. Person. *Ecol. Bull. Stockholm*. 32: 419-459.

Reddy, M.V. 1981. Micro-arthropod and the rate of litter disappearance in a pine plantation ecosystem in north eastern India. Pedobiologia. 22: 339-344.

Reddy, M.V. 1989. Mass loss and concentration of nutrients in relation to micro-arthropod abundance during needle decomposition in a pine plantation. *Proceedings of Indian Academy of Science (Animal Sciences)*. 98: 285-289.

Reddy, M.V. and B. Venkataiah. 1989. Influence of micro-arthropod abundance and climatic factors on weight loss and mineral nutrient contents of the Eucalyptus leaf litter during decomposition. *Biology and fertility of soils*. 8: 319-324.

Reddy, M.V. 1992. Effects of micro-arthropod abundance and abiotic variables on mass loss and concentration of nutrients during decomposition of Azadirachta indica leaf litter. *Tropical Ecology*. 33(1): 89-96.

Santos, P.E., E. Depree and W. Whitford. 1978. Spatial distribution of litter and micro-arthropods in Chihuahuan desert ecosystem. *J. Arid. Environ*. 1: 41-48.

Santos, P.F. and W.G. Whitford. 1981. The effects of micro-arthropods on litter decomposition in a chihuahuan desert ecosystem. *Ecology*. 62: 654-663.

Sestedt, T.R. and D.A. Crossley Jr. 1980. Effects of micro-arthropods on seasonal dynamics of nutrients in forest litter. *Soil Biology and Biochemistry*. 12:337-342.

Seastedt, T.R. and D.A. Crossley Jr. 1983. Nutrients in forest litter treated with naphthalene and simulated through fall. A field microcosm study. *Soil Boil. Biochem*. 15:159-165.

Seastedt, T.R. 1984. The role of micro-arthropods in decomposition and mineral processes. *Annual Review of Entomology*. 29:25-46.

Singh, K.P. and Chandra Shekhar. 1989. Concentrations and release pattern of nutrients (N.P and K) during decomposition of Maize and Wheat roots in a seasonally dry tropical region. *Soil Biol. Biochem*. 21(1): 81-85.

Strogan, C.L. 1978. Forest leaf litter decomposition in a vicinity of a zink smelter. *Oecologia*. 32: 203-212.

Swift, M.J., O.W. Heal and J.M. Anderson. 1979. Decomposition in terrestrial ecosystems. *Blackwell Scientific Pub*. Oxford.

Vats, L.K. and Shobha Handa. 1983. Proc. Sym. *Inst Ecol. and Resource Manage*. pp. 181-186.

Vossbrink, C.F., D.C. Coleman and T. Wooley. 1979. Abiotic and biotic factors in litter decomposition in a semi-arid grassland. *Ecology*. 62: 275-277.

Ward, D.J. and R.E. Wilson. 1973. Pesticide effects on decomposition and recycling of Avena litter in a monoculture ecosystem. *Am Midl. Nat*. 90: 226-276.

Witkamp, M. and J. Van Der Drift. 1961. Break down of forest litter in relation to environmental factors. *Plant and soil*. 15: 295-311.

Witkamp, M. and D.A. Crossley Jr. 1966. The role of micro-arthropods and micro-flora in break down of white oak litter. *Pedobiologia*. 6: 293-303.

Zlotin, R.I. 1971. Invertebrate animals a factor of the biological turnover. In 4th colloq. Pedobiol. Digon. Paris: *Ins. Nar. Rech. Agron*. 455-462.

Zlotin, R.I. and K.S. Khodashova. 1973. The role of animals in biological cycling of forest steppe ecosystems. Transl. (ed.) N.R. French. 1980. Stroudsburg, Pa: Dowden Hutchinsoa Ross: (from Russian) pp. 221.

5

Heavy Metal Pollution in Fresh Water and its Impact on Fisheries

K.L. Jain, R.K. Gupta and N.K. Yadava

Department of Zoology & Aquaculture, CCS Haryana Agricultural University, Hisar-125 004, India.

ABSTRACT

Heavy metals are one major component among various effluents which has attained alarming status as water pollutants over the last few decades due to heavy industrialization. Toxic metals are added in aquatic system from industrial processes, domestic sewage, discharge, steel dust, land runoff and fossil fuel burning. This envisages a great concern with the ecology of polluted waters and fisheries. Later being a natural food source for about half of the population of the world needs serious implications as the polluted food could draw not only nutritional complications in terms of diseases or other physiological disorders in humans and fish production. Liver and kidney mainly act for storage and transformation of such xenobiotics. Relationship between mercury content and weight of the fish and fish length have been reported by many workers. It has been said that the mercury content too increase with increase in the body weight. Out of several species of fishes analysed most of the species were found to contain equal amount of mercury in muscle than in liver and a few cases of increased level in liver than in muscle. Acute levels of metals in water, however, is said to stimulate "mucus" secretion which serve as a way of reducing their uptake in fishes. Differences in bio-accumulation of metals in body tissues also depend upon the animals' physiological activities. The accumulation of metals by various intercellular compartments is further a function of relative rate of metal binding and release, i.e. binding to Metallothione in (MT). Binding affinity of metal to MT is in the order: Hg> Cd> Cu> Zn. Cadmium toxicity depend upon intake of copper as copper displays cadmium from MT molecules resulting in cadmium elimination. On the other hand lead is synergistic to copper uptake and elimination. Most important are quantitative differences between the chemicals. Other major physiological aspects of metal toxicity is alterations in cellular enzyme activity. Cadmium along with mercury result in generally enzyme inhibition at high concentration. Various treatment techniques involving various chemicals and biological materials like bacteria, algae and other plant products can be used for the precipitation or elimination of various harmful metal toxicants. Heavy metals such as mercury, lead, cadmium and cobalt from waste-water can also be effectively sorbed by sulphurized activated carbon.

Water pollution today has become world's most serious problem, as water although an essential commodity for sustaining life, is being maximally exploited by the industrialists, civic bodies vis-a-vis, the water surface transporting units. Modern civilization with its rapidly growing industrial units and ever increasing population has led to an accelerated degradation of the fresh water resources. Today about 100 million people are known to severely suffer from these polluted waters, since a variety of causative agents are known to occur for various physiological disorders and water born diseases.

Water a valuable God's gift to man, is required in healthy state for every organism. Analysis of most of the natural water bodies revealed a critical loss of bio--diversity, frequent occurrence of pathogens and a variety of other noxious problems with aquatic plants and animals due to man's careless activity. In India textile and dying industries alone contribute 10 to 15% of about 1000 different chemicals in nature, leading to increased BOD and COD contents, thus simultaneously decreasing DO_2 and creating anaerobic conditions in the rivers. In Delhi about 200 million liters of untreated water is pumped into Yamuna. The major industries causing water pollution are the paint and the detergent industries, metal and surgical industries, steel and pulp industries and various other chemical industries. Fly ash is the major contributor of heavy metal toxicity in natural waters. Over 100 million tones of fly ash are produced each year and much of it is transported in the atmosphere before deposition. It is a major source of contamination of heavy metal such as cadmium in the rivers and estuarine sediments in the vicinity of such power stations. Even most of the wetlands in India including

Utter Pradesh today have become heavily polluted and unproductive.

The US Department of Health Education and Social Welfare has classified water pollutants in eight categories viz., 1. Sewage wastes 2. Infectious agents, 3. Plant nutrients, 4. Particulates 5. Radioactive substances, 6. Mineral and chemical substances, 7. Heat, 8. Organic chemical exotics. Heavy metals are one major component among various effluents which has attained alarming status as water pollutants over the last few decades due to heavy industrialization. Toxic metals are added in aquatic system from industrial processes, domestic sewage, discharge, steel dust, land runoff and fossil fuel burning. Traces of heavy metals such as Hg, Cd, Pb, As, Co, Mn, Fe and Cr have been identified as deleterious to aquatic ecosystem and human health. Excess inputs are of man made origin coming from industrial and mining effluents (Cd, Cu, Fe, Hg, Pb, Zn) presides (Cu, Hg, Zn), the burning of fuels and car exhausts (As, Ni, Pb, V, Zn). Since the rivers have become the natural dust-bins for discharging all sorts of city, industrial and other wastes, these destroy natural harmony between biotic and abiotic components of the aquatic ecosystem. This envisages a great concern with the ecology of polluted waters and its perspectives with regard to its fitness for biotic community including fisheries. Later being a natural food source for about half of the population of world needs serious implications as the polluted food could draw not only nutritional complications in terms of diseases or other physiological disorders in humans and fish production, but also produce more hazards to the living world on account of a variety of indirect interacting factors.

Mercury is a unique pollutant because of its apparent indestructibility and ubiquitousness. It attracted worldwide attention as a potential pollutant because of a tragic episode which occurred in the Japanese villages surrounding the Minamata Bay. The victims were struck with a mysterious nervous illness commonly called "Minamata disease" which was later shown to be caused by consuming fish and shell fish that has lethal accumulation of methyl mercury. It resulted in 121 cases of poising and 41 death. Another similar incident of mercury poisoning was reported again in Japanese villages near Niigata prefecture during 1969, which left 6 people dead with 47 poisoned cases.

The chlor-alkali industry in India has grown ten fold since 1947. Large quantities of caustic soda and soda ash produced are used in Paper and Rayon industries. Majority of the chlor-alkali factories adopt old process of using mercury as cathode. This has long since been discarded in Japan and in many other countries. The country imports around 200 tonnes of mercury every year of which 180 tonnes are used by the chlor-alkali industry alone and most of it finally goes out along with waste-water and sludge, and ultimately enters into the environment.

Central Inland Fisheries Research Institute and National Environment Engineering Research Institute in 1981 reported that Hooghly estuary received 28.5 ppm of chromium from tannery and 0.77 ppm to 129.7 ppm of mercury from various other industries. Kaviraj (1989) also surveyed Hooghly estuary during 1977-81 and reported that an accumulation 381.6 ppm of zinc and 9.6 ppm of chromium in the sediments, 120.4 ppm of zinc, 127.8 ppm copper and 20 ppm chromium in fishes and 112.2 ppm zinc, 10 ppm copper and 0.3 ppm chromium in shrimps.

The realization of the deleterious conditions of the rivers in the early nineteen century in Britain on account of disappearance of salmon from the Thomei and Mersey rivers prompted passing of certain acts to curb water pollution. In 1857, a royal commission on prevention of river pollution was set up. The federal water pollution central act of 1948 passed in America ascertained water quality and a new act passed in 1965 required the states to establish water quality standards. In India, the water (prevention and control of pollution) act came into existence in 1974 only and thereafter, the environmental act was passed on 23 May, 1986. As per WHO, permissible limits for toxic metals in drinking water are: cadmium, 0.01, lead 0.1 and chromium 0.05 mg per litre. As per the analysis of ground water quality as a measure of impact of Bhilai steel plant effluents, it has been reported that the post monsoon ground water samples representing entire water shed area are in general high in toxic elements.

Two kind of heavy metal occur in a animal body, one termed as essential metals and others are non-essential metals. Non-essential metals are present in intracellular compartment and appear toxic to various animal function because of their binding to Metallo Thione (MT). Experimental studies have proved that the role of heavy metal ions in living systems follows the pattern of natural availability. There is no life, that can survive without the participation of metal ions. Some of the major ions such as Na, K, Mg, C are essential to sustain life. In the same way some of the heavy metals are considered essential both for plants and animal nutrition, and they serve some useful biological functions in the body. Thus, Cu, Co, Mn, Mo, Se and Zn are essential to both plants and animals. However, some trace metals such as Cd, Hg, Pb and metalloids like As, Sb and Se are considered to be toxic, although some essential physiological roles have been inferred for As, Cd and Pb recently (Dara, 1993). Among the metals studies Cr, Ni, Cd and Be have been found to be potentially carcinogenic (i.e. causing cancer). Even the heavy metal although essential, on "acute exposure" or "chronic exposure", severely affect different body organs and produces various types of diseases and deformities in the human body.

Copper's necessary for the normal biological activity, excessive intake of however, can be toxic and can cause haemolysis and a disease of liver called Willson's disease. Generally liver, kidney, gills and intestine are the main site for MT formation in fishes which is either stored or excreted. Heavy metal residues Nickle, chromium, cobalt in combined waste-water from vegetable oil industries are generally below detectable limits (<0.01 mg/I). These waters however, are reported to be iron rich (Av. 1.5 mg/litre), and implying very poor DO_2 levels.

According to Haslan, (1978) the effect of sewage and industrial affluents on aquatic biotic system can be described as:

(a) A decrease and loss of the species, most sensitive to pollution.

(b) A decrease in species diversity at the sight of affluent industries.

(c) An increase in any species actually favoured by the polluted waters.

Heavy metals are important, not only as a source of phytotoxicity in plants but also these enter into food chain causing hazardous impact on animals, including the humans. Chromium accumulation in an animal body occur mainly in liver, whereas lead and cadmium accumulation do not vary significantly in different tissues. The major symptoms of heavy metal toxicity in mammals are hypertension, cirrhosis of liver, lung cancer, nausea, vomiting, salivation, diarrhoea, abdominal cramps etc.

Body size is closely related to fish growth and metabolism and also an important attribute to metal accumulation in fishes. Liver and kidney mainly act for storage and transformation of such xenobiotics (Heath, 1992). Maiti and Banerjee (2002) studied bio-accumulation of certain heavy metals in three fish species such as *Labeo rohita, Catla catla and Oreochromis nilotica* in waste-water fed wet lands in Calcutta. They reported considerable role of the age of fish in metal accumulation in different fish species. The bio-accumulation of metals in tissues is further species specific.

The expression 'uptake efficiency' has been used to refer to how readily a fish removes a contaminant from the water by the gills. The metal uptake is however, greatly influenced by pH and temperature. Copper uptake by the gills of feathered minnows were found to be©1 slower at low pH, cadmium appear to be taken up in direct correlation with pH and lead exhibit exactly the opposite i.e. the pH increases uptake (Camphell and Stokes, 1985). In terms of bio-magnification however, different metals behave differently for instance, mercury bio-magnifies easily, whereas cadmium and lead seemingly do not, nor do the body concentration of the later two metals increase with age and size of the fish, as dose mercury. The distribution of metals into body tissues and organs appear to vary greatly depending upon the fish species. Lead accumulates mainly in scales, gills, bone and liver and in an estuarine fish, liver and muscles accumulated almost no lead (Spry and Weiner, 1991).

Relationship between mercury content and weight of the fish and fish length have been reported by many workers. They found the mercury content to increase with increase in the body weight. Llyod (1977) measured Hg levels in the liver and muscle tissues of rainbow trout and brown trout from trout parks stream of southern Missouri and found a relationship between concentration and size. Few studies showed decrease in the mercury level with increase in the size.

Tariq *et al.* (1979) reported increased accumulation of mercury in muscle and liver with many cases of increased level in muscle than liver. Out of several species of fishes analysed by Greing *et al.* (1975) most of the species were found to contain equal amount of mercury in muscle than in liver and a few cases of increased level in liver than in muscle.

Acute levels of metals in water, however, is said to stimulate mucus secretion which serve as a way of reducing their uptake in fishes (Part and Lock, 1983). Accumulation of metals further depend upon the environmental conditions and their availability in water and the sediments. Differences in bio-accumulation of metals in body tissues also depend upon the animals' physiological activities. *O. nilotica* is said to have increased accumulation of copper in liver, than in *L. rohita or C. catla*. The accumulation of metals by various intracellular compartments is further a function of relative rate of metal binding and release, as determined by their relative metal binding affinity and competition among different metals for the binding sites i.e. binding to Metallo Thionein (MT). Low level of cadmium are known to induce MT formation but with chronic exposures cadmium binding capacity of MT decreases. *Catla* is, however, said to possess a strong tendency to retain lead in the kidney. Organic mercury shows strongest affinity to bind with MT, so much so that it displaces copper and zinc from the molecule. Methyl mercury however, do not bind with MT. Eaton (1985) reported binding affinity of metal to MT in the order: Hg>Cd>Cu>Zn.

Cadmium toxicity depend upon intake of copper as copper displays cadmium from MT molecules resulting in cadmium elimination. On the other hand lead is synergistic to copper uptake and elimination. There are also report for their *vice versa* effects. Zinc uptake is inhibited by low level of cadmium but zinc seems to have no effect on uptake of cadmium. These interactions suggest further complicate role of essential and non-essential metals.

The Tilapia fishes *(O. mossambicus)* generally accumulate higher amount of metal residue as compared to other fishes and shows higher susceptibility to metal toxicity. The metal in combination have been found to exhibit higher toxicity effect. It has been reported that zinc and cadmium alone cause 32% and 42% reduction in growth, respectively. Whereas these together at a sub-lethal concentrations exert 65% reduction. Likewise cadmium with copper cause 86% reduction in growth, whereas copper alone is 47% toxic. Even copper and zinc together could yield 73% reduction in growth in tilapia fishes. Increase in lead concentration also cause higher mortality. In common carps mortality or retarded growth occur at cadmium concentration ranging from 4-12 μg/l.

Bio-concentration factors have also evidenced greater bio-accumulation of metals by molluses than macro-phytes in waste-water fed wetlands (Bhattachayra and Chakraborty, 2002).

Physiological impact of heavy metals

Apart from acute toxicity, the heavy metals in water has proved dangerous and harmful because of their bio-accumulation and their impact on tissue degeneration. Beside these chemicals are also established to induce reproductive functions, thus influencing growth, survival and reproductive potential of animals/fishes. Loss of immunity, necrosis, inflammation, oedema are many other pathological conditions.

Chronic effect of copper adversely affected the growth and reproduction in fish. Precipitation of mucus on the gills of fishes exposed to copper leads to respiratory troubles. Fishes exposed to mercury exhibit neurological damage and histopathological changes in the visceral organs and gills. All metals are strongly chaleophilic (affinity for sulphur) in their mineralogy and chemistry and their affinity for sulphur is greater than oxygen. Hence they attack the sulphur bonds present in the enzymes and thus inhibit various metabolic activities that are taking place in the organism. They also exert toxic effect by combining with one or more reactive groups essential for normal physiological function.

Most important are quantitative differences between the chemicals i.e. chromium is not nearly as harmful to gill tissue as is nickel, when fish are immersed in solution of comparable toxicity (Hughes *et al.*, 1979). Zinc at a dose level of 71.5 μg/1 causes death due to hypoxia because of rapid drop of blood oxygen and pH. Cadmium causes both anaemia in a variety of fish species at low and high concentration as well as abnormally large number of malformed RBC and altering the leucocyte differential count. Cadmium is further reported to cause inhibition of serum antibody titers in cunners, but a six fold stimulation in stripped bass (Robohm, 1996). Newman and Maclean (1974) has also reported a dose dependent 3-4 fold increase in neutrophils and a dose dependent decrease in lymphocytes in the cunners (Marine teleost fishes). A study conducted by Verma (1999) from India has also evidenced significant increased in Hc content and RBC count in case of Indian major carps, exposed to copper and zinc while WBC counts were decreased. But in the group exposed to cadmium a significant decrease in HC content, RBC count along with WBC count, as in the case of other heavy metals was observed. In case of copper and zinc the lymphocytes were decreased significantly, while the granulocyte and monocyte population tend to increase and with cadmium monocyte and lymphocytes decreased significantly. A significant decrease in the antibody titers was observed in the group exposed to copper and zinc. The serum protein levels were also decreased in all the cases of heavy metals. Likewise, chronic exposure of the cyprinid *Barbus* to lead for up to 60 days at 47µg/l also resulted in severe reduction (12 to 31%) in RBC count, haematocrit, haemoglobin concentration and MCV (Tewari *et al.*, 1987). Other major physiological aspects of metal toxicity is alterations in cellular enzyme activity. For example lead is said to act as a stimulator of alkaline phosphatase, but a strong inhibition of both acid phosphatase and xanthine oxidase. Cadmium along with mercury result in generally enzyme inhibition at high concentration. There was a clear cut increase in carbonic anhydrase activity in flounders exposed to cadmium (Gould, 1977). Shaffi (1993) also reported significant inhibition in certain metabolic enzymes such as lactate dehydrogenase, succinic dehydrogenase and malate dehydrogenase in the fresh water fish species, following their 24 or 48 h exposure to mercury.

Prolonged cadmium exposure of male brook trout is reported to cause their hormone levels to continue to rise after spermiation at which time they normally decline and the testis further exhibit histopathological damage from cadmium possibly due to inhibition of the system responsible for normal metabolic breakdown of steroids. Hg both organic and inorganic forms also cause histological lesions in the Leydig's cells of the fresh water catfish *Clarius* (Kirubagaran and Joy, 1992) and lead at low concentration is also said to cause marked suppression of spermatogenesis in trouts. The metals further interfere with the dynamic cell surface measurements due to complexion of ions particularly the divalent cations like cadmium.

There are many other reports on the physiological and biochemical changes in animal tissue on account of heavy metal toxicity. The biochemical components like proteins, lipids, cholesterol and phospholipids in the liver of *Cyprinus carpio* were found significantly reduced following 60 days exposer to nickel (7.5 to 15.0 mg/l) and chromium (10.0 to 20.0 mg/l)

Heavy metal management

Today fishes have become the major diet and there have been attempts to devise ways to enhance fish production. The need of hour is to rationalize the use of industrial effluents on one hand and to minimize their interference with other biotic components including the fishes. Waste-water management technology has received considerable attention in recent years to meet the stringent regulation of water quality. Research in this field has resulted in development of various treatment techniques involving various chemicals and biological materials like bacteria, algae for the precipitation of elimination of various harmful toxicant. Organic loading in terms of biochemical oxygen demand (BOD) is found to be the limiting factor for ponds and rivers. Commonly around 9000 million cubic meters of waste-water is discharged in India, of which nearly 32% is currently treated with primary treatment systems. In eastern part of Calcutta alone a number of sewage fed ponds covering about 25000 hectare area are established which approximately produce about 7000 tonnes of fishes annually (Gupta and Misra, 1997).

Among various biological materials in use under waste-water management technology are also certain plant types being used for reducing heavy metal toxicity levels in water such as neem bark is used for the elimination of mercury, pine leaves for nickel, cypress leaves for lead and Ipomea leaves for cadmium. *Chlorele vulgaris* is reported to remove these metals up to 100 mg/l of the sewage waste (Shrivastava and Sohani, 2002).

Bio-accumulation method is effectively used for removal of toxic metals in waste-water treatment and industrial waste. This process is a component of commercial water treatment circuit and use a combination of flocculation and settling as a mechanism by which metals are removed. Micro-organisms and their extracellular products of metabolism play an active-role in flocculation. Majoritily, the metals present in the system seem to be removed from solution through adsorption onto the floes or onto the sludge by physico-chemical interactions involving the ion-exchange types of mechanism (Dahiya, 1990).

Absorption and adsorption of metals by micro-organisms involves more than one physiological activities. These mainly include-metal transport mechanism, which accumulate and store selective metals including Mg, Ca, Mn, Fe, Zn, Si, K, Cu, Co, Mo, The transport of metal in microbial cells has been studied in *Chlorella pyrenoidosa, Dunaliela tertiolecia,* Baker's yeast, *Necocosmospora vasinfectum* and *Escherichia coli* (Kelly *et al.*, 1979).

Failla *et al.* (1976) reported the accumulation and storage of Zn^{2+} by *Candida utilis*. Metal accumulation affected by chemical factors such as cations soluble amino acids and physical factors including temperature, conductance, pH etc. Mostly, the binding efficiency shows an acidic pH. When pH goes to alkaline the binding efficiency decrease (Gloaguen and Morvan, 1979). In general, binding ability of the barks for the metal is, Pb>Cu>Cr Ni>Fe.

Krishnan *et al.* (2002) have further stated that heavy metals such as mercury, led, cadmium and cobalt from waste-water can be effectively sorbed by sulphurized activated carbon. Sulphurized activated carbon was prepared from sugar industry waste, bagasse pith, by single-step steam pyrolysis in the presence of SO_2 and H_2S at 400°C. The maximum removal took place at the pH range of 4.0-8.0. Maximum removal of 98.2% for Pb(II), 97.5,% for Hg (II), 91.5% for Cd(II) and 73.6% for Co(II) was observed at an initial concentration of 100.0 mg/L at 30°C and pH 6.0.

REFERENCES

Bhattacharya and Chakraborty, S. (2002). In: *Ecology of Polluted Waters*, vol. 1 (Ed. Arvind Kumar), APH Publ. Corp., New Delhi. p. 287.

Campbell, P.G.C. and Stokes, P.M. (1985). *Can. J. Dish. Aquat. Sci.* 42: 2034.

Chatterjee, S. and Bhattacharya, S. (1984). *Toxicol. Lett.* 22: 187.

Dahiya, J.S. 1990. *Indian J. Microbiology* 30: 1-12.

Dara, S.S. (1993). S. Chand & Co. Ltd., , New Delhi, p. 210

Eaton, D.I. (1985). *Toxicol. Appl. Pharmacol.* 78: 158.

Failla, M.L., Benedict, C.D. and Weinberg, E.D. 1976. *J. of General Microbiology*. 94: 23.

Gloaguen, V. and Morvan, H. 1997. *IJEH*. A32: 901.

Gould, E. (1977). In: *Physiological response of Marine Biota to Pollutants* (eds.), Vermberg. F. J. and others), Academic Press, New York, p. 209.

Greig, R.A., Wengloff, D. and Shelpuk, C. 1975. *Pest. Monit.* J.9: 15.

Gupta, D.C., Peters E. and Yadava, R.N. (2002). In: *Ecology of Polluted Waters* (vol. 1) Ed. Arvind Kumar, APH Publ. Corp., New Delhi, p. 515.

Gupta, S.K. and Mitra, A. (1997). *J. Ind. Soc. Soil Sc.*, p. 52.

Haslan, S.M. (1978). *River plants*. Cambridge Univ. Press, Cambridge, p. 396.

Heath, A.G. (1992). Water Pollution and Fish Phyriology CRC Press, NC. Florida, USA, 1-245 pp.

Hughes, G.M., Perry, S.F. and Brown, V.M. (1979). *Water Res.* 13: 665.

Kelly, D.P., Norris, P.R., Brierley, C.L. (1979). *Symp. Soc. Gen. Microgeology*. 29: 263.

Kirubagaran, R. and Joy, K. (1992). *J. Fish Biol.* 14: 305.

Krishnan, K., Vinod, V.P. and Anirudhan, T.S. (2002). In: *Ecology of Polluted Waters* (vol. 1) Ed. Arvind Kumar, A.P.H. Publ. Corp., New Delhi. p. 549.

Lloyd, E.T. (1977). *Environ. Res.* 13: 62.

Maiti, P and Banerjee, S. (2002). In: *Ecology of Polluted Waters* (vol. 1) Ed. Arvind Kumar, APH Publ. Corp., New Delhi. p. 217.

Newman, M.W. and MacLean, S.A. (1974). Histopathology, in Physiological Response of the Cunner, *Tautogotabrus adspersus* to cadmium, NOAA Technical Report NMFS SSRF-681, 27.

Norey, C.G., Brown, M.W., Cryer, A. and Kay, J. (1990). *Comp. Biochem. Physiol.* 96C: 181.

Part, P. and Lock, R.A.C. (1983). *Comp. Biochem, Physiol.* 76C, pp. 259-263.

Robohn, R.A (1986). *Vety. Immunol. Immunopathol.* 12: 251.

Shaffi, S. (1993). *Physiol. Res.* 42: 7.

Shrivastava, V.S. and Sohani, D. (2002). In: *Ecology of Polluted Waters* (vol. 1) Ed. Arvind Kumar, APH Publ. Corp., New Delhi. p. 435.

Spry, D. and Wiener, J. (1991). *Environ. Pollut.*, 71: 243.

Tariq, W.K., M.D. George and R.S. Gupta (1979). *Mar. Pollut. Bull.* 10: 357.

Tewari, H., Gill, T. and Pant, J. (1987). *Bull. Environ. Contam.* Toxicol. 38: 748.

Verma, A.R. (1999). Research Thesis, Central Institute of Fisheries Education, Versova, Mumbai.

6

Aquatic Pollution and its Influence on Aquacultural Productivities - A Review

N.K. Yadava, R.K. Gupta and K.L. Jain

Department of Zoology & Aquaculture CCS Haryana Agricultural University, Hisar-125 004, India.

ABSTRACT

Due to population explosion, industrialization, lack of proper sanitary facilities and treatment of wastes, the water quality of Indian towns is deteriorating rapidly, all of which together put a great pressure on the existing water resources. It is felt that there is an urgent need to document the quality of water, productivity of water-bodies, and to identify the future trends, particularly in developing countries. The interest in the use of treated waste-waters in most of the countries is of comparatively recent origin, the concept of using sewage effluents for agricultural production is two thousand years old. World wide attention has been focused by recent events in many parts of the world on the problems of shortages and mal-distribution of food resources. The judicious use of waste-water is an alternative to meet the critical shortages of world food supplies. It is one of the methods of developing massive and economic production of protein from natural sources, which at present are adequate , though dwindling fast. Waste water has great potential in aquaculture. It provides tremendous scope for food production, using the biological system, as it is loaded with natural nutrients. One of the major achievements of recycled water in sewage-fed fish polyculture is that, it gives tremendous increase in the fish yield, if properly managed and monitored. Waste water reclamation, minimizes environmental pollution and its proper use can increase the economy and productivity.

Introduction

On account of indiscriminate discharge of urban, livestock, agricultural and industrial wastes, our aquatic system is increasingly becoming unstable. Untreated domestic sewage or industrial effluents when allowed to be discharged at an unchecked rate into water-bodies, create violent biological imbalances and may act as the causative factors leading to an ecological disaster. Impairment in the physico-chemical and other environmental conditions cause a heavy damage to the existence of aquatic flora and fauna. Though wastes disposal and pollution problems are at present comparatively less acute in India, than in many other developing countries, these, however, may attain a working proportions in the near future with an increase in industrialization and urbanization in the country, more so because the present ways of wastes disposal are neither good nor efficient for the judicious utilization of organic wastes. Fresh water ecology, since long, has been extensively investigated for fish culture (Jhingran, 1982). However, use of waste-water in aquatic system for fish culture is uncommon in India, except in a few states like, West Bengal and Tamil Nadu. It is not commonly exploited for fisheries. It is a useful alternative to conventional waste-water treatment system because, in this system, nutrients from domestic waste-water can be regarded as having a two-fold purpose of treating water and producing a useful product.

Domestic sewage is considered to be a good fertilizer for enhancing productivity in fish ponds, due to the presence of the nitrogen and phosphorus in adequate among. As it is well known that water content of domestic sewage is about 99%. Though, the rest consists of organic and inorganic constituents, the organic constituents form three quarters of total quantity. Despite of certain inherent drawbacks, sewage can gainfully be used for increasing biological productivity in the form of phytoplankton and zooplankton. Sewage from organized human settlement, corporate bodies, educational institutions, having residential facilities and hospitals, are being discharged in large quantities and ultimately it reaches the rivers close by. Fish culture forms one of the non-consumptive uses of sewage. The utilization of sewage as fertilizer for fish culture in fresh water ponds, has long been recognized in many countries. Several reports are available about the physico-chemical and micro-biological parameters of sewage-fed fish culture and large scale culture of Indian carps, exotic species and catfishes.

Although, production in waste-water fish ponds, depends largely on the amount of nutrients available in effluents, the application of sewage in ponds has certain limitations. Raw sewage is inimical to fish life due to high values of organic carbon (OC), BOD, CO_2, NH_3, H_2S and negative value of oxygen. However, these limitations can be overcome by primary treatment and dilution with fresh water. At present, there is need to enumerate and define the specific fertility limits, for an optimum production levels in an aquatic system. Our current knowledge regarding the maximum loading levels, rate of mineralization of such organic wastes, self purification processes, as well as the nutrient-productivity relationship, influence of pesticides and heavy metals of urban effluents on survival, growth, production and breeding of fish is very meagre.

The relevant information has been described under the following headings:

1. Pollution in aquaculture due to the organic wastes in excess.
2. Influence of physico-chemical factors on biological productivity of treated waters.
3. Survival, growth and production of fish in addition to some hydrological conditions of treated waters.

Untreated urban sewage effluents when allowed to be discharged at an unchecked rate into water bodies create violent biological imbalances in the aquatic system and may act as the causative factor leading to an ecological disaster. Thus impairment in the physico-chemical and hydro-biological conditions, cause a heavy damage to the existence of aquatic flora and fauna including fishes (Piccinetti and Mantrin, 1969). Therefore, with the continuous increase in industrialization and urbanization in our country, the problem of aquatic pollution is becoming more acute because, the present ways of disposal are neither good nor efficient for the judicious utilization of urban wastes for aquacultural production (Taylor, 1971); Jhingran, 1982). No work has been taken seriously and scientifically in details on the estimation of 'safe' levels of various urban sewage effluents which could successfully be utilized for fish culture without creating pollution problem in the treated waters (Yadava, 1988). The community waste-water or sewage is now a days looked upon as the residue of resources that the community has metabolized. For waste-water treatment and disposal, it is very important to look into recycling this residue through an ecologically balanced system of aquaculture and agriculture. The use of sewage effluent as fish pond fertilizer has been known in many countries but is of rather recent origin in India.

Pollution in aquaculture due to the organic wastes in excess.

Fish-culture forms one of he non-consecutive uses of sewage. The utilisation of sewage as fertilizer for fish culture in fresh water ponds has long been recognised in many countries (Nair 1944, Sarig 1956, Wolny, 1962).

Tsail (1970) reported that the increased pollution due to sewage in the little Petuxent River, Maryland resulted in fish migration and thus effecting the abundance of resident species and the fish community of the stream. Drastic effect was seen in the area lying immediately below the chlorinated sewage effluent.

Avery (1970) found that the rainbow trout, brown trout, brock trout and mountain white fish were found to occur above the sewage outfall while the rainbow trout was present below, where the complete mixing of the sewage had occurred. Above the outfall, the standing crops of rainbow trout were similar during summer and winter. Below the outfall, standing crop was lower during winter.

Pollution was recorded due to agricultural, domestic and industrial wastes in South River, Virginia (USA). It was further reported that these effluents depleted the dissolved oxygen and enriched the river by adding nutrients such as C, K and N which affected the life of fish and of other organisms adversely. Oliver (1973) reported that high concentration of heavy metal in the sediments of Ottawa and Ridean rivers appeared to be related pollution by the discharge from municipal and industrial wastes.

Tsail (1973) conducted comparative studies of water quality and fish species diversity in stream locations immediately above and below the out-falls of 149 secondary sewage treatment plants in Virginia, Maryland and Pennsylvania (USA). Sewage chlorine and turbidity increment resulting from sludge were the major causative factors for the high reduction in diversity of species below the outfall.

Dogadina and Chukhlebova (1973) studied the effects of urban sewage on the sanitary and biological regime of the Lopan river and found that sewage discharge generally exceeded the self-purifying capacity of the river.

Huang and Hwang (1973) also reported that domestic and municipal sewage contained a total phosphorus, inorganic phosphorus and organic phosphorus besides many nutrients required for the productivity of waters.

According to survey of water quality criteria of National Academy of Sciences, & National Academy of Engineering (1972), the concentration of heavy metals like Fe, Cu, Cd and Pb were found in toxic range at the main disposal point of Amritsar city effluents and if such effluents are discharged untreated in any water body, they may cause high damage to aquatic life, both plants and animals (Singh *et al.*, 1985) besides causing pollution problems.

Quasime *et al.*, (1969); Kaliyamurthy (1978) and Kapur (1981) observed the high values of available nutrients as, phosphates, silicates and chlorophyll contents were introduced in a limited amounts in the treated ponds.

Several reports are available about the physico-chemical and micro-biological parameters of sewage-fed fish cultures system (Chaudhari *et al.*, 1974; Balasubramanium *et al.*, 1992; Rajan and Raj 1994) and large scale culture of Indian carps, exotic species and catfishes (Ghosh *et al.*, 1973; Krishnamoorthi *et al.*, 1976; Naskar and Saha 1988; Govindan 1989; and Balasubramanian *et al.*, 1990).

Bhardwaj and Sharma (1998) studied the physico-chemical and biological conditions in a sewage-fed fish pond of Udaipur (Rajasthan). Ortho-phosphate values were found to be higher than the nitrate-nitrogen values. Cyanophycean algae were found to be dominant phytoplankton. Among the zooplankton, copepods were found to be contributing to 60% population.

Sewage waste-water of Dehradun valley was analysed on bimonthly basis (Augusthy *et al.*, 2000). The samples showed low dissolved oxygen and high BOD, throughout the year. TDS and TSS with carbonates and bicarbonates were high. Calcium, potassium, nitrate, phosphate, sulphate, chloride varied considerably during the study period. Heavy metals analysed for two months show concentration of lead, iron, chromium, zinc, mangesium. The sewage in untreated condition is toxic.

Influence of physico-chemical factors on biological productivity of treated waters

A water with a pH range from 6.5 to 9.0 before day break was considered to be the most suitable for pond culture (Swingle, 1957). He also stated that alkaline waters with 9.5 pH are unproductive because carbon-dioxide becomes unavailable in such alkaline waters and fishes are killed at about 11 pH.

Golachowska (1971) studied the pathways of phosphorus in a lake water and found that the production ability of a water body is determined not only by the absolute level of phosphorus contents but also the forms in which it occurred. In different types of lakes, total P during summer, consisted of 40-60 percent of dissolved orth-phosphates, 16-33 percent of dissolved organic P, 3-20 percent particulate organic P and 4.2 to 11.7 percent particulate mineral P. He further found that sources of dissolved organic phosphorus form were algae and bacteria.

Edwards *et al.*, (1972) studied the nutrient contents of barnlot run off water and reported nitrate-nitrogen contents. They further observed that most of the soluble nitrogen was present in the reduced form. While concentrations of phosphorus and potassium, were depending upon the environmental conditions favored oxidation. Shankar (1978) and Hegde (1981) have also observed lower NO_3-N values following fertilization, this reduction may be due to utilization of NO_3-N by phytoplankton or its transformation into lower-forms.

Some of the studies have indicated the nitrogen and phosphorus dynamics in three Central Texas impoundments and found that all lakes served as nutrient traps and based on N-F ratios, it was postulated that N is the nutrient factor which limited algal growth.

Egeland (1973), reported that phosphorus, a constituent of waste-water, if added to lakes and ponds, promotes algae growth, which in turn provided food for increased fish populations. Further, phosphorus could prompted excessive algae growth which might decrease the dissolved oxygen in water resulting in a corresponding decrease of desirable fish population.

Vasisht and Sharma (1975) have also indicated in their studies, the low production of plankton during summer rainy months (July-August) because of blanketing effect due to increased turbidity and thus decrease in photosynthesis rate, they further observed the cumulative effect of sulphates on primary production. There is always a direct and positive correlation between dissolved salts and conductivity, therefore, conductivity of any water body always depends upon the concentration of dissolved salts. When untreated organic wastes are discharged in water, they always increase the conductivity level very high in the treated waters, resulting pollution in water, thus deplete dissolved oxygen because conductivity has an inverse relationship with DO (Klein, 1960; Sreenivasan, 1964; Gill, 1974).

Tondon and Singh (1972) have discussed the relationship between plankton and various physico-chemical factors in a water body, they have concluded that no single factor can govern the production of a water body, on the other hand, the production is controlled by the cumulative or conjoint action of several factors in different degrees. The animal excreta, when recycled in a fish pond, is not only consumed directly by certain species of fish but also the nutrients obtained from these waters may lead to increase in autotrophic production which provides a substrate for heterotrophic production including fish production. Some of the heterotrophs may also form the good of certain fish species (Schroeder 1980; Kapur, 1984a).

Among the various combinations like fish-domestic sewage, fish-rice and fish-livestock integration is most obvious, with both the wasted food of the livestock, as well as the excrets finding its way into aquaculture system, where it is recycled and either utilized directly or indirectly, through the increased autotrophic/ heterotrophic production (Delmendo, 1980; Woynorovich, 1980a, b; Schroeder, 1980; Jhingran and Sharma, 1980; Soeder, 1981; Kapur, 1984a, 1987).

Aquaculture forms an essential component of all such integrated farming systems, as its products not only seems to be the cheapest in terms of quality and quantity of animal protein (Kapur, 1984b) but fish farming is one of the most efficient rearing systems (Bardach *et al.*, 1972; Edward 1980). Manure may enter the food web in aquatic ecosystem either directly consumed by fish or acting as a source of minerals and organic substances used for the production of autotrophic and heterotrophic organisms (Schroeder, 1980). Pande and Dass (1980) have stated that high levels of available potassium and manganese in the waters, depend upon the bottom sediments. They have further indicated that these metals were abundant in soil and rocks of the catchment area from where they got leached down in the water due to different kinds of organic or inorganic pollutants it receives especially, effluents from laboratory, hospitals, laundries etc., similar studies were also made by Hutchinson (1975) and Singh *et al.*, (1985).

Chapman and his group, in a series of papers, discussed various aquacultural approaches to recycling of dissolved nutrients in secondarily treated domestic waste-waters (Kawasaki *et al.*, 1982; Sliva *et al.*, 1982; Gordon *et al.*, 1982).

Microbes by digesting organic matter, liberate bound minerals and carbon dioxide, which become available for photosynthetic production (Anderson and MacFayden, 1976). In the presence of adequate nutrients, primary production reaches a maximum value set by the amount of solar energy penetrating the pond water.

According to Edwards (1980), economical methods to harvest phyto-plankton is to allow the fish to feed upon fish flies. Schroeder (1977), Hepher *et al.*, (1978) and Sandbank and Hepher (1980) have also tried phyto-plankton for fish culture.

The importance of application of organic manures in nursery ponds for a sustained production of fish food organisms have been emphasized by several workers (Alikunhi, 1957; Jhingran, 1982; Kapur, 1981, 1984a). Organic manures are composited in nature and contain almost all the nutrient elements required in the metabolic cycle.

During the use of organic Waste in aquaculture, it is imperative to maintain the adequate dissolved oxygen for the growth and survival of fish. The effect of treated water on dissolved oxygen depends on the added BOD and the temperature of the pond (Wolhfarth *et al.*, 1974).

Brown and Gratzek (1980) have observed that at level below 4 ppm of DO_2, fish may live but will not feed or grow. Fluctuations in DO_2 content is known to be characteristic of fertilized water bodies (Hepher, 1959). It has been reported that ponds with thick phyto-plankton population, sometimes reduce the oxygen concentration to less than 1 ppm in the early morning hours. He has further observed that low oxygen alone will not cause mortality of fish, except they suffer slight distress (Banerjea, 1967). Manjunatha (1977) observed that if an appropriate dose of organic manure is given to fish ponds then dissolved oxygen contents never go below critical level either in daytime or in night. His observations were also supported to Natarajan (1977), but it was further observed that when there was a large bloom of *Volvox* spp, as a result, the early morning dissolved oxygen levels during summer months were low, but after rainy season, DO_2, level less than 3 ppm were never recorded even during nights (Natarajan, 1985), also supported by Ashwathanarayanan (1979) and Chattopadhyay and Mandal (1982).

Wong and Cheung (1985) studied the two level aquatic food chain, simulated for the study of the transfer of heavy metals. The food chain involved animal manures and sewage sludges as nutrient sources. The results demonstrated that animal manures can yield fish and shrimps with higher level of productivity, but lower heavy metal contents, than sewage sludges. While Singh *et al.*, (1985) and Arora *et al.*, (1985) have also indicated the natural leaching of zinc metal at a very very low concentration in natural waters, either from the sediments or from dead shells of zooplankton at the bottom.

The rotifers are prime constituent of zooplankton and thus form major heterotrophic link in food chain. These organisms being hetero-trophic play a key role in cycling of organic materials in aquatic ecosystem. Very little information is available on the riverine rotifers. Notable contribution on the rotifers fauna of rivers are of Benerjea and Motwani (1962) Chacko and Rajgopal (1962), Prakash *et al.*, (1978), Sampath *et al.*, (1979), Mishra and Saksena (1990) and Mishra (1996).

Ahmad Md. Shamim (1996) studied ecology and survey of some algal flora of polluted habitats of Darbhanga. Observation was made on algae of polluted habitats of (1) Sewage-fed pond. (ii) Medical college waste-water (iii) Tannery factory. The distribution pattern was correlated with physico-chemical characteristics of the effluents. It was found that members of cyanophyceae were found dominant in all the polluted habitats. Total nitrogen, phosphate, potassium and chloride content played a vital role in their distribution pattern.

Basheer *et al.*, (1997) studied the seasonal variations in the primary productivity of a pond receiving sewage effluents. Their study illustrates data on primary productivity, chlorophyll content and total phyto-plankton biomass values of a pond receiving city sewage. Correlation coefficients were computed between NPP and GPP and other parameters. Total phyto-plankton and chlorophyll contents showed highly significant correlation with both GPP and NPP.

Venkataraman and Nandi (1998) studied the zooplankton of Damodar river and their importance in biomonitoring. A knowledge of biological aspects of pollution is not only of interest but also essential to study the problem of river pollution. It was found that Cladocera dominated with 28 species, 18 genera and 6 families. *Ceriodaphnia cornuta, Moina micrara and Macrothrix spinesa* were found to occur in most polluted region of the water.

Chaturvedi *et al.*, (1999) studied plankton community of 6 polluted sites in Sanganer; belonging to 3 dye waste-water pools and 3 sites along Amanishah drain carrying polluted water of the Jaipur city and Sanganer town, during spring and late rainy season of 1997. Zooplanktons were totally absent in the waste-water. Members of Bacillariophyceae such as *Amphora, Fragillaria, Frustulia, Navicula and Neidium;* of chlorophyceae such as *Oscillatoria* dominated the pools as wells as in Amanishah drain during the spring and the rainy season.

Saha *et al.*, (1999) studied the diversity of phytoplankton of a sewage polluted brackish water tidal ecosystem. Phytoplankton community of Jagannath canal, an important brackish water tidal ecosystem of north Sundarbans, West Bengal, India was examined for two years to study the diversity of phytoplankton of the canal at monthly intervals. Sixteen genera of phyto-plankton belonging to Chlorophyta (five genera), Bacillariophyta (seven genera and Cyanophyta (four genera) were recorded from the canal. Phyto-plankton counts varied from 47.07 to 89.70 × $10^5/m^3$ with the domination of Chlorophytes. Both density and diversity of phytoplankton was higher during the monsoon season (March to June) than those of pre-monsoon (November to February) and post-monsoon seasons (July to October).

Harikrishan and Aziz (2000) studied primary production in a fresh-water temple tank in Kerala. A year round study was carried out in a temple tank in Thiruvananthapuram, Kerala to investigate the primary productivity in relation to various physico-chemical components. The gross primary production in the temple tank ranged from 288 to 2843 mg Cm^{-3} day^{-1} and net production from 72 to 1188 mg $Cm^{-3}day^{-1}$. The mean gross production was found to be maximum during the post-monsoon period. It was found that temperature and nutrients had a direct effect on the production potential of the temple tank.

Jindal and Bhavleen (2000), studied the effect of heavy metal toxicity on the productivity of a fresh water pond at village Khuda Lahora (Chandigarh). The pond was eutrophic in nature with dominance of Cyanophyceae among phyto-plankton, and Cladocera among zooplankton. The water was alkaline and hard. The primary productivity ranged from 2.326 to 3.888 gcm^{-3} h^{-1}. Based on 50 percent reduction in gross primary production (EC_{50} Values) the order of toxicity of heavy metals to primary production was: Hg>Cu>Cd>Zn>Cr. Respiration like GPP and NPP also retarded from lower to higher concentrations of metals, but the average percentage retardation was generally more than that of production.

Shukla *et al.*, (2000) studied an impact assessment of Industrial waste on river Krishna with reference to the physico-chemical characteristics. The study was carried out at nine sampling stations fixed at 65 Kms stretch. The river received about 8.2×10^4 m^3/day of domestic and industrial waste-water having BOD between 3745 and 10378 mg/1 and COD between 12087 and 17565 mg/1. The factors like colour, temperature, turbidity, dissolved solids, volatile solids, nitrogen, sulphate, phosphate and chloride affects the chemistry of river water which subsequently altered the biological characteristics of the river to great extent.

Survival, Growth and Production of fish in relation to some hydrological conditions of treated waters

Schaperclaus (1959) gave impressive figures for carp production in ponds watered from the Berlin sewage fields. He found that the fish production ranged between 865 to 1.057 kg per hectare annually. It was further stated that the natural feeding conditions were created by the nutrients present in the sewage water. Similar type of studies were also made by Clark and George (1970) and Neelkanthan and Sondhi (1984). Hickling (1962) also stated that sewage contains abundant of salts, which give rise to a very rich population of planktons which are used as food by the fish. Low production of fish was also recorded in a sewage enriched tank by Ray *et al.,* (1979) though the phyto-plankton production was high than in the ordinary irrigation tank, it was probably due tough pH resulting in the production of unionized ammonia, which is toxic to fish and thus probably influenced the fish production.

Egri and Man (1972) studied the physico-chemical factors (colour, turbidity, temperature, alkalinity, Ca, Mn, chlorides, sulphates, H_2O, Fe, ammonia, nitrates dissolved oxygen and water oxidability) of Romanian pond waters with special reference to the elements of fishery importance. They concluded that the pond waters provide favourable ecological conditions for profitable fish production. Similar studies were also made by Respes *et al.,* (1972). Where as studies made by Singh and Bhowmic (1985) indicted that excess of chloride contents in a water body cause an indirect influences on the biotic life including fish. It was further observed that the deficiency in dissolved oxygen and high concentration of chloride contents due to addition of excess of sewage effluents in the Ganga river at Kanpur (U.P.), not only caused pollution in water but disappearance of some species of fishes was also recorded.

Muthuswamy (1974) conducted studies on the utilization of stabilization pond effluent for pisciculture and found that maximum weight of *Tilapia* was 127 g in about five months, *Cyprinus carpio* reached a length of 38 cm and 620 g by weight in about 6 months and the maximum weight of *Cirrhinus mrigala* and *Labeo rohita* were 420 g and 800 g respectively .The maximum primary production rate was observed to be 43.30 g oxygen m^2/ day.

Usually higher fish yield is obtained in the area where the primary productivity of water is very high and herbivore fishes are abundant, but primary productivity is not entirely sufficient to account for the high fish yields, therefore some other alternative pathways or food chains may be there in addition to autotrophic production which may be responsible for fish production in manure loaded ponds (Schroeder, 1987; Noreiga-Curtis, 1979; Kapur and LAL, 1986a; Lal and Kapur, 1987).

Schroeder (1980) has also recommended a manuring rate of 2-4 percent (as dry organic matter) of the fish biomass/d that could give a maximum fish yield of 30 kg./ha/d averaging over a growing season. Many organic manures like cow-dung, poultry, duck and pig excreta as well as sewage have been tried for fish production, as much work has been done on the productivity potential of cow-dung which is so far the most extensively used organic manure in aquacultural practices.

Hepher and Schroeder (1974) showed that organic wastes increased both fish yield and the efficiency with which the food was converted into fish flesh. The yield/ha was 70 percent greater than in ponds that did not receive organic wastes, further more 40 percent less supplemental feed was required.

If lower total yield per unit area of pond are acceptable, than organic wastes without supplemental feed may be the primary nutrient source for fish pond (Wohlfarh *et al.*, 1974).

Fish normally do not suffer from infection of *Salmonella spp.* However, they may carry these infections if caught from polluted waters (Shewan, 1962). There may be a public health problem in utilisation of animal wastes for fish farming, since pathogen organisms may occur in the manure, but no problems have been reported by Bryan (1977), Hovojee (1977) and Wohlfarth (1978). Janssen (1970) showed that these enterobacteriaceae can survive and multiply in the gut, mucus and tissue and thus, render the fish a potential vector for human diseases and serve merely as a passive vector. Allen and Hepher (1976) observed that fish grown in partially treated livestock wastes or other organic wastes, equal or sometimes are even better than to those fishes reared in other organic manure treated ponds as far as their tastes and odour are concerned. They have also reported that the off-tastes and odours in fish caught in sewage discharged areas are mainly due to odours arising from industrial hydro-carbons. In the effluent treated waters, high temperature of water certainly fastened the multiplication of micro-organisms and changed the taste and odour of the water (Silvery *et al.,* 1950).

Yadava and Kapil (1988) studied the ecology and chemical composition of domestic sewage waste-waters of Hisar city for fish culture. They studied the analysis of physico-chemical and biological characteristics of the domestic sewage waste-waters of Hisar city (29° 10′N; 75° 46′E) discharged in open fields which mainly consists of composite wastes waters from kitchen and households, laundries, hospitals, livestock farm and some soap factories, but major part of the waste-waters come from urban population. The study revealed an important relationship between physico-chemical and biological factors influencing the productivity in the treated waters if domestic sewage wastes were used for fish culture.

Specific studies on the effect of different BOD loadings of sewage on the instantaneous growth rate of fishes are totally wanting. In view of this, an attempt was made to assess the instantaneous growth rate of fishes grown in different BOD loaded sewage ponds (Rajan &Raj, 1997). The results showed that the initial mean instantaneous growth rate is higher and a steady decrease when the culture days are increasing irrespective of the BOD loadings.

Rajan (1997) studied the survival, feed utilization and growth of *Cirrhinus mrigala* (Ham.) in domestic waste-waters. The effect on the survival of fish *C. marigala* at different concentrations of domestic sewage showed that 60% of sewage was sublethal and 64% caused LC_{50} at 20 hr, exposure (10 days). The sublethal effect of sewage on feed utilization, growth and conversion efficiency in the fish were estimated. The feeding and absorption rates, absorption efficiency, growth rate and gross and net conversion efficiency were higher in 50 percent sewage than 25 percent.

The analytical work on sewage-fed fisheries was conducted to assess the water quality, production and potentiality in terms of fish prawn culture in Nalban sewage fed fisheries project for 90 days (Shobha Sondhia, 2000). It was observed that the physico-chemical parameters of the water body was within the normal range. The water exhibits alkaline nature, pH ranges from 7.1 to 8.4, dissolved oxygen concentration varies between 2.5 and 10.8 ppm. Water temperature was found between 17°C-32°C. Among the zooplanktons and phyto-planktons the cladocerans and cyanophyceae have the maximum percentage of 42.68 and 35.68 respectively. *Daphnia* and *Nostoc* were the dominated species, the macrophye vegetation found was *Eichornia, Lemna* sp., *Azolla sp., Pistia, Hydra sp., Spirodella sp.* The heavy metal concentration was in the normal range, including mercury, zinc, lead, arsenic and other metals except chlorides and sodium showed the higher value 42 mg/1 and 55 mg/1 respectively. The fishes produced were in a good health condition.

REFERENCES

Ahmad, Md. Sahamim (1996), Ecology and survey of some algal flora of polluted habitats of Darbhanga. *J. Env. & Pollut.* 3: pp. 147-151.

Alabaster, J.S. and R. Lloyd (1982). *Water Quality Criteria for Fresh water Fish.* Butterworth Scientific, London, Sydney, Wellington, Durban, Toronto, p. 361.

Alikunhi, K.H. (1957). Fish culture in India. Farm Bull. No. 20. Indian Council of Agricultural Research, New Delhi, p. 144

Allen, G.H. and B. Hepher (1976). Wastes and use of recirculating water in aquaculture. In: *Advances in Aquaculture,* pp. 478-487.

American Public Health Association, 1998. Standard methods for the examination of water and waste-water. Washington, D.C.

Anderson, J. and A. MacFayden (eds.) (1976). *The Role of Terrestrial and Aquatic Organisms in Decomposition Processes.* Blackwell Scientific Publ., London.

Arora, B.R., A.S. Azad, Bijay Singh and G.S. Sekhon (1985).

Pollution potential of municipal waste-waters of Ludhiana, Punjab. *Indian J. Col.* 12: 1-7.

Ashwathanarayana, C. (1979). Studies on the relative merits of three commonly available organic manures for fish culture. M.F. Sc. thesis, University of Agricultural Sciences, Bangalore.

Augusthy, P.O., S.P. Joshi and M.M. Srivastava (2000). Pollution Potential of the Sewage Waste Water of Dehradun. *J. Eniron. Pollut.* 7.(1): 3-75.

Avery, B.L. (1970) Effects of domestic sewage on aquatic insects and salmonids on the East Gallatin River, Montain, *Wat.* Res. 4: pp. 165-177 (Cited from *Biol. Abstr.* 52 (2) E. No. 7178).

Bailly, J.M. (1979). Aquaculture experiments in sewage effluents. In : *Fin Fish Nutrition and Fish Feed Technology.* Hawer, J.E. and K. Tiews (eds.), pp. 423-433.

Balasubramanium, S., M.R. Rajan and S.P. Raj. (1990) Sewage fed fish culture with reference to different BOD levels. Proc. on the National Workshop on Animal Biotechnology. Madras Veterinary College, July 26-28, Madras, pp. 266-274.

Balasubramanium. S., M.R. Rajan and S.P. Raj (1992). Microbiology of sewage grown fishes. *Bioresource Technology* 40 (1): 63-66.

Banerjea, S.M. (1967). Water quality and soil condition of fish ponds in some states of Indian in relation to fish production. *Indian J. Fish.* 14: 115-144.

Banerjea, S. and M.P. Motwani (1962). Some observations on pollution of the Suvaon stream by the effluents of a sugar factory, Balrampur (U.P.) *Ind. J. Fish.* 7: 107-128.

Banerjee, R.K., P. Ray, G.S. Singit and B.R. Dutta (1979). Poultry droppings-its manorial potentiality in aquaculture. *J. Inl. Fish. Soc. Indian,* 11: 94-108.

Bardach, J.E., J.H. Ryther and W.O. Mclarney (1972). *Aquaculture.* Wiley-Interscience, New York, London, Sydney, Toronto. pp. 868.

Basheer, V.S., Asif. A. and Aftab Alam (1997). Seasonal variations in the primary productivity of a pond receiving sewage effluents. *J. Inland fish* Soc. India 28: 76-82.

Bharadwaj, R. and L.L. Sharma (1999). Study of some physico-Chemical characteristics of a sewage fertilized seasonal pond of Udaipur (Rajasthan). *J. Environ. & Poll.* 6(4): 255-260.

Brown, E.E. and J.A. Gratzek (1980). *Fish Farming Handbook.* AVI Publishing Company, INC. Westport, Connectricut.

Bryan, F.L. (1977). Diseases transmitted by foods contaminated by waste-water. *J. Food Prot.* 40: 45-56.

Chacko, P.I. and A. Rajagopal (1982). Hydrobiology and fisheries of the Ennore river near Madras from April 1960 to March 1961. *Madras J. Fisher.* 1: 102-104.

Chatterjee, S.K. Arora and D.R. Gupta (1967). Some observations on utilisation of sewage for fish culture in oxidation ponds. *Environ Hlth.* 9: 156-161.

Chattopadyaya, G.N. and L.N. Mandal (1982). Influence of manure and fertilizers of some chemical and biological properties of soil and water of a brackish water fish pond. J. Inland Fish. Soc. India, pp. 190-200.

Chaturvedi, R.K., K.P. Sharma, Kamyani Sharma, S.M. Bhardwaj and Subhasini Sharma (1999). Plankton Community of Polluted Waters Around Sanganer, Jaipur. *J. Enuiron. & Pollut.* 6(1): 77-84.

Chaudhari, N., S. Niyoji and C. Chakraborty (1974). Investigation on biofileer plant at Gangulibagan, Calcutta. 1. Sewage characteristics. *Indian J. Environ.* 16(4): 12-15.

Clark, S.M. and G.R. George (1970). Limnological studies of lower Columbia River, 1967-78. US Fish Wild. *Serv. Spec. Sci. Rep. Fish.* 610: 1-15. (Cited from *Biol. Abstr.* 52 (1): E.No. 1280).

Curtis, P.N. (1979). Primary productivity and related fish yield in intensively manured fish fonds. *Aquaculture* 17: 335-334.

Dean, J.G. F.L. Bosqui and K.H. Lanoville (1972). Removing heavy metals from waste-water. *Environ. Sci. Technol.* 6: 518-522.

Delmendo, M.N. (1980). A review of integrated Livestock fowl-fish farming systems. In: Proc. ICIARM-SEARCA Conf. on Integrated Agriculture-Aquaculture Farming Systems, Manilla, pp.59-71.

Dogadiana, T.V. and N.A. Chukhelbova (1973). Effect of urban sewage on the sanitary and biological regime of the Lopan river. *Girdobiol. Zh.* 8: 93-97 (Cited from *Biol. Abstr.* 56 (11) : E. No. 63423).

Doudroff, P. and L.S. Dean (1967). Dissolved oxygen criteria for the protection of fish. *Amer. Fish. Soc. Spl. Publ.* 4: p. 37.

Edwards, P. (1980). A review of recycling organic wastes into fish, with emphasis on the tropics, *Aquaculture* 21: pp. 261-279.

Edwards, W.M., E.C. Simpson and M.H. Frero (1972). Nutrient content of barnlot run-off water. *J. Environ.* Qual. 1: 401-405. (Cited from *Biol.* Abstr. 55(11): E.No. 59653).

Egeland, D.R. (1973). Land disposal. 1. A. giant step backward. *J. Wat. Pollut. Cont. Fed* 45 pp. 1465-1475.

Egri, L. and C. Man (1972). A physico-chemical study of the water of small and large ponds and its effect on pisciculture. *Inst. Agron. Dr. Petru Grosa Gluj Luor. Stint. Ser. Zootech. 28:* 253-260 (Cited from *Biol.* Abstr. 54 (11): E. No. 58778).

Ellis, M.M. (1937). Detection and measurement of stream pollution in. U.S. *Bur, Fish, Bull.* 22 (58): 365-437.

Ghosh, A., M. K. Banerjee and H.L. Rao (1973). Some observations on the cultural prospects of silver carp. Hypopthalmichtyps molitrix (Valenciennes) in sewage-fed ponds, *J. Inland fish. Sci. India* 5: 131-133.

Ghosh, A., S.C., Banerzee and L.H. Rao (1974). Studies on hydro-biological conditions of a sewage fed pond with a note on their role in fish culture.

Gill, H.S. (1974). Effect of pollution upon the physico-chemical composition and fisheries of Punjab waters. Ph.D. Thesis, Punjab Agricultural University, Ludhiana, India, p. 147.

Gill, H.S. and H.S., Toor (1976). Bioassay studies on the toxicity to fish of distillery effluents. *Acta. Hydrobiol.* 18: 229-234.

Golachowska, J.B. (1971). The pathways of phosphorus in lake water. Pol. *Arch, Hydrobiol.* 18: 325-345. (Cited from *Biol. Abstr.* 54 (12) E.No. 65350).

Gopalkrishnan, V., P., Ray and B.B., Ghosh (1973). Present status of pollution in the Hooghly Estuary with special reference to the adverse effects abserved on fishery resources. *Symp. Environ. Pollut. Jan.* 17-19, Indian Ass. Wat. Pollut. Cont. And CPHERE, Nagpur.

Gordon, M.S., D.J., Chapman, L.Y., Kawasaki, E.T., Silva and D.P., Yu (1982). Aquacultural approaches to recycling of dissolved nutrients in secondarily treated domestic waste-waters-III. Conclusions, design and operational considerations for artificial food drains. *Water Res.* 16: 67-71.

Govind, D.V., K.V., Rajgopal and G.S., Singh (1978). Studies on the comparative efficiencies of organic manures as fish food producers. *J. Inl. Fish. Soc. India* 10: 101-106.

Govindan, V.S. (1989). Food and feed production from municipal waste-water treatment. Biological wastes 30: 169-179.

Harikrishanan, K. and P.K., Abdul Aziz (2000). Primary Production Studies in a Freshwater Temple Tank in Kerala. *Indian J. Environ. and Ecoplan.* 3(1): 127-130.

Hegde, R.N. (1981). Studies on the effect of fertilizing fish ponds with poultry manure and varying doses of NPK. M.Sc. Thesis, University of Agricultural Sciences, Bangalore.

Hepher, B. (1959). Chemical fluctuations of the water of fertilized and unfertilized fish ponds in a subtropical climate. *Bamidgeh* 11. (1): pp. 3-32.

Hepher, B. and G., Schroeder (1974). Waste water utilization in integrated aquaculture and agriculture systems. In *Waste Water use in Production of Food and Fiber-Proceedings*. U.S. Environmental protection agency, Environmental Protection Technology Series 660/2- 74-041. Washington, D.C., pp. 9-15.

Hepher, B., G., Sandbank and G., Shelef (1978). Alternative sources for warm water fish diets. Proc. Internatl. Symp. Fish Nutr. and Feed Tech. EIFAC. Symp. R/11.2, Hamburg.

Hickling, C.F. (1962). *Fish Culture.* Faber and Faber, London, p. 295.

Hora, S.L. and Pillay, T.V.R. (1962). Handbook on fish culture in the Indo-pacific fisheries region. FAO Fish. Biol. Techn. Paper. 14.

Hovojee (1977). Health effects from waste utilization. In: *Animal Wastes,* Taizanides, E.P. (ed.), Applied Science Publishers Ltd., London, pp. 105-109.

Huang, P.M. and C.P. Hwang (1973). Inorganic and organic phosphorus distribution in domestic and municipal sewage. *Wat. Sewage Works.* June, 1973: 82-83.

Hutchinson, G.E. (1975), Minor metallic elements in lake water. In: *A Treatise on Limnology* vol. 1 *part 2, Chemistry of the Lakes,* pp. 801-835.

Janssen, W.A. (1970). Fish as potential vectors of human bacterial diseases of fishes and shell fishes. *Amer. Fish. Soc. Special Publ.* 5: 284-290.

Jhingran, V.G. (1982). *Fish and Fisheries of India,* 2nd edn., Hindustan Publishing Corporation, Delhi.

Jhingran, V.G. and B.K., Sharma, B.K. (1980). Integrated livestock fish farming in India, Proc. ICIARM-SEARCA Conf. on Integrated Agriculture-Aquaculture Farming Systems, Manilla, Philippines, pp. 135-142.

Jindal, R. and Bhavlen Kaur (2000). Effect of heavy metal toxicity on the productivity of a fresh-water ecosystem. Indian. J. Ecol. 27(1): 27-32.

Kaliyamurthy, M. (1978). Organic production in relation to environmental features, nutrients and fish yield of lake Pulicat. *J. Inland Fish. Soc. India* 10: 69-75.

Kapur, K. (1981). The utilization of some organic wastes for fish culture, *Acta Hydrobiol.* 23: 95-102.

Kapur, K. (1984a). Effect of organic wastes on the primary and secondary levels of production in the treated waters. National Seminar on Organic Waste Utilization and Vermi-composting, Sambalpur. Dec. 1984, AVP. 12.

Kapur, K. (1984 b). The economic aspects of utilization of livestock waste for carp culture. National Seminar on Organic Waste Utilization and Vermi-composting, Sambalpur, Dec. 1984, AVP. 12.

Kapur, K. (1987). The role of organic wastes in composite fish culture in India. *Agric. Wastes,* U.K., 14 (In press).

Kapur, K. and H.S. Toor (1978). Effect of pollution on aquatic fauna with special reference to fish: a review. *Acta Hydrobiol. Krakow,* 1978, Poland, 20: 210-215.

Kapur, K. and K.K. Lal (1986a). Relative potency of certain livestock wastes for fish culture. In: McLean, J. (ed.), Proc. First Asian Fisheries forum, Manilla, Philippines (In press).

Kawasaki, L.Y., E., Tarifeno-Silva, D.P., Yu, M.S., Gordon and D.J., Chapman (1982). Aquacultural approaches to recycling of dissolved nutrients in secondarily treated domestic waste-waters-1. Nutrient uptake and release by artificial food chain. *Water Res.* 16: 37-49.

Klein, L. 91960). *River Pollution.* II. *Causes and Effects.* Butterworths, London, p. 456.

Krishnamoorthi, K.P., M.K., Abdulappa, R. Sarkar and R.H., Siddiwi (1975). Productivity of sewage fertilized fish ponds. *Water Res.* 9: 269-274.

Krishnamoorthi, K.P., M.K., Abdulappa and A.V.J., Rao (1976). Productivity of *Clarius batrachus* in sewage-fed fish ponds. *Indian J. Environ. Hlth.* 18 (1): 292-298.

Lal, K.K. and K., Kapur (1987). Temperature versus biochemical oxygen demand–its significance in quantitative and qualitative evaluation of certain livestock wastes for pisciculture. *Acta Hydrochem. Hydrobiol.* (In press).

Lyubumova, T.S. (1974). Effect of fertilization on zooplankton growth in carp nursery ponds in Southern Urals. *Hydrobiol. J.* 10: 1-95.

Manjunatha, M. (1977). Studies on the use of different combinations of organic manures in relation to fish culture. M.F. Sc. Thesis, University of Agricultural Sciences, Bangalore.

Mara, D. (1976). Sewage Treatment in Hot Climates. Chichester, New York, Brisbane, Toronto: John Wiley and Sons. p. 168.

Michael, R.G. (1969). Seasonal trends in physico-chemical factors and plankton of a fresh water fish pond and their role in fish culture. *Hydro-biologia.* 33: 145-160.

Mishra, S.R. (1996). The zooplankton of Indian rivers: A Review. In: Assessment of Water Pollution (Ed : S.R. Mishra), APH. Publishing Corporation, New Delhi, pp. 810-92.

Mishra, S.R. and D.N., Saksena (1990). Seasonal abundance of the zooplankton of waste-water from the industrial complex at Birla Nagar (Gwalior). India. *Acta. Hydrochim. Hydro-biol.*18: 215-220.

Mishra, S.R. and D.N., Saxena (1998). Rotifers and their seasonal variation in a sewage collecting Morar (Kalipi) river, Gwalior, India, *J. Environ. Biol.* 19 (4): 363-374.

Muthuswamy, S. (1974). Studies on the utilisation of stabilization pond effluent for pisciculture. M.Sc. (PEE) Thesis, University of Madras (Cited from *Indian Ass. Wat. Pollut. Cont. News 1.*, Feb., 1974).

Nair, K.K. (1944). The effects of Calcutta sewage on fish life. *Proc. Natn. Inat. Sci. India.* 18: p. 465.

Naskar, K. and S.K., Saha (1988). Utilisation of domestic wastes for integrated farming in the tropics. *J. Aqua. Trop.* 3: 37-41.

Natarajan, M. (1977). Studies on the relative merits of three locally available organic manure for fish culture. M.F. Sc. Thesis, University of Agricultural Sciences, Bangalore.

Natarajan, M. (1985). Recycling of poultry droppings through tilapia fish culture. Effect of monosex and mixed sex culture on production. *Agricultural Wastes* 12: 29-36.

National Academy of Sciences and National Academy of Engineering (1972). Water Quality Criteria EPA-R_3-73-003. Lis. Govt. Printing Press, Washington, D.C.

Neelkanthan, S. and H.S. Sondhi (1984). Renewable biomass waste management and utilization for energy, feed and manure. National Seminar on Organic Waste Utilization and Vermi-composting, Sambalpur, Dec., 1984.

Neelkanthan, S and H.S., Sondhi (1984). Renewable biomass waste management and utilization for energy, feed and manure. National Seminar on Organic Waste Utilization and Vermi-composting, Sambalpur, Dec., 1984.

Noble, R. (1975). Growing fish in sewage. *New Sci.* 67: 259-261.

Noreige-Curtis, P. (1979). Primary productivity and related fish yield in intensely manured fish ponds. *Aquaculture* 17: pp. 73-105.

Oliver, B.C. (1973). Heavy metal levels of Ottawa and Ridean River Sediments. *Environ. Sci. Tech.* 7: 125-137 (Cited from *Indian Ass. Wat. Pollut. Cont. Newsl.*, Jan., 1974).

Oswald, W.J., A.M., Asce, H.B., Gottas and M., Asce (1957). Photosynthesis in sewage treatment. *Trans. Am. Soc. Civil Engg.*, 1 and 2: 73-105.

Pandey, A.C., Krishnan Gopal and A.K., Pandey (2000). Pollution and fish physiology: A Review. *Aquacult* 1(1): 1-8.

Pande, J. and S.M., Das (1980). Metallic Contents in water and sediments of Lake Nanital. *Indian J. Ecol.* 7: 114-118.

Philipose, M.T. (1959). Fresh-water phyto-plankton of Inland fisheries. Proc. Symp. on Algalogy, pp. 272-291.

Piccinetti, C. and G. Manfrin (1969). Observations on the mortality of fishes and other organisms established in the Adriatic. *Net. Lab. Biol. Mar. Pescafano.* 3: 78-93 (Cited from *Biol* Abstr. 51 (2): E.No. 111441).

Prakash, C., D.C. Rawat and P.F., Grover (1978). Ecological study of the river Jamuna. IAWPC Tech. Annual- V, pp. 32-45.

Quasim, S.Z., W., Wallershans, P.M.A., Bhattathiri and S.A.H., Abidi (1969). Organic production in tropical estuary. *Proc. Indian Acad. Sci.* 69: 51-94.

Rajan, M.R. (1977). Survival, Feed utilization and growth of *Cirrhinus mrigala* (Ham.) in domestic waste-waters. *Environ. Ecol.* 15: 291-295.

Rajan, M.R. and S.P., Raj (1994). Sewage recycling through fish culture. *Environ. Ecol.* 12 (2): 247-251.

Rajan, M.R. and S.P., Raj (1997). Instantaneous growth rate of sewage grown fishes. *J. Environ. Biol.* 18 (2): 127-130.

Ray, P. and A, David (1966). Effects of industrial wastes and sewage upon the chemical and biological composition and fisheries of the River Ganga at Kanpur (U.P.). *Environ. Hlth.* 8: 307-339.

Ray, P., N.G.S., Rao and R.K., Banerjee (1979). Fertility and productivity status of a sewage enriched tank compared to a general tank in Karnataka State. Symposium on Inland Aquaculture, CIFRI, Barrackpore (W.B.), Feb. pp. 12-14.

Respes, R.O., C.E., Wood, J.M., Inglis and W.J., Clark (1972). Seasonal variations in selected pysico-chemical conditions of a small lake in Brazos Country, Texas, Southwest Nat. 17: 240-263 (Cited from *Biol. Abstr.* 55 (9), E.No. 48096).

Shyamalendu Bikash Saha, S.B. Bhattacharyya and Amalesh Choudhury. (1999). Diversity of phyto-plankton of a sewage polluted brackish water tidal ecosystem. *J. Enuron. bil.* 1: 9-14.

Saha, K.C., D.P., Sen, P.C., Mukherjee and S.K., Chakravarty. (1958). Physico-chemical qualities of Calcutta sewage from the view point of pisciculture and the danger of feeding raw sewage to confined fisheries. *Indian J. Fish* 5: 35-59.

Saleem, A. (1963). Preliminary studies on the population and causes of mortality of fish in the Soar Rivers. *Arric Pakistan* 13: 551-562.

Sampath, V., A., Sreenivasan and R., Ananthanarayanan (1979). Rotifers as biological indicators of water quality in Cauvery river. *Proc.*

Symp. Environ. Biol. The Academy of Environmental Biology, Muzaffarnagar (India). 441-452.

Sandbank, E. and B. Hepher (1980). Microalgae growth in waste-water as an ingredient in the diet of warm water fish. In: *Algal Biomass.*

Sarig, S. (1956). Determining the influence of various fertilizers in fish ponds. Bamidgeh 7: 3-9.

Sarkar, R. and K.P., Krishanmoorthy (1979). Diurnal variation studies of zooplankton in sewage fertilized fish ponds of Nagpur. *Indian J. Environ. Hlth.* 20 (4): 366-389.

Schaperclaus, W. (1959). Die Karpfenteichswirtschaf in der Deutsche Demokratischen Republik. *Sitz. Ber. Bd. VII fift. 7 Deutsche Akad. Landwirtschaftwiss Berlin,* 1958 (Cited by C.F. Hickling).

Schroeder, G.L. (1975). Night time material balance for oxygen in fish ponds. *Bamidgeh* 7: 65-75.

Schroeder, G.L. (1977). Agricultural wastes in fish farming: A commercial application of the culture of single celled organisms for protein production. *Water Res.* 11: 419-420.

Schroeder, G.L. (1978). Autotrophic and heterotrophic production of microorganisms, intensely manured fish ponds and related fish yields. *Aquaculture* 14: 303-325.

Schroeder, G.L. (1980). Fish farming in manure loaded ponds. In: Pullin, R.S.V. and Z.H. Shehadeh (eds.). Proc. ICIARM-SEARCA Conf. on Integrated Agriculture Aquaculture Farming Systems, Manila, pp. 73-86.

Sewan, J.M. (1962). Food poisoning by fish and fishery products. In: *Fish as food.* vol. 2, Borostom, G.P. (ed.), Academic Press, New York, pp. 443-466.

Shankar, K.M. (1978). Studies on the effect of fertilizing fish ponds by combining super-phosphate with cattle dung or sewage sludge cake. M.F. Sc. Thesis, University of Agricultural Sciences, Bangalore.

Shelef, G., C.J., Soeder and M., Balaban (eds.). Proc. Symp. on the Production and Use of Microalgal Biomass, Akka, Sep. 1978, pp. 697-709.

Shobha Sondhia (2000), Status and study of physico-chemical parameters and potentiality of nalban sewage fed fisheries, salt lake, Calcutta, *Aquacult,* 1 (1): 9-12.

Shukla, G.R., Rekha Rani Sharma and Ram Manohar (2000). Impact assessment of Industrial pollution on Krishna River. *Aquacult.* 1 (1): 69-90.

Silva, E. T., L.Y., Kawasaki, D.P., Yu, M.S., Gordon and D.J., Chapman (1982). Aquacultural approaches to recycling of dissolved nutrients in secondarily treated domestic waste-waters. II. Biological productivity of artificial food chains. *Water Res.* 16: 51-57.

Silvery, J.K.G., J.C., Rassel, D.R., Rodden and W.C., McCormick. (1950). *J. Am. Wat. Woks. Ass.* 42: 1038 (Cited by Ray and David 1966).

Singh, A.K. and B.N., Bhowmick (1985). Effect of sewage on physico-chemical characteristics and bacterial population of river Ganga at Patna. *Indian J. Ecol.* 12: 17-19.

Singh, J., B.D., Kansal and R.S., Panesar (1985). Chemical composition of waste-water of Amritsar city. *Indian J. Ecol.* 12: 12-16.

Soeder, C.J. (1981). Towards enforced aquaculture. *U.O.F.S. Publ. Series* C. No. 3: 187-194.

Sreenivasan, A. (1964). Limnological features of and primary production in a polluted moat at Vellore, Tamil Nadu, *Environ. Hlth,* 6: 237-245.

Srivastava, C.P. (1998). Pollutants and Nutrient status of Raw Sewage. *Indian J. Environ. Protection* 18(2): 109-111.

Swingle, H.S. (1957). Relationship of pH of pond water to their suitability for fish culture. 9th Pacific Sci, Congr., Bangkok -Thiland (Cited by C.F. Hickling, 1962).

Tondon, K.K. and H., Singh (1972). Effect of certain physico-chemical factors on the plankton of the Nagal lake. *Proc. Indian Acad. Sci.* 76: 15-25.

Tsai, C.F. (1970). Changes in fish population and migration in relation to increased sewage pollution, little Patuxent River, Maryland. *Chasspeake Sci.* 11: 34-41 (Cited from *Biol. Abstr,* 51 (33): E.No. 128910).

Tsai, C.F. (1973). Water quality and fish life below sewage out-falls. *Trans, Am. Fish, Soc.* 103: 281-292 (Cited from *biol. Abstr.* 56 (10) E.No. 53867).

Vaas, K.F. (1948). Notes on the fresh water fish culture in domestic sewage in the tropics. *Landboug Batavia* 20: (Cited by C.F. Hickling, 1962).

Van Schoor, D.S. (1959). Fertilization of fish ponds. *Rco Dep. Hat. Concer. Cape Prov.* 16: 65-68.

Vasisht, H.S. and B.K., Sharma (1975). Ecology of a typical urban pond in Ambala city of the Haryana States *Indian J. Ecol.* 2: pp. 79-86.

Vasisht, H.S. and C., Sheikher (1983). Ecology of waste-waters of the Postgraduate. Institute of Medical Education and Research and Punjab University Campus, Chandigarh. *Indian J. Ecol.* 10: 1-6.

Venkataraman, K. and N.C., Nandi (1998). Zooplankton of Damodar river and their importance in bio-monitoring. Proc. Zool. Soc., Calcutta. 50: 3-11.

Verde, R.S. and S.N., Ajwani (1964). Industrial pollution of Ulhas river. *Environ. Hlth.* 6: 13-17.

Vlasov, V.A. (1972). Effects of organic and inorganic fertilizers on the growth and development of juvenile carp in spawning ponds. *Dokl. Mosk. Skh. Akad. Im. K.A. Timiryazeva.* 381: 220-223 (Cited from *Biol. Abstr.* 56 (1): E.No. 60124).

Welch, P.S. (1952). *Limnology.* McGraw Hill Book Co. Inc., New York, p. 538.

Wilber, C.G. (1971). *The Biological Aspects of Water Pollution.* Charles Thomas Publishers, U.S.A., p. 296.

Wohlfarth, G.W., V., Moar and G., Hylata (1974). Fish and Aquaculture Research Station. *Dor. Israel Private Communication. Wastes,* London, pp. 10-70.

Wonly, A. (1962). The use of purified town sewage for fish rearing. *Roczn. N. ink. rol* (B). 81 (2): 231-249.

Wong, M.H. and Y.H., Cheung (1985). Heavy metal contents of *Macrbrachium hainanense* fed with waste-grown *Chlorella kyrenoidosa. Agric. Wastes* 13: 1-14.

Woynarovich, E. (1980a). Utilization of piggery wastes in fish ponds. In: Pullin, R.S.V. and Z.H., Shehadeh (eds.). proc. ICIARM-SEARCA Conf. on Integrated Agriculture-Aquaculture Farming Systems, Manilla, pp. 125-128.

Woynarovich, E. (1980b). Raising ducks in fish ponds. In: Pullin, R.S.V. and Z.H., Shehadeh (eds.) Proc. ICIARM-SEARCA Conf. on Integrated Agriculture-Aquaculture Farming Systems, Manilla, pp. 129-135.

Yadava, N.K. and R.P., Kapil (1988). Ecology and chemical composition of domestic sewage waste-waters of Hisar city used for fish culture. Proc. Nat. Sem. Recent Trend Fish Biol. pp. 51-59.

7

Feeding Ecology of Rohu Larvae in Relation to Light and Dark Conditions

Seema Langer, Sushma Khajuria and Tajinder Kour

Department of Zoology, University of Jammu, Jammu

ABSTRACT

The studies reveal that in spite of being visual feeder, rohu larvae are capable of feeding in total darkness and prey capture success is comparable to that observed in illuminated conditions. Visibility of prey to predator is an important factor that increases the capture success. The vulnerability of prey among other factors is dependent upon the brood condition (presence or absence of eggs) of prey organisms. Presence of brood because of their opacity make them prominent feeding targets as compared to translucent preadults (without brood). In addition to visibility, presence of brood is an indicator of prey being superior in terms of stored energy and mostly fish larvae are energy maximizers.

Keywords: Feeding, rohu, light, dark.

Introduction

Fish larvae generally feed on diverse zooplankton organisms with selection for a specific plankter at a particular growth phase (Malhotra and Langer, 1993). Various theories of prey selection have been proposed for planktivore fishes.

Birge and Juday (1922) reported calorific value of prey to be an important criterion in prey selection while studying the calorific value of daphnids. Birge and Juday (1922) reported that daphnids carrying many embryos have about 20% of their dry wt. in fat while immature adults with fewer eggs are lean with only about 3.9% fat.

Another important factor contributing to the prey selection is the visibility of prey to predator. Several reviews of the mechanism of prey selection by planktivore fishes (Eggers, 1977; Confer *et al.*, 1978; O. Brien, 1979) have identified that differential visibility due to prey size, shape, colour and motion, differential evasive ability among prey and behavioural preference on the part of predator for certain prey, collectively contribute to prey selection.

Material and Methods

The experiment was conducted to test fish larval feeding under light and dark conditions. The feeding trial was carried out in the petridish containing food organisms (*Simocephalus*) at a density of 5 individuals/ml. To ensure the absolute darkness, the test containers in replicate were placed in glass aquarium covered on all sides with black drawing sheets. Four larvae were preconditioned under dark to acclimatize under these conditions and were not given food for about 12 hours. Two larvae were put in test containers with its replicate and allowed to feed. An equal number of fish larvae were introduced into petridish in replicate containing food organisms with same density and were allowed to feed under natural daylight conditions to compare the feeding of larvae under light and dark conditions.

Results and Discussion

To observe the feeding ecology of rohu larvae in relation to light and dark conditions, the larvae were offered live food organisms *(Simocephalus)*. In the present experiment it was observed that feeding success increased progressively with increasing age which is an effect to meet the energy demands of the growing larvae which also increases with age, and visibility of prey to predator is also an important factor that increases the capture success.

Table 7.1 Feeding of rohu larvae under light and dark conditions (Density: 5 individuals/ml: Volume of water: 20 ml; Prey offered- *Simocephalus*)

Larval Age (Days Post-hatching)	Feeding in Light	Feeding in Darkness
12	20%	17%
16	23%	19%
20	26%	20%

Griffith (1975) while studying prey consumption by *Salmo,* indicated that trout selectively search for prey and are energy maximizers which means that they tend to select only those animals which are superior in terms of energy (Emlen, 1966 and 1968) and/or nutrient value (Pulliam, 1975). Like *Salmo Tor* has been reported to choose and capture zooplankton with large clutches (Malhotra and Langer, 1990) and high calorific values (Galibraith, 1967) which obviously means that both these species (*Salmo* and *Tor*) are visual feeders besides being energy maximizers. That the *Salmo* is definitely a visual feeder and it is also supported by the view of Mellore (1975) who has noted that ephippial daphnia were selected over, similarly sized non-ephippial females. Visual feeding in an anthernid fish has also been suggested by Zaret and Kerfoot (1975) who noted that fish keyed on eye pigmentation of ceriodaphnia selecting those individual which possessed large eyes. Besides, being a visual feeder rohu is capable of feeding in darkness as well. In an experiment, when larvae of different age groups (12, 16 and 20 days post hatching) were fed in total darkness even then they were able to capture prey and surprisingly the prey capture success was also comparable with that observed in light conditions. Present observation indicate ability to feed in total darkness as has already been reported in *Xenomelaniris venezuelae* (Unger *et al.*, 1984), *Cyprinus carpio* (Khadka, 1986) and *uclpae harangues* (Batty *et al.*, 1986; Batty, 1987; Blaxter and Batty, 1990). Blaxter (1986), however contradicted some of earlier claims suggesting that in absence of actual light measurements one cannot be sure of total darkness reaction mentioned above. Laboratory studies (Townsend and Risebrown, 1982); Mills *et al.*, 1986; Batty, 1987; Batty *et al.*, 1990), showed that some are capable of feeding at extremely low light intensities. Therefore, in field studies, moonlight alone can provide threshold level illumination. (Unger *et al.*, 1984).

In laboratory studies (where light is altogether denied), however phenomenon needs further explanation. In this regard mech-noreception (tactile), chemoreception in certain species and the electro-reception are the possible mechanisms that could aid the larvae in prey detection and capture when vision is not of much help (Luenzkovich, 1988). According to Blaxter (1986), in many fish species, free neuromasts are present and are able to detect mechanical disturbances caused by the prey in vicinity (rheostatic). Further, there are reports that chemoreception may also play a role in the foraging behaviour of larval fishes (Blaxter, 1986; Bres, 1986; Lucnzkovich, 1988). Bres (1986) reported individual prey selection in silver carp and attributed this variability (individual selection) to chemical cues (such as taste and odour).

Acknowledgements

The authors are thankful to The Head, Department of Zoology, University of Jammu, for providing necessary facilities.

REFERENCES

Birge, E.A. and Juday, C. (1922). The inland lakes of Wisconsin. The plankton, Part I. Its quality and chemical composition. *Bull wisc Geol. Nat. Hist. Surv*, 64: 1-22.

Batty, R.S., Blaxter, J.H.S. and Libby, D.A. (1986). Herring (clupea harengus) filter feeding in the dark. *Mar. Biol.*, 91: 371-375.

Batty, R.S., Blaxter, J.H.S. and Richard, J.M (1990). Light intensity and the feeding behaviour of herring, clupea harengus. *Mar. Biol.*, 107: 383-388.

Blaxter, J.H.S. and Batty, R.S. (1990). Herring behaviour in light and dark. In: P.J. Herring, A.K. Campbell, M. (eds.), *Whitfield in the sea*. Cambridge University Press, U.K. pp. 209-220.

Bres, M. (1986). A new look at optimal foraging behaviour, rule of thumb in the rainbow trout. *J. Fish. Biol.*, 29: 25-36.

Blaxter. J.H.S. (1986). Development of sense organs and behaviour of teleost larvae with special reference to feeding and predator avoidance. *Trans. Amer. Fish. Soci.*, 115: 98-114.

Confer, J.L., Howick, G.L., Corgette, M.H., Kramer, S.L., Fit Ggilition, S. and Landesberg (1978). *Visual predation by planktivores* 31: 27-37.

Emlen, J.M. (1966). The role of time and energy in food prefernece. *Am. Nat.*, 100: 610-611.

Emlen, J.M. (1968). Optimal choice in animals. *Am. Nat.*, 102: 385-389.

Eggers. D.M (1977). The nature of Prey selection by Planktivorous. *Fish. Ecology*, 58: 46-59.

Griffith, D. (1975). Prey availability and the Predators. *Ecology*, 56: 1209-1214.

Gali braith, M.G. (1967). Size selective predation on Daphnia by rainbow trout and yellow perch. *Trans. Amer. Fish. Soc.*, 96: 1-10.

Luenzkovich, J.J. (1988). The role of prey detention in the selection of prey by pinfish *Lagodon rhomboides* (L). *J. Exp. Biol. Ecol.*, 123: 15-30.

Malhotra and Langer, S. (1990). Culture of simocephalus vetulus (Schodler, 1858) a food organism of fish larvae, in different nutrient media. *Recent Trends in Limnology*, pp. 41-48.

Mills, E.L., Confer, J.L. and Kretchmer, D.W. (1986). Zooplankton selection by young yellow perch: The influence of light, prey density and predator size. *Trans. Amer. Fish. Soci.*, 115: 716-725.

Mellors, W.K (1975). Selective predation on ephippial Daphnia and the resistance of epippial eggs to digestion. *Ecology*, 56: 974-980.

Malhotra Y.R. and Langer, S. (1993). Effect of temperature on growth and fecundity of selected species of cladocera. *J. Ind. Inst. Sci.* 73: 335-345.

O'Brien, O.K. and Riessen, H. (1979). Helmets and invisible armour: Structures reducing predation from tactile and visual planktivores. *Ecology*. 60(2): 287-294.

Pulliam, H.R. (1975). Diet optimization with nutrient constraints. *Amer. Nat.*, 109: 765-768.

Townsend, C.R. and Risebrow, A.J. (1982). The intluence of light level on the functional response of zooplanktonivorus fish. *Oecologia*, 53: 293-295.

Unger, P.A. Lewis, W.H. Jr. and Mc. Clearn, D.H (1984). Non usual feeding in visual planktivore, Xenomelaniris venezuelae. *Oecolgia*, 64: 280-283.

Zaret, T.M. and kerfoot, W.C. (1975). Fish Predation on Bosmina longirostris, body size selection. *Ecology* 56: 232-237.

8

Seasonal Incidence of *Telorchis* (Trematoda: Digenea) in the Fresh Water Turtle, *Kachuga* (Family: emydidae) in Jammu

B.R. Pandoh[1], Anil K. Verma[2] and V. K. Gupta[3]

1.Govt. College for Women Gandhi Nagar, Jammu (J & K).
2. Govt. Degree College (Boys), Udhampur (J & K).
3. Division of Pharmacology, Regional Research Laboratory, (CSIR), Jammu-180001

Introduction

The genus *Telorchis* was erected by Looss and Luhe in the year 1899 independently without one knowing about the work of the other. Of the 57 species described under the genus from different reptilian hosts from all over the world, three have been described from India alone. Two species, *T. dhongokii* and *T. pellucidus* were first of all described by Mehra and Bukhari (1932) from a fresh water turtle *Kachuga dhongoka* from U.P and Chatterji (1948) added the third *T. gorakhpuri* n.sp. from the same host.

From Jammu province of J & K state, Sudan (1979) described two species *T. longicirrus* and *T. brevicirrus* from *Kachuga tectum tectum* and *K. smithi* as new species under the genus. Subsequently, Sharma (1990) redescribed the species *T. brevicirrus* of Sudan (1979) but considered *T. brevicirrus* conspecific with *T. pellucidus* (Mehra and Bokhari, 1932;. Wharton, 1940). In the present investigation, large number of specimens of *T. longicirrus* and *T. pellucidus* were recorded from the intestines of *Kachuga tectum tectum* and *Kachuga smithi* collected from Ghomanasa-

Khul-Chenab fed stream (32° 71′ N 74°8′E and elevation 280 msl) situated about 15 km North West of Jammu city.

Previous studies on various trematode parasites (*Telorchis* included) recorded from different vertebrate hosts from Jammu province (Sudan, 1979; Khar, 1983; Verma, 1988; Singh, 1989; Janak Raj, 1990 and Sharma, 1990) dealt mainly with the taxonomic descriptions. In this study the seasonal distribution of the trematode *Telorchis* in the reptilian hosts *Kachuga tectum tectum* and *K. smithi* (Family; Emydidae) has been reported.

Methodology Followed

The host specimens were collected and identified by the methods as described by (Gupta, 1978). Physiography and topography of the collecting site, recovery of the parasites, their fixation and staining has already been described elsewhere (Pandoh, 1992). Seasonal and sex-wise distribution of *Telorchine* trematodes was recorded and statistically analysed (Sokal and Rohlf, 1973). Identification of worms was done after Yamaguti (1971) and Pandoh (1992). The data analysis was done by using following formulae:

$$\textit{Percent Prevalence} = \frac{\text{Hosts infected}}{\text{Hosts examined}} \times 100$$

The intensity was calculated as:

Number of individuals of the parasite species in each infected host.

The Mean Intensity was determined by using following equation :

$$= \frac{\text{Total number of individuals of a particular parasite species in a sample of hosts}}{\text{Total number of infected individuals of the hosts species}}$$

The Abundance was calculated as:

$$= \frac{\text{Total number of individuals of a particular parasite species in a sample of hosts}}{\text{Total number of individuals of the hosts (infected and uninfected) in a sample}}$$

The parasite load was determined by the following formula:

$$= \frac{\text{Total number of parasites recovered from infected hosts}}{\text{Number of infected hosts}}$$

The usage of the terms prevalence, intensity, mean intensity, abundance, etc. followed here is according to the recommendation of an ad hoc committee of the American Society of Parasitologists (Margolis *et al.*, 1982).

Observations and Discussion

The two telorchine trematodes (*Telorchis pellucidus* and *T. longicirrus*) showed an interesting pattern of infection in the two turtle hosts. While *T. pellucidus* was found parasitic in *K. t. tectum* over most part of the year reaching peak load in March, July and December (Tables 8.1 and 8.2), in *K. smithi*, the worms were recorded only once during the present investigation that too from only a single host specimen in the month of July (Table 8.3). On the other hand, *T. longicirrus* was found infecting *K. t. tectum* during the months of January through May, reaching the peak intensity load in March and *K. smithi* in the months of February, March and April with abnormally low parasitic load (Tables 8.1 and 8.2). These observations not only indicate the greater resilience of *K. t. tectum* for the infection of *Telorchis* spp. but also the occurrence of *telorchine* infection in *K. smithi* as only instance of accidental infections or variations in host susceptibility about the ecological factors affecting the organization of endohelminth communities as Price (1980) has stressed, not much information is available.

However, what is interesting is the ecological adjustment shown by the two telorchine trematodes *vis-à-vis* their host *K. t. tectum* in having established dispersed temporal patterns at interspecific level carefully reached as not to overlap in time and concur in existence within the same microhabitat (i.e., in the same host specimen) so as to avoid the detrimental results of competitive exclusion; a factor amongst so many that affects the parasite-host inter-relationship (Malhotra, 1989).

Although both the species of *Telorchis* exhibited an over-dispersed distribution amongst the hosts examined (Table 8.4) which is a normal phenomenon seen in parasitic distribution

Table 8.1 : Showing data of *Telorchis* recovered from *Kachuga tectum tectum* collected from Ghomanasan Jammu province

Month of Collection	Jan.	Feb.	Mar.	Apr.	May	Jun.	Jul.	Aug.	Sep.	Oct.	Nov.	Dec.
No. of Hosts examined	2	2	11	15	5	1	3	2	2	1	1	2
No. of Hosts infected	2	2	11	14	3	×	1	1	1	×	1	2
% infected	100	100	100	93.33	60.0	×	33.33	50.0	50.0	×	100	100
Taxa for which data compared:												
1. *Telorchis pellucidus*	×	1	70	9	×	×	31	10	10	×	7	65
No. of Hosts infected	×	1	5	1	×	×	1	1	1	×	1	2
Parasitic load	×	1	14	9	×	×	31	10	10	×	7	32.5
2. *T. longicirrus*	28	2	127	84	2	×	×	×	×	×	×	×
No. of Hosts infected	2	1	5	10	1	×	×	×	×	×	×	×
Parasitic load	14.0	2.0	25.4	8.4	2.0	×	×	×	×	×	×	×
Total number of parasites	28	3	197	93	2	×	31	10	10	×	7	65
Overall parasitic load	14.0	0.66	19.70	8.45	2.0	×	31.0	10.0	10.0	×	7.0	32.5

Table 8.2 : Showing data of *Telorchis* recovered from *Kachuga smithi* collected from Ghomanasan Jammu province

Month of Collection	Jan.	Feb.	Mar.	Apr.	May	Jun.	Jul.	Aug.	Sep.	Oct.	Nov.	Dec.
No. of Hosts examined	×	2	3	3	×	1	2	3	×	1	×	×
No. of Hosts infected	×	2	2	2	×	1	1	×	×	×	×	×
% infected	×	100	66.66	66.66	×	100	50.0	×	×	×	×	×
Taxa for which data compared :												
1. *Telorchis pellucidus*	×	×	×	×	×	×	23	×	×	×	×	×
No. of Hosts infected	×	×	×	×	×	×	1	×	×	×	×	×
Parasitic load	×	×	×	×	×	×	23.0	×	×	×	×	×
2. *Telorchis longicirrus*	×	1	3	1	×	×	×	×	×	×	×	×
No. of Hosts infected	×	1	2	1	×	×	×	×	×	×	×	×
Parasitic load	×	1.0	1.5	1.0	×	×	×	×	×	×	×	×
Total number of parasites	×	1	3	1	×	×	23	×	×	×	×	×
Overall parasitic load	×	0.5	1.5	0.5	×	×	23.0	×	×	×	×	×

Table 8.3 : Seasonal distribution of infection and parasitic load per month of *Telorchis pellucidus* in the chelonian hosts *K. tectum tectum* and *K. smithi*

	K. tectum tectum				K. smithi			
Months	n	ni(%)	PN	PL	n	ni(%)	PN	PL
March	11	6(54.54)	70	11.66	2	*(00.00)	*	*
April	15	1(6.66)	9	9.00	2	*(00.00)	–	–
May	5	*	–	–	–	–	–	–
June	1	*	–	–	1	*(00.00)	–	–
July	3	1(33.33)	31	31.00	2	1(50.00)	23	23
August	2	1(50.00)	10	10.0	3	*(00.00)	–	–
September	2	1(50.00)	10	10.00	–	–	–	–
October	1	*	–	–	1	*(00.00)	–	–
November	1	1(100.0)	7	7.0	–	–	–	–
December	2	2(100.0)	65	32.50	–	–	–	–
January	2	*	–	–	–	–	–	–
February	2	1(50.00)	1	1.0	2	*(00.00)	–	–
Total	47	14(29.78)	203	113.16	13	1(7.69)	23	23

Where, $\overline{X}$ 14.02 ± 10.67 (1.0-32.50) $\overline{X}$ 23.0 ± 0.0 (23)

n= number of hosts screened for infection; ni = number of hosts infectd (% prevalence of infection in parentheses); PN = total number of parasites recovered from infected hosts; PL = parasitic load (PN/ni), * = hosts are not infected and – = hosts not available.

Table 8.4 : Seasonal distribution, prevalence of infection and parasitic load per month of *Telorchis longicirrus* in the chelonian hosts *Kachuga tectum tectum* and *Kachuga smithi*

Months	K. tectum tectum				K. smithi			
	n	ni(%)	PN	PL	n	ni(%)	PN	PL
March	11	5(45.45)	127	25.4	2	2(100.0)	3	1.5
April	15	10(66.66)	84	8.4	2	1(50.00)	1	1.0
May	5	1(6.66)	2	1.0	–	–	–	–
June	1	*	*	*	1	*	*	*
July	3	*	*	*	2	*	*	*
August	2	*	*	*	3	–	–	–
September	2	*	*	*	–	–	–	–
October	1	*	*	*	1	*	*	*
November	1	*	*	*	–	–	–	–
December	2	*	*	*	–	–	–	–
January	2	2(100.0)	28	14.0	–	–	–	–
February	2	1(50.00)	2	1.0	2	1(50.00)	1	1.0
Total	47	18(40.42)	243	49.8	13	4(30.76)	5	3.5

Where 12.78 ± 1.37 (1.0 – 25.4) 1.25 ± 0.25 (1.0 –1.5)

n =number of hosts examined; ni = number of hosts found infected (% prevalence of infection in parentheses); PN = total number of parasites recovered; PL = Parasitic load (PN/ni); – = hosts not available; * = parasites not recovered.

(Kennedy, 1970; Crofton, 1971), yet of the two turtle hosts *K. t. tectum* was found more often parasitized and consistently more heavily by the telorchine worms than *K. smithi* – a finding which points to some sort of difference in infection process and is consistent with one of the special features of over-dispersed distribution.

Tables 8.5 and 8.6 reveal an interesting finding on the differential parasitic pattern of *Telorchis pellucidus* and *T. longicirrus vis-à-vis* the sexes of the turtles, *K. t. tectum* and *K. smithi*. While *T. pellucidus* emerged as parasite found more commonly in male members of the two turtles, *T. longicirrus*, on the contrary, emerged to be more frequent in the females of the two turtle hosts.

Apart from indicating possibly the involvement of some intrinsic preference between the sexes of the host and the parasitising species of *Telorchis*, the data in Table 8.5 also reveals how the phenomenon of competitive exclusion seems to operate in this instance at the same time.

Perusal of the data in Tables 8.5 and 8.6 indicate an interesting finding with regard to the variation in host susceptibility of the two emydine turtles to parasitic pattern of *T. pellucidus* and *T. longicirrus*. While *K. t. tectum* shows significantly high (55.0%) parasitic intensity in its females than in its male members (29.62%), the relative abundance of *T. longicirrus* infection in the two sexes of the host *K. t. tectum* is, however, comparable suggesting thereby more or less equal degree of susceptibility of the host *K. t. tectum* at the level of the sexes for the infection of *T. longicirrus*. Contrarily, however, female *K. smithi* are significantly more susceptible to the infection of *T. longicirrus* than are its male members although from the point of view of intensity and the parasitic load *K. smithi* appears to be far less susceptible than *K. t. tectum* to the infection of *T. longicirrus* (Table 8.7).

This is apparent from the observations (Table 8.6) that of a total number of 248 worms of *T. longicirrus* collected during the present investigations, 234 (98%) were recovered from *K. t. tectum* turtles and not even 2 percent from *K. smithi*.

The infection of *T. pellucidus*, on the other hand, seems to skew more in favour of the male members of the two turtle hosts.

Table 8.5 : Showing sex–wsie distribution, prevalence, intensity, parasitic load and abundance of *Telorchis pellucidus* in the host *K. tectum tectum* and *K. smithi*

Host (s)	Host sex (n)	Infected (ni) (%)	Uninfected	PN	PL	Intensity (range)	Abundance
K.t. tectum	Male (27)	11(40.74)	16	169	15.36	(1–60)	6.25
	Female (20)	3(15.00)	17	34	11.33	(1–31)	1.70
K. smithi	Male (7)	1(14.28)	6	23	23.00	–	3.28
	Female (6)	–(00.00)	6	00	00	–	–

Table 8.6 : Showing sex-wise distribution, prevalence, intensity parasitic load, abundance of *Telorchis longicirrus* in the host *K. tectum tectum* and *K. smithi.*

Host (s)	Host sex (n)	Infected (ni) (%)	Uninfected	PN	PL	Intensity (range)	Abundance
K.t. tectum	Male (27)	8(29.62)	19	125	15.62	(2-29)	4.62
	Female (20)	11(55.00)	9	118	10.72	(1-70)	5.90
K. smithi	Male (7)	1(14.28)	6	1	1.00		0.14
	Female (6)	3(50.00)	3	4	1.33	(1-2)	0.66

Table 8.7 : Showing infelection of *Telorchis* in Emydine turtles from Ghomanasan from Jammu province.

	K. tectum tectum				K. smithi			
	M (n=27)	%	F (n=20)	%	M (n=7)	%	F (n=)	%
Telorchis pellucidus	11	(40.74)	3	(15.0)	1	(14.28)	0	–
T. longicirrus	8	(29.62)	11	(55.0)	1	(14.28)	3	50.0

Where, M = Male and F = Female.

While around 90% of the total number of worms of *T. pellucidus* were recovered from *K. t. tectum,* a pattern rather similar to that observed in *T. longicirrus,* but significantly different in respect of sex-related distribution pattern of the parasites (Table 8.5). Moreover, the relative abundance of *T. longicirrus* in the two sexes of *K. t. tectum* is not significantly different when compared with the relative abundance of *T. pellucidus* which on the contrary in male *K. t. tectum* is significantly higher than that in the females of the host. This difference is also apparent at the level of the two sexes of the other turtle host, *K. smithi* in which all the worms of *T. pellucidus* recovered were collected only from males and none from females. Variations in sex-related host susceptibility, particularly in regard to infections by cestodes is documented (Semenova, 1975; and Nama and Parihar, 1976) and has been attributed to variability in infections in different sexes.

These findings therefore, lead one to conclude that the telorchid infection found in the turtles of J & K state is predominately *K. t. tectum* oriented and that of the two species known so far *T. pellucidus* shows a positive sex-related distribution, being more frequently found in males than in females of the host, although no biological significance can be placed on this difference at this time.

Acknowledgements

We are grateful to Head P.G. Deptt. of Zoology, University of Jammu, Jammu for providing necessary facilities at the department for conduction of the work. Senior author (BRP) is indebted to the Principal, Govt. College for Women, Gandhi Nagar, Jammu for moral support and encouragement.

REFERENCES

Chatterji, P.N. 1948. On some heterophyid trematodes of the genus *Haplorchis* Looss, 1988. *Prco. 34th Ind. Congr. Abstr.* 7: 175.

Crofton, H.D. 1971. A quantitative approach to parasitism. *Parasitology,* 62: 163–179.

Gupta, V.K. (1978) Studies on the female reproduction cycles of some fresh water chalones from Jammu. Ph. D. Thesis Unpublished-University of Jammu, Jammu, p. 378.

Janak Raj. 1990. Studies on *Clinostome metacercariae* from fishes in Jammu. M. Phil. Dissertation (Unpublished), University of Jammu, Jammu. pp. 1-93.

Kennedy, C.R. 1970. The population biology of helminths of British fresh water fishes – *Proc. British Soc. Parasit, 8:* 145-159.

Khar, H.K. 1983. Studies on the larval ecology of trematodes in some water bodies of Jammu Ph. D. Thesis, University of Jammu, Jammu: 1-386.

Malhotra S.K. 1989. Ecological ichthyioparasitology. *Ind. Rev. life sciences,* 9: 253–296.

Margolis *et al.* (1982). The use of ecological terms in parasitology (report of an ad hoc committee of the American Society of Parasitologists). *J. Parasitol.,* 68: 131–33.

Mehra and Bokhari (1932) c.f. Yamaguti, 1971.

Nama, H.S. and Parihar, A. 1976. Quantitative and qualitative analysis of helminth fauna in *Rattus rattus rufescens.* J. Helminth. 50: 99–102.

Pandoh, B.R. 1992. Ecological Studies on Trematodes of some aquatic vertebrates of Jammu province Ph. D. Thesis (Unpublished), Univ. of Jammu, Jammu: 1–544.

Price, E.W. 1980. The evolutionary biology of parasites. Princton Univ. Press, Princton, New Jerseay.

Semenova, N.N. 1975. Affect of some ecological factors on formation of helminth fauna of rodents. *Mater. Nauch Knof. Vses. Obs. Gelmint.* 27: 126–134.

Sharma, S. 1990. Telorchid infection in Emydid turtles in Jammu. A dissertation submitted of Univ. of Jammu for M. Phil Degree (Unpublished): 1-100.

Singh, G. 1989. Studies on Lecithodendriid infection in frogs in Jammu. M. Phil Dissertation submitted to Univ. of Jammu (Unpublished): pp. 1-99.

Sokal, R.R. and Rohlf, F.J. 1973. Introduction to biostatistics. A series of books in Biology, W.H. Freeman and Co. Publ. San. Francisco, p. 368.

Sudan, O.S. 1979. "Studies on trematode fauna of Jammu province". Thesis submitted for the degree of Doctor of Philosophy to Univ. of Jammu, pp. 1-453.

Verma, A.K. 1988. "Polymorphism in *Paradistomoides gregarinum*". M. Phil. Dissertation, Univ. of Jammu, pp. 1-99.

Wharton, G.W. 1940. The genus *Telorchis, Protenes* and *Auridistonum.* (Trematoda; Reniferidae). *J. Par.,* 26(6): 497-518.

Yamaguti, S. 1971. Synopsis of Digenetic Trematodes of Vertebrates. Keigaku Publishing Co., Tokyo, Japan, 1 and 2: 1-1074.

9

Seasonal Population Dynamics of *Paradistomoides gregarinum* (Trematoda: Digenea: Dicrocoelidae) in the Lizards *Caltoes versicolor* Daudin and *Hemidactylus flaviviridis* Rupell from J & K State

***Anil K. Verma and P.L. Duda**

**Department of Zoology, Govt. Degree College (Boys), Udhampur.*
P.G. Department of Zoology, University of Jammu, Jammu – 180006.

ABSTRACT

Seasonal population dynamics of *Paradistomoides gregarinum* – a discrocoeline trematode in the two lacertilian hosts viz. *Calotes versicolor* Daudin and *Hemidactylus flaviviridis* Rupell has been studied. The prevalence of infection, parasitic load; relative frequency and maturation stages of the parasite in the two hosts has been discussed in this communication.

Key word : *Paradistomoides*, Seasonal dynamics, *C. versicolor*; *H. flaviviridis*.

Introduction

Paradistomoides gregarinum (Tubangui, 1929) Travassos 1944 is the dicrocoeline fluke that most frequently infects the lizards and is parasitic in the gall-bladder of squamate reptiles. The present communication reports on the seasonal dynamics of this parasite in the two very common lizards namely, *Calotes versicolor* (family : Agamidae) and *Hemidactylus flaviviridis* (family : Gekkonidae), in Jammu (J & K state) the place where the present work was carried out. The two lizards are distinctly different in their habits and habitat, one (*H. flaviviridis)* is a compulsive indoor species and the other (*C. versicolor*) is given to outdoor field life. The difference in their ecological preference appears to strongly influence the incidence, load, recruitment period and prevalence of the fluke in the two lizards.

Materials and Methods

The lizards used in this study were mostly hand collected between March, 1997 to February, 1998, from same locality in Jammu city (J & K state) (Fig. 9.1), ninety five specimens of *Calotes versicolor* and 73 of *H. flaviviridis* were consumed. After capturing the lizards in their natural habitats, the specimens were brought to the Departmental laboratory and kept in well aerated and lighted cages for few days before dissection.

During this interval, the lizards were fed grubs, snails, and insects. Water was provided *ad libitum*. The hosts were dissected within 3-4 days of their capture. Gall-bladder, liver and bile-duct were excised out of the body and placed separately in petri-dishes containing normal physiological saline. A scrutiny for parasites was made in these organs. The parasites were studied alive as far as practicable. A few of them were fixed, stained and mounted by techniques discussed elsewhere in details (Verma, 1988; Duda and Verma, 2002). Identification and systematics followed are mainly after Yamaguti (1971). Numerical, micrometrical and morphometrical data was taken mostly from the preserved specimens, only after ascertaining their taxonomy when alive. Measurements are given as the range followed by the mean in parentheses.

Observations

This description of *paradistomoides gregarinum* is based upon study of both living as well as fixed specimens and measurement of over 200 stained specimens recovered from *Calotes versicolor* and *Hemidactylus flaviviridis.* Body elongated rarely slender or ovoid, 0.61-5.27 mm (2.40) long by 0.29-0.59 mm (1.05) wide at acetabular level. Tegument without spines. Oral sucker spherical

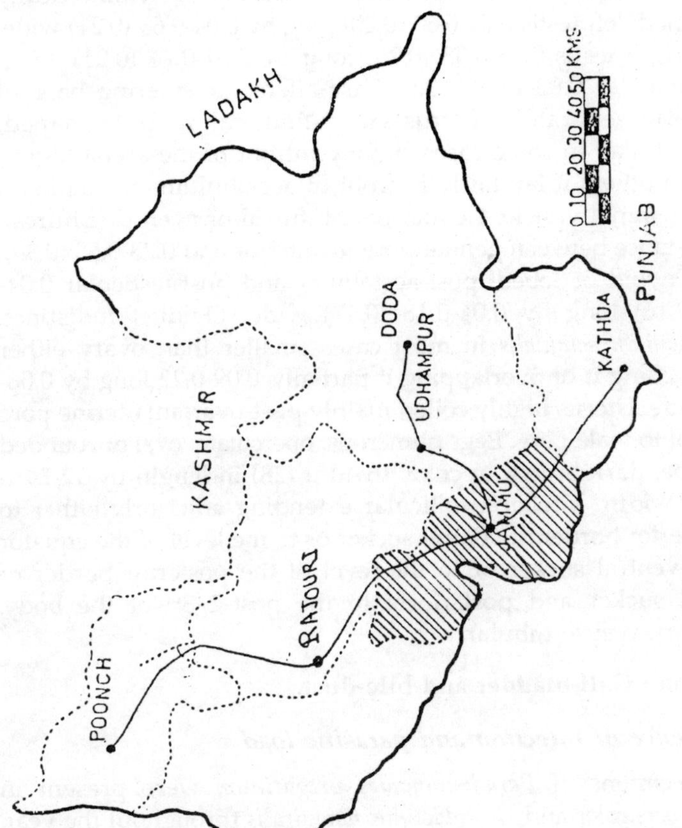

Fig 9.1: Map of Jammu Division (J & K)

to subspherical 0.07-0.37(0.22) long by 0.09-0.39 (0.24) wide. Prepharynx absent; pharynx well developed, 0.01-0. 14 (0.06) long by 0.01-0.15 (0.08) wide. Oesophagus short but conspicuous 0.01-0.26 (0.06) long by 0.01-0.06 (0.02) wide. Caecal bifurcation immediately preacetabular. Caeca extending nearly to posterior extremity. The oro-acetabular distance varied from 0.15-0.95 (0.45). Testes rounded or oval either asymmetrically or symmetrically positioned, left testis 0.04-0.60 (0.20) long by 0.04-0.63 (0.21) wide, while right testis 0.03-0.53 (0.20) long by 0.04-0.74 (0.21) wide. Intertesticular 0.03-1.00 (0.32). Vasa-efferentia entering base of cirrus-sac separately. Cirrus-sac cylindrical to pear-shaped, preacetabular, in some cases slightly infront of the acetabulum , while in others it lay fairly in front of acetabulum, containing a coiled seminal vesicle. Genital pore bifurcal or even prebifurcal. The distance between genital pore to anterior end 0.23-0.58 (0.39). Ovary ovoid or lobed, post-acetabular and post-testicular 0.04-0.36 (0.16) long by 0.04-0.48 (0.17) wide. Oviduct indistinct. *Receptaculum seminalis* in most cases smaller than ovary either just touching it or overlapping it partially, 0.09-0.22 long by 0.06-0.15 wide. Uterus highly coiled mainly post-ovarian; uterine pore adjacent to male pore. Eggs numerous, operculate; oval or rounded in shape, dark-brown in color 16-40 μ (28) in length by 12-24 μ (16) in width. Vitellaria follicular extending anteriorly either to the anterior border of ventral sucker or to the level of the equator of the ventral sucker or to the level of the posterior border of ventral sucker and posteriorly to the post-2/3rd of the body. Excretory vesicle tubular.

Location : Gall-bladder and bile-duct.

Prevalence of infection and parasitic load

Specimens of *Paradistomoides pregarinum* were present in *Calotes versicolor* and *Hemidactylus flaviviridis* throughout the year. The present study reveals a definite seasonal variation in the prevalence of infection and parasitic load of *Paradistomoides gregarinum* in both the lizards hosts (Table 9.1) In the gekkonid lizards, the prevalence of infection because of this parasite was relatively higher during the months from September and October than at any other time of the year. In the Agamid hosts, on the other hand, the prevalence of infection due to this trematode over most part of the year is more less the same. Moreover, these data also indicate that the prevalence of *Paradistomoides* infection in the agamid host touches the lowest in November and December months when in the case of the gekkonid host, it is significantly reduced only during the months from June to August and again in December and January months though the latter reduction is comparatively of a lesser order. Parasitic load in the agamid host is higher in comparison to that in the gekkonid host during most part of the year (Table 9.1).

The overall frequency of the parasite, seasonal index and parasitic load of the parasite in the two hosts stood highest in the month of May (Table 9.1), whereafter, both the frequency and the load exhibited a sudden drop during June and July followed by a further yet milder drop from September onwards to December, and thereafter, the number of parasites recovered per month from comparable number of infected hosts progressively increased to touch the peak value in May. When this data is compared with the temperature curve of the said year, the two curves more or less follow a similar contour; the maximum number of parasites having been apparently obtained during the hotter and humid period of the year than during relatively colder period.

Maturation of *Paradistomoides gregarinum*:

In order to assess the rate of growth and development of *P. gregarinum*, specimens of this parasite were assigned to one of the four growth phases (G_1 to G_4) on the basis of body-size and relative degree of maturity of the reproductive organs.

Phase G_1: Body length 1.01-3.25; width 0.33-1.34. Reproductive system consisting of undifferentiated germinal cells; distinct reproductive organs not present. This phase represented the smallest and least developed worms recovered. Phase G_1 worms occurred throughout the year but were found most frequently in January and February when the prevalence and parasitic load were (40%, 3.00; 55.5%, 3.20) and (71%, 3.00; 62.5%; 3.00) in Agamid and Gekkonid hosts (Table 9.1; Fig. 9.1). The relative density in two months was 1.20 and 2.14 in Agamid and 1.77 and 1.81 in Gekkonid.

Table 9.1. Prevalence of Infection, Relative density and Parasitic load per month of *Paradistomoides gregarinum* in the Agamid host *Calotes versicolor* and gekkonid host *Hemidactylus flaviviridis*.

Host	Calotes versicolor					Hemidactylus flaviviridis			
Months	Pr (%)	P_n	RD	PL	Months	Pr (%)	Pn	RD	PL
Mar 1997	78(9)	30	3.33	4.29	Mar 1997	62.5(8)	7	.87	1.40
Apr	75(12)	41	3.42	4.55	Apr	66.6(9)	16	1.77	2.66
May	81(16)	53	3.31	4.42	May	66.6(9)	16	1.77	2.66
Jun	60(10)	29	2.90	4.83	Jun	33.3(12)	9	.75	2.25
Jul	62(8)	21	2.64	4.20	Jul	42.8(14)	10	.71	1.66
Aug	60(10)	28	2.80	4.67	Aug	40.0(10)	5	.50	1.25
Sep	60(10)	16	1.60	2.67	Sep	68.7(16)	25	1.56	2.27
Oct	57(7)	14	2.00	3.50	Oct	75.0(8)	10	1.25	1.66
Nov	40(5)	6	1.20	2.00	Nov	60.0(10)	9	.90	1.50
Dec	33(3)	4	1.33	4.00	Dec	50.0(6)	7	1.16	3.20
Jan 1998	40(5)	6	1.20	3.00	Jan 1998	55.5(9)	16	1.77	3.20
Feb	71(7)	15	2.14	3.00	Feb	62.5(8)	15	1.87	3.00
Overall	X=63.7 (8.5)	X=21.9	X2.57	X=4.05±.76 (2.67-4.67)	Overall	X=56.3% (9.92)	X=12.25	X=1.23	X=2.19±.65 (1.25-3.20)

Where,
X = Mean values
Pr = Prevalence of Infection
P_n = The total number of parasites recovered
RD = Relative density i.e. total No. of parasites obtained to the total No. of hosts dissected out.
PL = Parasitic load i.e. the number of the parasite obtained per infected host.
Figures in brackets represents the number of hosts specimen examined for the parasite.

Phase G_2: Body length 0.61-4.29; width 0.24-1.69. Differentiation of reproductive systems apparent. Testes and ovary separate from one another and histologically distinct. Vitellaria in the process of develovpment, seminal receptacle, and seminal vesicle present or absent. When present, seminal vesicle and Seminal receptacle lacking sperms; vitelline follicles immature not closely packed. *P.gregarinum* specimens of phase G_2 marking early stages of development were most commonly encountered in April collections. Prevalence of infection and parasitic load of these worms during this month was (75%; 4.55) and (66.6%; 2.66) respectively in Agamid and Gekkonid hosts. The relative density was 4.55 in Agamids and 2.66 in Gekkonids in this month.

Phase G_3: Body length 2.02 – 5.27; width 0.43-2.59; well developed and differentiated male and female reproductive systems present. Seminal vesicle and seminal receptacle usually containing sperms. Vitelline follicles abundant, mature. Uterus packed with eggs. Prevalence of infection and parasitic load of these mature, non-gravid worms was highest in July, when the temperature was 41°C and decreased again in August. The prevalence of Infection, parasitic load and Relative density in Agamid and Gekkonid host in July was (62%, 4.20 and 2.64) and 42.8%, 1.66 and 0.75) respectively.

Phase G_4: Body length 1.32 – 4.34, width 0.64 – 1.92. Male and female reproductive systems somewhat regressed. Eggs very rare. These specimens were abundant in Septmeber but prevalence decreased in November.

Discussion

The study reveals that the lizards first became infected with *Paradistomoides gregarinum* in Septmeber and the period of recruitment continued into November or December. Maturation of *P. gregarinum* started in March to May, as indicated by an increase in the body size, gonad-proportion and parasitic load of phase G_2 specimens. Fully mature specimens (Phase G_3) were found in May to August prior to sharp decline in the temperature of environment which occurred with the onset of autumn. As non-gravid specimens of *P. gregarinum* completed egg production at the end of summer, their frequency and load decreased as senescent worms were lost from their hosts. Concurrent with summer maturation of *P. gregarinum* was an overall decline in both the prevalence and the load. In May an average of 3.8 mature worms (Phase G_3) occurred per infected lizard, whereas in July and August each infected lizard harboured an average of 3.1 and 3.7 mature (Phase G_3) worms respectively. While the exact reason for the decline is not known, intraspecific competition for the limited resources of space and nutrients within the gall-bladder of the hosts is suspected. A more or less similar view has been maintained by Kuntz and Font (1984) in their study on the trematode *Allopodocotyle boleosomi*.

Although most individuals of *P. gregarinum* completed egg production during the warm summer months as is indicated by gravidity of the uterine coils. Some specimens, however, had not commenced or finished sexual reproduction prior to the onset of colder air temperature in autumn.

On the basis of the rather constant parasitic load of G_1, G_2, G_3 and G_4 phase worms throughout the fall and winter, it is probably that growth and development was retarded until the return of warmer air temperature in the follwing spring. A modest rise in the number of mature specimens in May apparently supports this hypothesis. A similar pattern of summer egg production by the remainder of the population is also documented (Font, 1983).

The data presented above clearly shows that of the two host lizards, the agamid lizard *Calotes versicolor* not only hosts higher incidence of infection but also carries a higher parasitic load of the worm *P. gregarinum*, in comparison to the gekkonid host lizard, *Hemidactylus flaviviridis*. While it cannot be immediately stressed whether or not this differential parasitic load and prevalence (incidence) is due to different degrees of tolerance of the two hosts for the said worm, apparently, the capacity of the gekkonid host for sustaining *P. gregarinum* is comparatively lesser than that of the agamid host.

One possible explanation for this difference in parasitic relationship between the two lacertilian hosts for *P. gregarinum* could be the vastly different habits exhibited and the habitats preferred by these two hosts. *Calotes versicolor* being given to an outdoor life has available to it comparatively larger areas for foraging without any spatial constraints from which the other host lizard, *Hemidactylus flaviviridis* suffers. As an excepted consequence, far greater amounts of food must be coming available to the agamid host than to the gekknoid one. The greater the uptake, the greater the energy reserves at the disposal of the agamid lizard for partitioning and hence greater the capacity of this lizard for the consort parasite species.

This view gets fortified by the observations (i) that the parasitic load touches a low in these hosts during hottest season of the year which precisely coincides with the period of reduced foraging activity due to aestivation by this lizard being poikilothermic; (ii) that during the winter months, the load in the agamid host is not as low as in the gekkonid host, could also be attributed to the foraging-activity. Beacause of its living out door in the fields, *Calotes versicolor* during winter months is often seen basking on sunny days and therefore, has access to some food that may be spotted by it. This way, its food or energy reserves do keep augmenting and hence only little reduction in the values of the parasitic load. A similar view has been maintained by Gerald (1967) in his study on the incidence of nematode infection in the painted turtle *Chrysemys picta*.

The adaptation of *P. gregarinum* to a more restricted number of definitive host species seems to be occurring in this case and presents the possibility for study of the intriguing phenomenon of the evolution of host-specificity. Such an investigation, like the understanding of the mechanisms underlying the seasonal population dynamics of *P. gregarinum*, is dependent upon additional knowledge of the life cycle of this parasite.

REFERENCES

Duda, P.L. and Verma, A.K. 2002. Incidence of gall-bladder infection in some lacertillian hosts from Jammu Province. In: *Perspectives in Animal Ecology and Reproduction*, V.K. Gupta, A.K. Verma and J.P. Sharma (Eds.). Daya Publishing House, New Delhi, vol. 1, pp. 41-48.

Font, W.F. 1983. Seasonal population dynamics of five species of intestinal helminthes of the brook stiddeback, *Culaėa inconstans*. *Can . J. Zool.* 61: 2129-2137.

Gerald, W.E. and Gibbons, J.W. 1967. Seasonal incidence of parasites in the painted turtle, *Chrysemys picta*. *J. Parasitology*, 53(4): 818-821.

Kuntz, S.M. and Font, W.F. 1984. Seasonal dynamics of *Allopodocotyle boleosomi* (Pearse, 1924) n. comb. (Digenea: Opecoelidae) in Wisconsin darters (Etheostomatinae). *Can. J. Zool.*, 62: 2666-2672.

Travassos, L. 1944. Revisao da familia dicrocoeliidae. Odhner 1910. *Mem, J. Inst. Osw. Cruz.*, 2: 1-357.

Tubangui, M.A. 1929. *Paradistomoides gregarinum*, a new name for trematode Paradistomum magnum. *Philipp. J. Sci.*, 38(4): 443.

Verma. A.K. 1988. Polymorphism in *Paradistomoides gregarinum*. M.Phil. Dissertation, University of Jammu, Jammu, p. 99

Yamaguti, S. 1971. *Synopsis of the Digenetic Trematodes of the Vertebrates.* Keigaku Publ. Tokyo, Japan, vol. I and II: pp. 1-1074.

10

Ecological and Behavioural Studies on Rhesus Monkey, *Macaca mulatta* (Zimmermann) – A Review

S.K. Gahlawat, R.K. Gupta and R.C Gupta[1]

Department of Zoology and Aquaculture, CCS Haryana Agricultural University, Hisar-125004

[1]Professor, Department of Zoology, K.U. Kurukshetra – 136119

ABSTRACT

The present studies were carried out on certain behavioural and ecological aspects of free-ranging rhesus monkey, *Macaca mulatta* (Zimmermann) at Saraswati and Veer Sonti forests in Haryana. The geographical position of Saraswati forests is 29° 58′ N latitude and 76° 32′ E longitude and Veer Sonti forests is 30° 0′ N latitude and 77° 1′ E longitude. The climate of these forests is typical to that of plain of northern India. Typical vegetation of these forest sare characterized by *Eucalyptus* sp., *Dalbergia sissoo, Prosopis juliflora, Morus alba* and *Capparis sepiaria,* etc. A total of 6 troops have been identified at both the sites (three each). All the troops are pure forest dwellers. The major topics dealt within this study are: (i) group size and composition, (ii) home-range, (iii) daily life schedule, (iv) feeding habits and (v) quantitative behavioural observations.

Introduction

Present work concerns rhesus macaque monkey [*Macaca mulatta* (Zimmermann)] in respect of its certain selected ecological and behavioural aspects viz. group size, group composition, home-range, daily life schedule, feeding and quantitative behavioural observations. Rhesus monkey (*Macaca mulatta*) belongs to order primates; suborder Anthropoidea; superfamily Cercopithecoidea (Catarrhini); family Cercopithecidae.

The study sites have been confined to two natural forests nearby Kurukshetra University campus. One of these two study sites hails from Saraswati forests. The Saraswati forests are located on Kurukshetra-Pehowa-Cheeka road about 35 km away from Kurukshetra University campus in the West. Its geographical position is 29° 58′ N latitude and 79° 32 E longitude. Its total area is 11231 acres.

The other study site refers to Veer Sonti forests situated on Kuruksthera-Saharanpur link road about 18 km east of Kurukshetra University. It total areas is 474.50 acres. Its geographical position is 30° 0′ N latitude and 77° 1′ E longitude.

In the present study attention has been focused on six troops of rhesus monkeys. Out of these six troops, one troop is purely road side, another troop is semi-forest dweller and the rest four troops are purely forest dweller.

In order to acquaint one with the lineage an evolution of rhesus monkey (*Macaca mulatta*) in time and space, it would be worthwhile to briefly understand its ancestry and distribution. According to Moody (1962) primates appeared on the horizons of our planet some 63 million years back in Eocene period and Coenozoic era. Gradual cum sustained evolutionary trends diversified the original stock of primates into two major streams viz. Catarrhini (old world monkeys) and Platyrrhini (new world monkeys).

The chief taxonomical characteristic features of Catarrhini are: cheek pouches, raised nose and a narrow internal septum. Old world monkeys also have two premolars and bony external auditory meatus. *Macaca mulatta* (Rhesus monkey) and *Macaca radiata* (Bonnet monkey) are two examples.

On the other hand the distinguishing characteristic features of Platyrrhini are: absence of cheek pouches and a broad internasal septum. The new world monkeys also have three premolars and do not have any external auditory meatus. *Ateles* - the spider monkey is the prominent example. As the name indicate, the Platyrrhini primates are confined to the Central and South America.

Interestingly enough, monkeys and langurs are restricted in their distribution to South Asia. Approximately 60 species of various monkey and langurs have been reported. Some of these are:

Macaca mulatta (Zimmermann)	–	Rhesus monkey
M. radiata (Geoffroy)	–	Bonnet monkey
M. silenus	–	Lion-tailed monkey
M. arctoides		
M. assamensis		
M. fascicularis	–	Crab-eating macaque
M. fuscata fuscata	–	Japanese monkeys
M. nemestrina	–	Pigtail monkeys
M. specoiosa		
M. sylvanus	–	Barbary ape
M. irus	–	Philippine macaque
M. maura		
Presbytis entellus	–	Hanuman langur
P johnii	–	Nilgiri langur
P. phayrei	–	Phayreis leaf monkey
P. pileatus	–	Capped langur
Erythrocebus patus or *Circopithecus patus*		The hussar or African red monkey

A number of species of monkey and langurs have been reported from Indian subcontinent. Some of the prominent species are:

Macaca mulatta	-	Rhesus monkeys
Macaca radiata	-	Bonnet monkeys
Macaca silenus	-	Lion-tailed macaque
Presbytis entellus	-	Hanuman langur
P. johnii	-	Nilgiri langur
P. phayrei	-	Phayrei's leaf monkey
P. pileatus	-	Capped langur

As far as the distribution of free ranging rhesus monkey (*Macaca mulatta*) in Indian subcontinent is concerned, evidences point out towards the fact that it displays a relatively higher level of tolerance as it frequents habitats of wide-ranging varieties. For example, rhesus monkey range extends from 70° E longitude in Afghanistan to 120° E coast of China (Lindburg, 1971). The distribution of rhesus monkey also extends northward as far as

27° N latitude in some places. It must be mentioned that Himalayan foothill population is also found at altitudes of at least up to 8000 feet. Workers like Pocock (1932), Miller (1933), Fooden (1964), Southwick *et al*. (1965), Hill (1965), Valeric *et al*. (1969), Singh (1969), Lindburg (1971) and Roonwal and Mohant (1977) have worked on investigating distribution pattern of free-ranging rhesus monkey (*M. mulatta*) in Asia. In post-partition India two prominent species viz. rhesus monkey (*M. mulatta*) and bonnet monkey (*M. radiata*) have been studied from different angles by different scientists (Bernstein and Mason, 1963; Southwick *et al*., 1965; Altmann, 1968; Neville, 1968; Lindburg, 1971; Pirta *et al*., 1978; Singh, 1981; Mukherjee, 1983; Seth and Seth, 1983; Guerra, 1988).

According to these scientists, the distribution of rhesus monkey is confined to northern India whereas that of bonnet macaque to southern India. Precisely speaking rhesus monkey (*Macaca mulatta*) is found in mixed-deciduous forests of Northern India (Seth, 1983). It co-exists with hanuman langur (*Presbytis entellus*). On the other hand, bonnet monkey (*Macaca radiata*) is found in evergreen forests of Western ghats to dry areas of Deccan plateau. It co-exists with lion-tailed macaque (*Macaca silenus*) in the evergreen forests of South India. It also co-exists with hanuman langur (*Presbytis entellus*) in the dry deciduous forests of Western Ghats (Sugiyama, 1964, 1968, 1971). According to Poirier (1970), bonnet monkey (*Macaca radiata*) also co-exists with Nilgiri langur (*Presbytis johnii*) in Nilgiri hills.

It must be mentioned that the northern boundary of *Macaca radiata* is formed by Vindhyachal ranges and, therefore, these ranges also constitute the southern boundary of *Macaca mulatta*.

Here it is relevant to briefly understand popular themes concerning ethology of rhesus monkeys viz., rhesus ontogeny, age-sex classification and difficulties in age-sex classification.

Rhesus Ontogeny

Infant: Pre-natal and neonatal stage – post zygotic organisms develop characteristics which themselves counsel further development. Rhesus females carry the foetus of either male or female for about six lunar months or the gestation period in rhesus females is of 6 months (Lancaster and Lee, 1965; Brandt and Mitchell, 1971). Sensory-motor nerves regulate the behavioural development during the pre-natal stage (Sperry, 1971; Tobach *et al.*, 1971). Development of various reflexes and demonstration of different capacities can be specified as individual behaviour.

Free ranging rhesus females give birth in the dark places, which makes observation studies difficult, however, cases of day time delivery and behaviour have been reported (Brandt and Mitchell, 1971; Lindburg, 1971). After six months, the neonate separates from mother anatomically and physiologically and interwined behaviourally with mother and the immediate social and physical environment.

Primate mothers ordinarily do not help the infant to find the breast and support continuously for contact. As a result, infant is equipped with "rooting reflex", innate "suckling" movement and "grasping" as a traditional postural adjustment ventral to mother. Sensory motor behaviour of infants develop as fast as they release grip for tactile exploration at 5 days, locomotion at 9 days, eye-hand coordination at 18 days, and eat solid food at the age of 21 days (Hinde *et al.*, 1964). Few behaviours appeared after birth have been reported to disappear within first month, e.g., following head up orientation and auditory response to lipsmack. Jensen (1961) found that prehension and discrete picking gradually develop after one month to six months. So, many changes in behaviour do occur in early weeks of life among rhesus infants.

Normally, at the age of three months, the infants become independent from mother but for security purposes, infants approach mothers even after attaining independent stage. Males become independent within few weeks. Males are careless by exhibiting less of their behaviour towards their mothers and appear to be more responsive to their environment.

The "infants play behaviour" shifts from solitary to social play with peers, after attaining mutual independence. After 10-12 weeks, infants rarely play without partner. Male infants spend more time in play than females because female infants try to remain with mothers and siblings even after weaning. Juveniles of both the sexes do not include infants in play groups.

The infants are generally mishandled in social interactions by the juveniles. Adolescent and adult females play an 'aunt' like role with new born infants (Rowell *et al.*, 1964; Spencer-Booth, 1968; Handy-Neely, 1978). Adult and sub-adult males protect infants from external danger and show some extent of 'Parental behaviour' (Mitchell, 1977). At the age of twelve moths, behavioural repertoire of infants is virtually complete (Sade, 1971). Simpson and Howe (1980) reported a high degree of individual differences at the age of twelve months.

Juvenile: The weaning process, which changes infants to juvenile status, appears to be gradual. The juveniles become more independent to mother in feeding, movement and daily routine. The birth of a new sibling is the most abruptive event in the weaning process.

Handen and Peter (1980) analyzed behavioural data of bonnet juveniles and it revealed several patterns:

(1) Either sex shows continuous decline in association with age mates;

(2) Solitary time of females peaks at 1.5 years after which they begin to associate with adult females;

(3) Solitary time males peaks at 2.5 years after which they associated predominantly with adult males;

(4) Both males and females showed higher association with age mates of same sex than to opposite sex.

Altmann (1968) and Clark (1978) characterized juveniles as least interacting class.

This so-called solitary period is mostly spent in play behaviour with same sexed peers. Gard and Meier (1977) suggested several factors responsible for variation in form and frequency of play behaviors, e.g., age, sex, status, temperature, activity and vocalizations, etc. Drickamer (1976) also accounted grooming activities as high as 25 percent for juveniles rhesus monkeys of total grooming behaviour of a free-ranging group. Sexual behaviour of juveniles resemble with adults, but physical and physiological maturity differ between them. Menarche in juvenile females occur at about 2 years of age (Van Wagenen, 1952) and testarche among males (spermatogenesis) appear at about 3.5 years (Van Wagenen and Simpson, 1954).

Adolescent/Sub-adult: One year gap exists normally in between puberty and maturity in case of females. It starts at about 2 years, when a juvenile female enters adolescent class. This event occurs in birth season and as a result solitary juvenile female starts taking interest in new born infants and mothers (Spencer-Booth, 1968; Quiatt, 1979). In this way adolescent female increases her social interactions to get admission in central female assembly. A female, probably, starts taking interest in breeding activities at 2.5 years (Neville, 1968).

Menarche is manifested with slight vulvar and pinkish swelling of perineal skin. At the age of 2.5 years, pinkish swelling spreads to entire hind quarters. Adolescent is a period of endocrine instability, as evidenced by irregular menstruation, preponderance of menstruation without ovulation, follicular artresia, excessive swelling of skin, and comparative sterility (Ashley-Montagu, 1957). The first birth recorded at 4 years of mean age (Drickamer, 1974). This reveals adolescent females are incapable of breeding and maternal care in behavioural sense.

Sub-adult males are lacking red pigmentation of scrotum and a fully red face. Sub-adults are smaller than fully adult male and occupy peripheral position relative to group (Neville, 1968). Sub-adult males are often aggressive towards females and immatures. As they mature, tend to threaten the lowest ranking males to get admission in adult hierarchy. According to Hamburg (1971), this period is the most aggressive period.

During breeding season, the testicular size increases and continuously weight increases (Glick, 1979). The sub-adult males ejaculate less frequently with longer latencies. These also mount more frequently making more thrusts per mount than fully adult males (Michael and Wilson, 1973). Thus males between 4 to 5 years, sexually mature in physiological sense are not mature behaviourally.

Adults: Adults members in group take part in breeding activities. At adult stage, the sexual skin reddens and becomes bright red during breeding season. A period beyond 3 years among females and 5 years among males upto death (20 years approximately) is considered as a single stable class.

Adults of both the sexes sit for long periods, engage in less locomotor activity, but increased head movements (Drapper, 1966).

Adult females account for 60 percent grooming activities of a group (Drickamer, 1976), while adult males rarely groomed, other but females in breeding season.

A development view among adults have been followed in relation to maternal behaviour and parity of mothers (Seay, 1966), which effects and sharpens mothers' infant care tricks. Grewal (1980) accounted on nulliparous and parous females among free - ranging Japanese monkeys. Grewal observed 50 females of four classes: (1) nulliparous females (3 to 4 years), (2) primiparous (4 to 7 years), (3) multiparous young adult females (7 to 10 years) and (4) multiparous old females (10 to 22 years). According to him multiparous young females spend more time in allo-grooming, non-specific body contact and proximity than do females of other classes and most of their social interactions are centered around their own off-springs. Social interactions of multiparous females are fewer than those of the young mothers. Old females stop taking part in breeding activities, but old males continue to mount females. Evidently, old males loose rank in central hierarchy and often assume peripheral position in the group. Most of the old males leave group in this process.

Some examples of age-sex classification

A very common practice among researchers interested in development is to divide the time course of ontogeny into stages or periods. In this way the researchers assigned the animals by the age and by their appearance and behaviour. However, it has certain limitations. Some examples of classification used by different authors for rhesus monkey are shown in Tables 10.1 and 10.2 (after Singh, 1981).

Problems of age-sex classification

Different workers adopt different criteria within a species for age-sex classification. Hinde (1971) concludes that age-sex classification are generally not helpful. He states that one major problem of the application of stages to development is that their use tends to take away from the fact that development is a continuous, ongoing process. In order to make a general comparison between different species and situations within species, a comparison of age limits of certain stages may be useful.

Table 10.1 Examples of age sex classification for *Macaca mulatta* with characteristics and age. (After Singh, 1981)

Source	Age-sex class	Characteristics
Bernstein and Mason (1963)	Juvenile male	18 months to 4 years; weight 7 lb; canine growing; and no sex colour.
	Juvenile female	2 years; 5-7 lb weight; milk teeth and permanent incisors; no sex colour.
	Subadult male	6 years; 19 lb weight; dentition complete; red sex skin colour.
	Subadult female	4 years of age; canine still growing; 7.5 lb weight; red sex skin colour.
	Adult male	More than 7 years; 22-32 lb weight; completed dentition; red sex skin colour.
	Adult female	More than 5 years; 10-12 lb weight, complete dentition and red colour.
Southwick *et al.* (1965)	Infants	Born during study year.
	Immatures	From 1 to 3 years old.
	Mature	At least 4 years old.
	Full Adults	Six years of age; when they have full dentition and growth.
Drapper (1966)	Age only	Yearlings.
	Age only	2-3 years.
	Age only	4-5 years.
	Young and older	6 years and over.
Altmann (1968)	Infants	0 to 1 year.
	Juvenile males	1 year to less than 4 years.
	Juvenile females	1 year to less than 3 years.
	Adult males	More than 4 years.
	Adult females	More than 3 years.
Neville (1968)	Infant	With mother.
	Juvenile	Intermediate between infant and an adult or subadult.
	Subadult male	Lack pigmentation of sex skin and smaller than adult male.
	Adult male	5.5 years.
	Adult female	Takes part in breeding at 2.5 or 3.5 years.
Lindburg (1971)	Infants	Less than 1 year of age.
	Juvenile-I	1 to 2 years.
	Juvenile-II	2 to 3 years.

Source	Age-sex class	Characteristics
	Juvenile-III (male)	3 to 4 years.
	Subadult male	4 years and old.
	Adults	Young, middle and aged.
Drickamer (1976)	Infants	Less than one year.
	Juveniles	1 to 3 years.
	Adults	4 years and older.

Table 10.2 A finer gradation of age-sex classes stage approach to rhesus ontogeny (After Singh, 1981)

Age-sex class	Characteristic
Infants	Normally less than one year. Period before weaning. May be extended if sibling is not born at 1 year of age.
Juvenile	More than one year and smaller than adolescent/subadult in size.
Juvenile female	More than one year and less than two years. Period between weaning and menarche. A vulvar perennial skin. Spend much time in same sexed peers.
Juvenile male-I	More than one year and less then two and half years. In a process from centre of the group to peripheral sub-group of male juveniles. Normally the process terminates in mating season. Exception may continue in centre of group (e.g., son of a high ranking female).
Juvenile male-II	More than two and half years and less then four years. Live in same sexed peers group at periphery, in the process from subgroup to peripheral position alone to the group. Receive agonistic interactions from group members. Testes size increased without colour.
Adolescent female	More than two years and less than three years. Display pinkish swelling on entire hindquarters. Associate closely with adult females. May be conceived even at 2.5 years to reproduce at 3 years, normally it does not occur. Try to form consort relation with adult males, but they show disinterest, though mount them.
Subadult juvenile	Between 4 and 5 years of age. Bigger than

Contd.

Age-sex class	Characteristic
male	male-II and smaller than adult male. Scrotum starts becoming Pinkish. Assume peripheral position. In the process to enter group and often repelled back by adult members. Submission to adult male only. Not allowed to form consort during mating period.
Young adult female	More than three years. May be nulliparous or primiparous. As a central member of female assembly.
Young adult male	More than five years. Often peripheral male and lower in male hierarchy. Red scrotum. Show submission to mature males only. Form consort in mating season.
Mature male	Member of central male hierarchy. Never emit submissive behaviour and form consort relations during mating period.
Mature Female	From disparous to reproductive age.
Old female	Non-reproductive. Reduces to juvenile size. Loss of fur and twisted tail. Also lost teeth. Increased passivity. Live in close association with adult members in centre.
Old male	Though takes part in breeding activities. But consort relations are disturbed. Becomes submissive and assume peripheral position. Loss of teeth and tail twists.

A tentative approach

There are problems to know how of rhesus ontogeny and related problems. In the field study, it is very difficult to identify all the animals by age and sex.

Table 10.3 summarizes tentative age-sex classification for the present study, may be called hypothesis. It is only to identify the individuals by outlook.

Due to a combination of several pressures caused by modern man at the population of rhesus monkeys, the overall declining trends set in thereby has brought this primate on the brink of attaining the status of an endangered species. Large scale export of rhesus monkeys in fifties, sixties and early seventies to United States and many European countries for biomedical research dealt a heavy blow to the rhesus monkey population. Beside, this extensive destruction of forests also became a threatening measure to a widespread population across northern India. The sum total of these factors warranted the Government of India to ban the export of rhesus monkey in 1977-78 and take appropriate measures to rehabilitate this primate. In this given situation, it should become obligatory on scientists to provide "state of the art" in respect of distribution, factual ecological and ethological information so as to enable the concerned government agencies to take appropriate steps for the rehabilitation and conservation of rhesus monkey in its natural habitats. As such various centres of primate research have been emerged in India viz. Zoological Survey of India, Calcutta; Department of Anthropology, Delhi University, Delhi; Department of Psychology, University of Mysore, Mysore; Department of Zoology, Aligarh Muslim University, Aligarh; Department of Psychology, Meerut University, Meerut; Department of Psychology, Utkal University, Bhubaneswar; Jodhpur University Jodhpur, and Department of Psychology, Jaipur University, Jaipur.

Table 10.3: Age-sex classification of rhesus monkeys

Age-sex class	Characteristics
Infants	Normally less than one year. Period before weaning. May be extended if sibling is not born at 1 year of age.
Juveniles	More than one year and small than adults. Not allowed to take part in breeding activities. Oftenly found on sub-periphery and periphery. At later stages pinkish swelling on the sexual skin. Receive agonistic interactions from group members.
Adults	Bigger than juveniles in size.
Adult female	More than three years. As a central member of female assembly. Take parts in breeding activities. Sexual skin becomes bright red.
Adult male	More than 5 years. As a member of central male hierarchy. Never emit submissive behaviour. Take part in breeding activities during mating period. Scrotum becomes bright red.

Here, it must be mentioned that no research work has been done on free-ranging rhesus monkey (*Macaca mulatta*) troops residing in various forests of Haryana. The present work is a humble attempt in this direction. In the present study various ecological and ethological aspects like group size, group composition, home-range, daily life schedule, feeding and quantitative behavioural observations have been looked into.

The past record of research work on rhesus monkeys and their related species available in varied journals of Primatology, Ecology and Ethology, etc., along with few master books on primates prove convincingly that a number of scientists across the continent have stirred their minds to understand the overall existence of the apes and monkeys in the present and past. The information is being summarised here in a chronological order as far as possible in respect of the certain selected topics viz., distribution, group size, group composition, home range, daily life schedule, feeding, sexual behaviour and quantitative behavioural observations. Each one of these topics is being catalogued in textual form below.

Distribution

Pocock (1939) observed that the monkeys are distributed from 70° E longitudinal in Afghanistan to about 120° E Coast of China and 35° N latitude in eastern Afghanistan to 40° N latitude in northern China and within India the Tapti river in the west and the Godavari river in the east which mark the effective barrier of the rhesus monkey species (Fielder, 1956; Prater, 1965), and also some population is found in the south of the Godavari river (Southwick *et al.*, 1965; Spillett, 1966). However, Fooden *et al.* (1981) redefined the boundary of rhesus and bonnet macaques in Peninsular India. According to them, the Manjra plateau line which joins the northern end of the Velikonda range, eastern ghat is the effective barrier and not the Tapti and Godavari rivers as previously reported.

Southwick (1961) found that the rhesus monkeys are also residing in villages, towns and temples of north India. Pirta *et al.* (1978) observed the population of two non-human primates (*Macaca mulatta* and *Macaca radiata*) resident of Dehradun forests in north Indian and urban areas of Mysore in South India.

Seth and Seth (1983) in their brief report on population dynamics of free-ranging rhesus monkeys in different ecological conditions in India described that rhesus groups are residing in tropical forests, villages and urban habitats, in both tropical and arid biomes. Wada (1983) observed that the rhesus monkeys are widely spread through South Asia, in various habitats ranging from dry forests to humid forests and from tropical to temperate or sometimes alpine zones vertically. The map of distribution of rhesus monkeys was prepared (Tan *et al.*, 1965; Fooden, 1971; Puget, 1971; Koyama and Shekar, 1981; Shang *et al.*, 1981; Wada, 1983).

Koyama and Shekar (1981) found that the hanuman langur (*Presbytis entellus*) inhabited moist deciduous, dry deciduous, even thorn forests and very rarely in evergreen or semigreen forests along the western ghats. According to them bonnet monkeys occurred in moist deciduous, the dry deciduous forests from 21° 05′ N to 73° 35′ E near Songadh in Gujarat but never in the dry scrub or thorn forests. They also found that rhesus monkeys inhabited along the Satpura range in the north of Tapti. Its western limit was 21° 20′ N to 73° 30′ E in Kherwads forests of Gujarat. Rhesus monkeys also inhabited in east Maharashtra and in north Andhra Pradesh.

Mukherjee (1983) in his survey of non-human primates of Tripura, India found that 7 species are inhabited in forests of the State out of which rhesus monkey (*Macaca mulatta*), phayrei's leaf monkey (*Presbytis pharyrei*) and capped langur (*P. pileatus*) are less in abundance. The phayrei's leaf monkeys are found only in Tripura, India.

Group size

The number of individuals of a particular species living in a particular area constitute the size of a group. The variation in the group sizes in forests, urban and semi-urban areas have been reported earlier by various workers (Parkash, 1962; Jay and Lindburg, 1965; Mukherjee and Gupta, 1965; Southwick *et al.*, 1965; Singh, 1969; Lindburg, 1971; Mukherjee and Mukherjee, 1972). Mukherjee (1969) observed the group size of macaques as 33 and 30 in two different villages of Uttar Pradesh.

Due to different ecological conditions, the variation in the group size had been observed 8 to 98 (Jay and Lindburg, 1965); 6 to 90 (Makwana, 1978) and 18 to 102 (Seth *et al.*, 1983).

Seth and Seth (1983) presented the data on group size, group composition and reproductive rates of 9 free-ranging rhesus monkey groups inhabiting different habitats and suggested that these was a increase in group size in respect of the habitat occupied. However, Southwick *et al.* (1961, 1965) observed the group size of rhesus monkeys as 50 and Neville (1968) obtained the mean size of a group as 28 near Haldwani forests.

Pirta *et al.* (1978) found a significant difference in group size as well as in adult male and female ratio of two species viz. *Macaca mulatta* and *M. radiata.* Their energy requirement also play a major role for the variation in group size of the two species.

Group composition

Infants, juveniles, sub-adults/adolescent as well as adults constitute the composition of a group. To distinguish them, various parameters like scars, hair patterns, colour, tail and body profile were taken into consideration by Lindburg (1971). Koyama and Shekar (1981) found three groups of hanuman langur (*P. entellus*) along the Western Ghats. The mean value of the socio-economic 'sex ratio' i.e., the number of adult females per adult male for the 24 troops of hanuman langur was observed c.6.9 and this ratio among bonnet monkeys was c.1.9.

Koford (1966) observed a 'sex-ratio' of 96 males per 100 females in Cayo Santitago colony. Rahaman and Parthasarathy (1967, 1978) observed the adult male/female ratio as 0.526 in Hasan district and for Bangalore monkey it was 0.785, the same for South Karnataka and Dharwar were 0.94 and 0.828 respectively (Simonds, 1965; Sugiyama, 1971). The data from F.R.I showed 3:1 ratio of adult females and males in a group and which was considerably different from the Aligarh temple group in which the female/male ratio was almost equal (Southwick *et al.*, 1965).

However, Pirta *et al.* (1978) concluded that a significant higher percentage of females in the rhesus and the males in bonnet affected the socio-economic 'sex-ratio' of the two species.

Koyama and Shekar (1981) also observed the percentage of adults, sub-adults, juveniles and infants; the adults constitute about 59 percent of the total population in hanuman langur (*P. entellus*), however, in bonnet monkeys (*M. radiata*) accounted for about 48.3 percent of the total number.

Home-range

In primates, the members of one group live in a circumscribed area of a particular niche for their co-operation with other. This area is known as home-range and except minor seasonal fluctuations, it remains stable for years (Altmann, 1962; Simonds, 1965; Lawick-Goodall, 1967; Pirta *et al.*, 1977, 1978). The stereotyped movements of the monkeys were mainly influenced by the availability of food (Chance, 1956; Mukherjee, 1969). Southwick (1962) observed that for such movements, the sufficient number of animals moved together as a unit.

The home-range size had been found to be an environmentally labile behavioural system and varies both within and between the species of non-human primates (Southwick and Siddiqui, 1974; Cutton-Brook and Harvey, 1977). Pirta and Singh (1978) observed that the home-range increased with age and rearing experiences. However, addition of more individuals did not cause a change in the home-range. Pirta and Singh (1980) observed that the decrease in home-range size was either due to emigration or due to trapping of alpha male. They also found a slight increase in home-range due to disturbance caused by elephants, trapping and conspecific group. However, this increase was for a very short period.

Neville (1968) concluded that the monkeys living in less favourable areas had larger home-ranges. Makwana (1978) observed the variation in home-ranges from 1.3 km^2 to 13.4 km^2 and it also overlapped the home-range of other groups. Rahaman and Parthasarathy (1978) observed that the dominant troops occasionally invaded the neighbouring range and after a short vocal display, returned back to their own range. They also found that, the mean area covered in a day by the troop was about 1 km^2 and the linear extent was about 3 km^2. They further concluded that this all was mainly dependent on food and physiographic conditions and the troop size had no influence on it. They also observed that the fluctuations in the home-ranges were influenced by season and human interferences. Pirta and Singh (1982) observed the average home-range size in 3 habitats to be 5.18 km^2, 1.52 km^2 and 0.017 km^2, respectively.

The removal or loss of a alpha male, his experience and personality had been found to be crucial for the changes in the home-range size (Carpenter, 1942; Bernstein, 1966; Southwick and Siddiqui, 1967; Neville, 1969; Rowell, 1972; and Pirta and Singh, 1980).

Daily life schedule

Lindburg (1971) observed the variation in the daily activity cycles of the forests group with respect to the pattern, frequency, and intensity and he also, suggested to follow the natural cycle of the warm-wet season (July-October), the cool season (November-February) and the hot-dry season (March-June) to describe such differences. During the warm-wet season, the first activity starts with the feeding which easily began at down. However, he also observed grooming at this time but within a relatively short period. In the cool season, the lateness of the start of morning foraging which persisted irregularly throughout the day due to less food sources at this time of the year. Similarly in the hot-dry season foraging became again an early morning activity and for rest of the day foraging occurred intermittently with long periods of inactivity or with periods of grooming, play and dozing.

Nolte (1955) observed that the daily routine among bonnet monkeys started with mutual grooming, defecation and urination. He also observed that the first quart of an hour was noiselessly. He further found that the mutual grooming occupied by a considerable part of the day without influence of the dominance. Maslow (1936) also observed the same pattern in the laboratory animals. Except dogs, the other domestic animals and birds of different sizes did not disturb the routine behaviour of the monkeys. Neville (1968) divided the daily activities of the forests monkeys into 3 major classifications viz., the overall troop movements, foraging of the troop and resting of the troop. He also observed that there was rarely uniformity of the activity among the individuals of the given group.

Bernstein (1972) observed the maximum frequencies of different activities viz., travel, feeding, play, sex and agonism at the time of sunset. He further observed that such activities were also influenced by the temperature, weather and replacement of alpha male.

Khalid and Qaza (1986) studied the diurnal dispersion pattern of free-ranging rhesus monkeys using three dimensional version of Clark-Evans model. They found two peaks of dispersion, one in the morning and the other in the evening. The clumping of the animals occurred in the late hours of the day.

Feeding habits

Lindburg (1971, 1976) observed the dietary habits of rhesus monkeys in north India which revealed that the diet was largely frugivorous and also included the variety of leaves, stems, flowers and insects. He found no evidence of feeding on animal matter other than insects. Roonwal (1956), in a similar study observed that the rhesus monkeys consumed wild mushrooms during the month of August. Mukherjee (1969) also observed that the main food of the rhesus monkeys in northern Uttar Pradesh composed of leaves of mango, pakar and neem trees, mango fruits, grasses and the food grains left over in the village fields. These monkeys also consumed insects as a part of their food. Makwana (1978) also observed the dietary habits of hanuman langur which consisted of leaves, fruits and flowers only. Rahaman and Parthasarathy (1978) also reported the diet of bonnet monkeys which included fruits, flowers, leaves, stems, bark, pith, gum and stilt roots. They also observed that the females and infants caused the greatest damage to the plants and fruits of coffee, orange, paddy and cardamom especially in the morning hours. The feeding and other activities of these monkeys were also influenced by rainfall. However, they were less active in inclement weather. The selection of food by these monkeys was by smelling and testing.

The diet of the monkeys also include certain food plants having toxic properties. The seeds of *Abrus precatiorious* which cause death in humans and animals (Gunn, 1969) were regularly eaten by the monkeys without any apparently adverse effect. The seeds are also used in India for poisoning cattle and humans (Chakravarthy, 1969). Lindburg (1976) observed the monkeys consuming the leaves of *Casearia gravcolens* Delz. and *C. tomentosa* Rxb. The fruits of these are used to poison fish (Gupta, 1962).

Siddiqui and Southwick (1979) observed a remarkable adjustment of the food habits among the rhesus monkeys of India with respect to local environment. The peak feeding in these rhesus monkey groups occurred in the morning as well as in evening with an intervening period of rest and reduced feeding.

In India most of the rhesus monkey groups, both in rural and urban habitats being fed artificially is a common practice. Such feeding results in a significant increase in aggressive competition in the rhesus monkey groups (Southwick *et al.*, 1976).

Sexual behaviour

The sexual behaviour starts with the onset of puberty. Females beyond the age of 3 years and the males attaining the age of 5 years are considered to be sexually mature. These adult members become an integrated unit and take part in breeding activities. At this stage, the sexual skin becomes red and during breeding season, it is bright red and that serves as a criteria of specification. The females are closely associated with males or with several males successively. At some time, during her period of receptivity, these associations (consorts) vary in intensity and duration. However, the characteristic pattern includes proximity, reciprocal grooming, copulation, restlessness and excitement (Carpenter, 1942a, b; Lindburg, 1967; Loy, 1970). In some consort pairs, there was a weak bond and the relationship was easily broken up for some other activities (Neville, 1966; Kaufmann, 1967; Lindburg, 1967).

Draper (1966) observed an increased movement of head, sittings for long periods, less locomotory activities and infrequently change in the postures in the adults. Drickamer (1976) found 60 percent grooming activities in adult females and the males groomed the females in breeding season. Post and Baulu (1978) observed the high social contacts of adult females than males. During this period males show a more aggressive behaviour and dominance display (Parkar and Parkar, 1979).

Carpenter (1942) found the incidence of the sexual behaviour correlated positively and highly with the dominant status. Wilson

(1981) concluded that the social rank among females also influenced their pattern of sexual activity. He observed that the high ranking females were more engaged in the copulatory series than the lower ranking females, also the lower ranking females copulated with as many as available males as did the high ranking females. The results of Drickamer (1974) related to sexual behaviour of rhesus monkey could not support the conclusion that the higher ranking male monkeys engaged in a disproportionately higher number of sexual act in a sexual group.

Rahaman and Parthasarathy (1978) observed the peak of the copulation frequency in bonnet monkeys during August to October while for Darbar monkeys, the peak copulation frequency was observed from April to September (Sugiyama, 1971) and from October to November in Bangalore monkeys (Rahaman and Parthasarathy, 1969). They all suggested the influence of environmental factors on the onset of breeding. Jay (1963) and Hall and DeVore (1965) reported the maximum reproductive success in a troop for dominant males. However, Simonds (1965) did not observe this in bonnet monkeys.

In monkeys, the copulation is restricted within the troop but, the Gibbons mated with the suitable members of the other troop also (Carpenter, 1940). Rahaman and Parthasarathy (1978) observed a trend of mounting among themselves in babies and juveniles. This led to more familiarity and more stronger bond among them and fewer chances of inbreeding among the members of different age groups and consequently between their parents and off-springs and between older and younger siblings.

Lindburg (1967) observed the masturbating activity while groomed by females. However, Rowell (1963) observed this activity in non-mating season. The adult males who were associated with pregnant females did masturbate for ejaculation (Carpenter, 1942a; Altmann, 1962). The females ignored the grooming of the male after his ejaculation (Carpenter, 1942b). The masturbation in males was also observed during breeding season when rejected by females (Neville, 1966). The females were not found to be involved in masturbating. However, the females might have experience the sexual pleasure while mounting juveniles or pulling and sucking their nipples (Carpenter, 1942; Neville, 1966).

Loy (1970) observed the homosexual behaviour between the females which occurred during the mating season. Southwick *et al.* (1965) observed the homosexual behaviour in males. Kaufmann (1965) also witnessed the homosexual behaviour in laboratory males.

In general, the behaviour of adults is much more affected by reproductive seasonality and ecological changes.

Mother-infant relationship

Among primates, maternal care, infant development and socialization has been thoroughly studied in rhesus monkeys and man (Lindburg, 1971). The development of the infants had been affected with the quality of maternal care (Seay, 1966; Mitchell and Sterens, 1968). Kaufmann (1966) provided a descriptive data on mother-infant interactions under free-ranging conditions. Swartz and Rosenblum (1981) found that the mother-infant relationship is the first and probably the long lasting and most intense relationship. The early social interactions between mother-infant greatly determine the developing infants ability to deal adequately with the social encounters and the degree to which it is engaged in social interactions. They also observed the effect of individual differences in mother and infant on shape and nature of relationship within species and across species. They further observed that the birth prccess paid a major role in the establishment of mother-infant bond.

Neville (1966) observed the maximum births in April and May with a second peak of births in September to October in Rajasthan, however, Koford (1966), Vandenbergh and Vessey (1966) and Hausfater (1972) observed the births in January to June. As soon as the infants are born, the mother starts licking and clinging the birth fluid from infants and this helps the infants to make the physical contact with the mothers. However, after the placenta had been delivered, mother ignored the infant and licks and eat the placenta and after it, the mother again attend the infant. Mother spent much of her time in grooming, licking, patting and making eye contact during the first few days of the infant life (Swartz and Rosenblum, 1981). At the time of danger in a group or a slight sign of dis-interest, mothers quickly retrieve the infants (Jay, 1963; Hinde *et al.*, 1964; Bertrand, 1969; Struthsaker, 1971).

Only in few species it had been observed that the mother allowed the other female to hold or carry the new born (Jay, 1963; Struthsaker, 1971), however, mother allowed the handling of slightly older infant in most of the species (Rowell *et al*., 1964; Van Lawick-Goodall, 1968; Bertrand, 1969; Roseblum, 1971a). The old males of the several species including savanna, babbons (De Vore, 1963) and barbary apes (Burton, 1972; Lahiri and Southwick, 1973) showed active interest in infants for long periods of time.

The physical contact between mother and infant was influenced by the mothers feeding behaviour (Johnson; 1986). Guerra (1988) worked on the maternal transport and locomotory development in infant of rhesus monkeys. He found that as the infant grow older, mothers stopped carrying the infants. He also observed an increase in the incidence of transport in either jockey position or ventro-dorsal with increasing age, for this a higher motor development of the infant is required to climb up.

Guerra (1988) also carried out study on withdrawal and retrieval activity of mother in rhesus monkey infants from birth upto 60th day of age. They observed an increase in the number of withdrawals with age. They also concluded that the increasing motor development in the infants was being reflected by with-drawls and the gradual dissolution of mother-infant relationship.

Sub-adults and non-pregnant adult females had also been found to contribute for infants care (Jay, 1963; Rowell *et al*., 1964; Poirier, 1968; Lan-Caster, 1971; Struthsaker, 1971). Spencer-Booth (1968) observed that the siblings showed more behaviour towards infant than the animal of the same category present in a group. In other species, males had also been found to be caretakers (De Vore, 1963; Ransom and Ransom, 1971; Burton, 1972).

Natal coat had been observed in infants of many species which distinctly differ from the coat of an adult in langurs (Jay, 1963, Poirier, 1968); babbons (DeVore, 1963); stump tailed macaques (Bertrand, 1969); vervets (Struthsaker, 1971). The infants of monkeys and apes are very active at the time of birth. Clinging was strong at the time of birth in many species but some mothers had to support for the first few days (Hinde *et al*., 1964; Bertrand, 1969; Ransom and Rowell, 1972).

Lindburg (1971) observed that sometimes the adult females approached the mother with a new infant placed their one hand

and 'uttered' as a vocalization and as an 'exclusive cough' (Rowell and Hinde, 1962). However, the mother did not resist such kind of approaches (Lindburg, 1967, 1971). Mothers also gave the appearance of being 'short tempered' during birth season. There were also more female deaths and wounds at this time which could be attributed to the complications associated with pregnancy or birth (Wilson and Boelkins, 1970). Herbert and Trimble (1967) observed the aggression in adult males during the birth season that might be related for the protection of females and their off-springs. The males also protected the infants from the attacks by the intruders (Redican and Mitchell, 1973).

The suckling behaviour has been found to be universal and characteristic among mammals which is important for survival and psychosexual development (Blass and Teicher, 1980). The mothers in the lactating group who sucked their infant frequently and with a great number of individual nipple contact in each bout fail to conceive in the next breeding season (Monterrat, 1989). He also suggested that the difference in the suckling pattern at the early stages might reflect the difference in the maternal conditions and this difference during the breeding season also affected the quality of ovulation. Except maternal parity, all other variables like previous experience and social status, sex of the infant and the presence of peers had been found to affect the mother-infant relationship.

REFERENCES

Altmann, S.A. (1962). A field study of the sociobiology of rhesus monkeys, *Macaca mulatta*. *Ann. N.Y. Acad. Sci.*, 102: 338-435.

Altmann, S.A. (1968). Sociobiology of rhesus monkeys, III. The basic communication net work. *Animal Behaviour*, 32: 17-32.

Altmann, J. (1974). Observational study of *behaviour*: Sampling methods. Animal *Behaviour*, 149: 227-267.

Ashley-Montagu, M.F. (1957). *The Reproductive Development of Female*. New York: The Jullian Press, Inc.

Bernstein, I. (1966). A field study of the male role in *Presbytis cristatus* (Lutong). *Primates*, 7: 402-403.

Bernstein, I.S. (1972). Daily activity cycles and weather influences on a pigtail monkey group. *Folia primatol.*, 18(5-6): 390-415.

Bernstein, I.S. and Draper, W.A. (1964). The behaviour of juvenile rhesus monkeys in groups. *Animal Behaviour*, 12: 84-91.

Bernstein, I.S. and Mason, W.A. (1963). Activity patterns of rhesus monkeys in a special group. *Animal Behaviour,* 11: 455-460.

Bertrand, M. (1969). *The behaviour repertoire of stumptail macaque.* Switzerland and New York: In press.

Blass, E.M. and Teicher, M.H. (1980). Suckling behaviour. *Science,* 210(4465): 15-21.

Brandt, E. and Mitchell, G. (1971). Patutition in primates: Behaviour related to birth. In: L. Rosenblum (ed.), *Primate Behaviour: Developments in Field and Laboratory Research,* vol. 2, Academic Press, New York.

Burton, F.D. (1972). The integration of biology and behaviour in the socialization of *Macaca sylvanus* of Gibraltar, In: *Primate Socialization,* F.E. Poirier (ed.), Random House, New York, pp. 29-62.

Carpenter, C.R. (1940). A field study in Siam of the behaviour and social relations of gibbon (Hylobates. lar). *Comparative Psychological Monograph,* 16: 1-212.

Carpenter, C.R. (1942). Sexual behaviour of free ranging rhesus monkeys (*Macaca mulatta).* 1. Specimens, procedure and behavioural characteristics. *Journal of Comparative Psychology,* 33: 113-142.

Chakravarthy, R.S. (1969). More about *Abrus precatorius. Science,* 166: 44.

Chance, M.R.A. (1956). Social structure of a colony *Macaca mulatta. British Journal of Animal Behaviour,* 4: 1-13.

Clark, M.R. (1978). Social interactions of juvenile female bonnet monkeys, *Macaca radiata. Primates,* 19: 517-524.

Cutton-Brook, T.H. and Harvey, P.H. (1977). Primate ecology and social organization. *J. Zool., Lond.,* 183: 1-39.

DeVore, I. (1965). *Primate Behaviour: Field Studies of Monkeys and Apes. Holt, Rinhart and Winston,* New York.

Draper, W.A. (1966). Free ranging rhesus monkeys: Age-sex differences in individual activity patterns. *Science,* 151: 476-478.

Drickamer, L.C. (1974). A ten year summary of reproductive data for free ranging *Macaca mulatta. Folia Primatologica,* 21: 61-80.

Drickamer, L.C. (1976). Quantitative observations of grooming behaviour in free-ranging *Macaca mulatta. Primates,* 17: 323-335.

Fiedler, W. (1956). Ubersicht uber das system der Primates. *Primatologia,* 1: 1-266.

Fittinghoff, N.A., Jr., Lindburg, D.C., Gomber, J. and Mitchell, G. (1974). Constistency and variability in the behaviour of mature, isolation-reared, male rhesus macaques. *Primates,* 15: 111-139.

Fooden. J. (1964). Rhesus and crab eating macaques: intergradation in Thailand. *Science,* 143: 343-365.

Fooden, J. (1971): Report on primates, collected in Western Thailand January-April 1967. *Fieldiana Zoology,* 59(1): 1-62.

Fooden, J., Mahabal, A. and Saha, S.S. (1981): Redefinition of rhesus macaque – bonnet macaque boundary in Peninsular India (Primates: *Macaca mulatta, Macaca radiata*). *J. Bom. Nat. Hist. Soc.,* 78(3): 463-474.

Gard, G.C. and Meier, G.W. (1977). Social and contextual factors of play behaviour in sub-adult rhesus monkeys. *Primates,* 18: 367-377.

Glick, B.B. (1979). Testicular size, testosterone level, and body weight in male *Macaca radiata*: Maturational and seasonal effects. *Folia Primatologica,* 32: 268-289.

Grewal, B.S. (1980). Changes in relationships of nulliparous and parous females of Japanese monkeys at Arashiyama with some aspects of troop organization. *Primates,* 21: 330-339.

Guerra, R.F. (1988). Withdrawal and retrieval by mother in rhesus monkey infants (*Macaca mulatta*). CIENC CULT. SAO PAULO, 40(7): 703-705.

Guerra, R.F. (1988). Maternal transport and locomotor development in infant rhesus monkeys (*Macaca mulatta*). *Rev. Bras. Biol.,* 48(2): 227-236.

Gunn, C.R. (1969). Abrus precatrious: Pretty but poisonous. *Science,* 164: 245-246.

Gupta, B.L. (1962). *Forest flora of the Chakrata, Dehradun and Saharanpur Forest Divisions, Uttar Pradesh.* Third ed. Govt. of India Press, Calcutta.

Hall, K.R.L. and DeVore, I. (1965). Baboon social behaviour. In: (Ed) DeVore (ed.), *Primate Behaviour, Field Studies of Monkeys and Apes,* Holt, Rinehart and Winston, New York, pp. 53-110.

Hamburg, D.A. (1971). Psychobiological studies of aggressive behaviour. *Nature,* 230: 19-23.

Handen, C.E. and Peter, S.R. (1980). Social development of bonnet macaques from six months to three years of age: A longitudinal study. *Primates,* 21: 350-256.

Harlow, H.F., Rowland, G.L. and Griffin, G.B. (1964). The effect of total social deprivation on the development of monkey behaviour. Recent Research on Schizophrenia, 19: 116-135.

Hausfaster, G. (1972). Intergroup behaviour of free-ranging rhesus monkey (*M. mulatta*). *Folia Primatol.,* 18(1/2): 78-107.

Hauty, G.T. (1972). Activities cycles of the monkey (*M. mulatta*) under different environmental conditions. *J. Interdisciplinary Cycles Res.*, 1(2): 187-197.

Hendy-Neely, H. and Rhine, R.J. (1977). Social development of stumptail macaque (*M. aretoides*): Momentary touching and other interactions with adult males during infants' first 60 days of life. *Primates*, 18: 589-600.

Hill, W.C.O (1966). *Primates Comparative Anatomy and Taxonomy*. Vol. VI, Catarrhini, Cercopithecoidea (Cercopithecinae). Wiley, New York

Hinde, R.A., Rowell, T.E. and Spencer-Booth, Y. (1964). Behaviour of social living rhesus monkeys in their first six months. *Proc. Zool. Soc., London*, 143: 609-173.

Hinde, R.A. and Spencer-Booth, Y. (1971). Towards understanding individual differences in rhesus mother-infant interaction. *Animal Behaviour*, 19: 165-173.

Itani, J. (1975). Twenty years with Mt. Takasaki monkeys. In: *Primate Utilization and Conservation*, G. Bermant, D.G. Lindburg (eds.), Wiley, New York, pp. 101-126.

Jay, P. (1963). Mother infants relations in langurs, In: *Maternal behaviour in Mammals*, H.L. Rheingold, (ed)., Wiley, New York, pp. 282-304.

Jay. P.C. (1965). The common langur of North India. In: *Primates behaviour: Field study of monkey and Apes*, DeVore (ed.), Holt, Rinhart and Winston, New York, pp. 197-249.

Jay. P.C. and Lindburg, D.G. (1965). *Indian primate ecology project*. Unpublished report to the National Centre for Primate Ecology, Davis, California.

Jenson, G.D. (1961). The development of prehension in a Macaque. *Journal of Comparative and Physiological Psychology*, 54: 11-12.

Johnson, R.L. (1986). Mother-infant contact and maternal maintenance activities among free-ranging rhesus monkeys. *Primates*, 27(2): 191-203.

Kaufmann, J.H. (1966). Social relations of infant rhesus monkeys and their mothers in a free-ranging band. *Zoological*, 51: 17-28.

Kaufmann, J.H. (1967). Social relations of adult males in a free-ranging band of rhesus moneys. In S.A. Altmann (ed.), *Social Communication among Primates*, University of Chicago Press, Chicago.

Kaufman, I.C. and Rosenblum, L.A. (1966). A behavioural taxonomy for *Macaca nemestrina* and *Macaca radiata*: Based on longitudinal observation of family groups in the laboratory. *Primates*, 7: 205-258.

Khalid, M. and Qaza, J.L. (1986). Diurnal dispersion pattern of a group of free-ranging rhesus monkeys *(M. mulatta). Pak. J. Zool,*. 18(4): 385-394.

Koford, C.B. (1966). Population changes in rhesus monkeys: Cayo Sanitago, 1960-1964. *Tulane Stud. Zool.,* 13: 1-17.

Koyama, N. and Shekar, P.B. (1981). Geographic distribution of rhesus monkeys and the bonnet monkeys in West Central Asia. *J. Bomb. Nat. Hist. Soc.,* 78(2): 240-255.

Lahiri, R.K. and Southwick, C.H. (1966). Parental care in *Macaca sylvana. Folia Primatol.,* 4: 257-264.

Lancaster, J.B. (1971). Play mothering: The relations between juvenile females and young infants among free-ranging vervet monkeys (*Cercopithecus aethiops*). Foila Primatol., 15: 161-182.

Lancaster, J.B. and Lee, R.B. (1965). The annual reproductive cycle in monkeys and apes. In I. DeVore (ed.), *Primate Behaviour: Field studies of monkeys and apes*. Holt, Rinchart and Winston, New York.

Lawick-Goodall, J.V. (1967). *My friends wild chimpanzees* (Washington: National Geographic Society).

Lindburg, D.G. (1967). A field study of the reproductive behaviour of the rhesus monkey (*Macaca mulatta*). Ph.D. Thesis, University of California, Berkeley, p. 132.

Lindburg, D.G. (1969). Rhesus monkeys: mating season mobility of adult males. *Science,* 166: 1176-1178.

Lindburg, D.G. (1971). The rhesus monkey in North India: An ecological and behavioural study. In: *Primate Behaviour,* L.A. Rosenblum (ed.), Academic Press, New York, Vol 2, pp. 1-106.

Lindburg D.G. (1976). Dietary habits of rhesus monkeys *(Macaca mulatta)* (Zimmermann) in Indian forests. *J. Bomb. Nat. Hist. Soc.,* 72: 261-269.

Loy, J. (1970). Behavioural responses of free-ranging rhesus monkeys to food shortage. *Amer. J. Phys. Anthrop.,* 33: 263-272.

Makwana, S.C. (1978). Field ecology and behaviour of the rhesus macaque (*Macaca mulatta*): group composition, home-range roosting sites, and foraging routes in the Asarori forest. *Primates,* 19(3): 483-492.

Maslow, A.H. (1936). The role of dominance in the social and sexual behaviour of infra-human primates: IV. The determination of hierarchy in pairs and in a group. *J. Gened. Psychol.,* 49: 161-198.

Michael, R.P. and Wilson, M. (1973). Changes in the sexual behaviour of male rhesus monkeys, *Macaca mulatta,* at puberty: Comparison with the behaviour of adults. *Folia Primatologica,* 19: 384-403.

Miller, G.S., JR. (1933). The groups and names of macaques. In: C.G. Hartman and W.L. Straus, Jr. (eds.), *The Anatomy of the Rhesus monkey* (*Macaca mulatta*). Williams and Wilkins, Baltimore.

Mitchell, G.D. (1968). Attachment differences in male and female infant monkeys. *Child Developments*, 39: 611-620.

Mitchell, G. (1970). Abnormal behaviour in primates, In: *Primate Behaviour*, L.A. Rosenblum (ed.), ALI pp. 195-249. Academic Press, New York.

Mitchell, G.D. (1977). Parental behaviour in non-human primates. In: *J. Money and H. Musaph (eds.), Handbook of Sexology*. North-Holland Biomedical Press, Elservier.

Montserrat, G. (1989). Suckling behaviour and fertility in rhesus monkeys. *J. Zool. (Lond.)*, 217(3): 449-469.

Moody, P.A. (1962). *Evolution of Man*. Harper and Row, New York, Evanston and London and John Weather Hill, Inc. Tokyo, pp. 221-223.

Mukherjee, A.K. and Gupta, S. (1965). Habits of rhesus macaque, *Macaca mulatta* (Zimmermann) in the Sunderbans, 24-Perganas, West Bengal. *Journal of the Bombay Natural History Society*, 62: 145-146.

Mukherjee, R.P, (1969). A field study on the behaviour of two roadside groups of rhesus monkeys *Macaca mulatta* (Zimmermann) in northern U.P. *J. Bomb. Nat. Hist. Soc.*, 66(1): 47-56.

Mukherjee, R.P. (1983). Survey of non-human primates of Tripura, India. *J. Zool. Soc. India*, 34(1/2) 70-81.

Mukherjee, R.P. and Mukherjee, G.D. (1972). Group composition and population density of rhesus monkey *Macaca mulatta* (Zimmermann) in North India. *Primates*, 13(1): 65-70.

Neville, M.K. (1968). Ecology and activity of Himalayan foothill rhesus monkeys. *Ecology*, 49: 110-123.

Nolte, A. (1955). Field observations on the daily routine and social behaviour of common Indian monkeys with special reference to the bonnet monkey (*Macaca mulatta* Geoffry). *J. Bomb. Nat. Hist. Soc.*, 53: 177-184.

Parker, S. and Parker, H. (1979). The myth of male superiority: Rise and Demise. *American Anthropologist*, 81: 289-309.

Parkash, I. (1962). Group organization, sexual behaviour and breeding season of certain Indian monkeys. *Japanese Journal of Ecology*, 12: 83-86.

Pirta, R.S. and Singh, M. (1978). Establishment of home-range, intraspecific and interspecific relations in Rhesus monkey (*M.*

mulatta) under infant-infant rearing conditions. *Proc. Ind. Acad. Sci.* 387(9): 267-278.

Pirta, R.S. and Sing, M., (1978). Some factors affecting the size and location of home is natural groups of Rhesus monkeys (*M. mulatta*). *Proc. Ind. Acad. Sci.* (Animal Science).

Pirta, R.S. and Singh, M. (1980). Changes in home-range of rhesus monkeys (*M. mulatta*) groups living in natural habitats. *Proc. Ind. Acad. Sci.*, 89: 515-525.

Pirta, R.S. and Singh, M. (1982). Differences in home-range of rhesus monkey (*M. mulatta*) groups living in three ecological habitats. *Proc. Ind. Acad. Sci. (Animal Science)*, 91: 13-26.

Pirta, R.S., Parkash, P. and Singh, M. (1977). A population study of two species of non-human primates: *M. mulatta* and *M. radiata*. *J. Bomb. Nat. Hist. Soc.*, 77: 429-434.

Pocock, R.I. (1932). The rhesus macaques (*Macaca mulatta*). J. Bomb. Nat. Hist. Soc., 35: 530-551.

Pocock, R.I. (1939). *Information on the Geographical range of the rhesus monkey*.

Poirier, F.E. (1968). The Nilgiri langur (*Presbytis johnii*) mother-infant. *Privates*, 9: 45678.

Poirier, F.E. (1970). The Nilgiri larger (*Presbutis johnii*)of South India. In: Primate Behaviour, L.A. Rosenblum (ed.), Academic Press New York, Vol 4, pp. 251-383.

Post, W. and Baulu, J. (1978). Time budgets of *Macaca lulatta Primates*, 19: 125-140.

Prater, S.H. (1965). *The book of Indian Animals*. Bombay Natural History Society, Bombay, p. 324..

Puget, A. (1971). Observations on rhesus macaque, *Macaca mulatta* (Zimmermann), en Afghanistan. *Mammalia*, 35(2): 199-203.

Quiatt, D. (1979). Aunt and Mothers: Adaptive implications of allomaternal behaviour of non-human primates. *American Anthropologist*, 81: 310-319.

Rahaman, H. and Parthasarathy, M.D. (1967). A population survey of the bonnet monkey (*Macaca radiata* Geoffroy) in Bangalore, South India. *J. Bombay Natural History Society*, 64(2): 251-225.

Rahaman, H. and Parthasarthy, M.D. (1969). The home-range, roosting places and the day ranges of the bonnet macaque (*M. radiata*). *J. Zool. (London)*, 157: 267-276.

Rahaman, H. and Parthasarathy, M.D. (1978). Behavioural variants of bonnet macaque (*Macaca radiata*) inhabiting cultivated gardens. *J. Bombay Natural History Society*, 75: 406-425.

Ransom, T.W. and Rowell, T.E. (1972). Early social development of baboons. In: *Primate Socialization*, F. Poirier (ed.), New York, pp. 105-144.

Redican, W.K. and Mitchell, G. (1973). A longitudinal study of parental behaviour in adult male rhesus monkeys. I. Observations on the first dyad. *Psychol.*, 8: 135-136.

Roonwal, M.L. (1956). Macaque monkey eating mushrooms. *Journal of Bombay Natural History Society*, 54: 171.

Roonwal, M.L. and Mohnot, S.M. (1977). *Primates of South Asia: Ecology, Sociobiology and Behaviour.* Harvard University Press, Cambridge (Mass.) and London.

Rosenblum, L.A. (1971). Kinship interaction patterns in pigtail and bonnet macaques. *Proc. 3rd Int. Congr. Primatol.* Zurich.

Rowell, T.E. (1972). Female reproductive cycles and social behaviour in primates. In: D.S. Lehrman, R.A. Hinde, and Evelyn Shaw (eds.), *Advances in the study of Behaviour*, (Vol. 4), Academic Press, New York.

Rowell, T.E. and Hinde R.A. (1962). Vocal communication by the rhesus monkey (*Macaca mulatta*). *Proceedings of the Zoological Society of London*, 138: 279-294.

Rowell. T.E., Hinde, R.A. and Specner-Booth, Y. (1964). 'Aunt' – infant interaction in captive rhesus monkeys. *Animal Behaviour*, 12: 219-226.

Sade, D.S. (1971). *Life cycle and social organization among free-ranging rhesus monkeys*. Paper read at the meeting of American Anthropological Association, New York.

Seay, B. (1966). Maternal behaviour in primiparous and multiparous rhesus monkeys. *Folia Primatologica*, 4: 146-168.

Seth, P.K. and Seth, S. (1983). Population dynamics of free-ranging rhesus monkeys in different ecological conditions in India. Am. *J. of Primatology*, 5: 61-67.

Seth, P.K. and Seth, S. and Sukla, A.K. (1983). Sociobiology of free-raning rhesus monkeys. In: *Prospective in Primate Biology* P.K. Seth (ed.), pp. 37-44.

Siddiqui, M.F. and Southwick, C.H. (1979). Feeding behaviour of rhesus monkeys (*M. mulatta*) in northern Indian plains. *Proc. Zool. Sco. (Calcutta)*, 31(1/2): 53-67.

Simonds, P.E. (1965). The bonnet macaque in South India. In: I. DeVore (ed.), *Primate Behaviour: Field Studies of Monkeys and Apes*. Hot, Rinehart and Winston, New York.

Simpson, M.J.A. and Howe, S. (1980). The interpretation of individual differences in rhesus monkey infants Behaviour, 72: 128-155.

Simpson, M.J.A., and Simpson, A.E. (1977). One-zero and scan methods for sampling behaviour. *Animal Behaviour*, 25: 726-731.

Singh, R. (1981). Study of behavioural patterns of age-sex classes during transitional periods in free-ranging rhesus monkey (*Macaca mulatta*). Ph.D. Thesis, University of Rajasthan, Jaipur.

Southwick, C.H. (1962). *Patterns of inter-group social behaviour in primates, with special reference to rhesus and howling monkeys*. Ann. N.Y. Acad. Sci., 102: 436-454.

Southwick, C.H. and Siddiqi, M.F. (1974). Population dynamics of rhesus monkeys in northern India. In: *Primate, Conservation*, Prince Rainier III of Monaco and C.H. Bourne (eds.), Academic Press, New York, pp. 339-362.

Southwick, C.H., Beg, M.A. and Siddiqi, M.R. (1961). A population survey of rhesus monkeys in villages, towns and temples of Northern India. *Ecology*, 42: 538-547.

Southwick, C.H., Beg, M.A. and Siddiqi, M.R. (1961). A population survey of rhesus monkeys in Northern India: II. Transportation routes and forest areas. *Ecology*, 698-710.

Southwick, C.H., Beg, M.A. and Siddiqi, M.R. (1965). Rhesus monkeys in North India. In: I. DeVore (Ed.), *Primate Behaviour: Field studies of Monkeys and Apes. Holt, Rinehart and Winston*, New York.

Southwick, C.H., Siddiqi, M.F., Farroqui, M.Y. and Pal, B.C. (1976). Effect of artificial feeding on aggressive behaviour of rhesus monkeys in India. *Animal Behaviour*, 24: 11-15.

Spencer-Booth, Y. (1968). The behaviour of group companions towards rhesus monkey infants. *Animal Behaviour*, 16: 541-557.

Sperry, R.W. (1971). How a developing brain gets itself properly wired for adaptive function. In: E. Toback, L.R. Aronsone and E. Shaw (eds.), *The Biopsychology of Development*. Academic Press, New York.

Struthsaker, T.T. (1971). Social behaviour of mother and infant of vervet monkeys (*Cercopithecus aethiops*). *Animal Behaviour*, 19: 233-250.

Sugiyama, Y. (1964). Group composition, population density and some ecological observations on hanuman langur (*Presbytis entellus*). *Primates*, 5: 7-48.

Sugiyama, Y. (1968). Ecology of the loin-tailed macaque (*Macaca silenus* Linnaeus): A pilot study. *J. Bomb. Nat. Hist. Soc.*, 65(2): 283-293.

Sugiyama, Y. (1971). Characteristics of the social life of bonnet macaques (*Macaca radiata*). *Primates*, 12(3-4): 247-266.

Swartz, K.B. and Rosenblum, L.A. (1981). *The social context of parental behaviour: a perspective on primate socialization*. pp, 417-454.

Tan, S.T., Ma, Y., Ban, T.S., Ban, T.Y. and Chow, H.Y. (1965), Birds and mammals in the Chin-Chau district, Shansi. *Acta Zoological Sinica*, 17(1): 86-102 (In Chinese).

Tobach, E., Arnson, L.R. and Shaw, E. (1971). *The biopsychology of Development*. New York: Academic Press.

Valverio, M.J., Miller, R.L., Innes, J.R.M., Courtney, K.D., Pallotta, A.J. and Gittmacher, R.M. (1969). *Macaca mulatta* (*Management of a laboratory breeding colony*). Academic Press, New York.

Vandenberg, B. (1978). Play and development from an ethological perspective. American Psychologist, 33: 724-738.

Vandenberg, J. (1966). Rhesus monkey bands. *Natural History*, 75: 22-27.

Van Lawick-Goodball, J. (1968). The behaviour of free-living chimpanzees to Gene strut Reserve. *Animal Behaviour Monogr.*, 1: 161-311.

Van Wagenen, G. (1952). Age at menarche of the laboratory rhesus monkey. *Anatomical Record*, pp 112-436.

Van Wagenen, G. and Simpson, M.E. (1954). Testicular development in the rhesus monkey. *Anatomical Record*, 118: 239-252.

Vessey, S.H. (1968). Interaction between free-ranging groups of rhesus monkeys. *Folla Primatol.*, 8: 228-239.

Wada, K. (1983). Rhesus monkey distribution in the lower Himalayas and secondary forest succession. *J. Bomb. Nat. Hist. Soc.*, 81: 335-363.

Wada, K. (1983). Ecological adaptation in rhesus monkeys at the Kumaon Himalaya. *J. Bomb. Nat. Hist. Soc.*, 80(3): 469-498.

Wilson, M.E. (1981). Social dominance and female reproductive behaviour in rhesus monkeys (*M. mulatta*). *Animal Behaviour*, 29(2): 427-482.

Wilson, A.P. and Boelkins, R. (1970). Evidence for seasonal variation in aggressive behaviour by *Macaca mulatta*. *Animal Behaviour*, 18: 719-724.

White, L.E. and Hinde, R.A. (1975). Some factors affecting mother-infant relations in rhesus monkeys. *Animal Behaviour*, 23: 527-542.

Zhang, Y.Z., Wang, S. and Quan, G.Q. (1981). On the geographical distribution of primates in China. *Journal of Human Evolution*, 10: 215-256.

11

An Ecological and Behavioural Study on *Macaca mulatta* (Zimmermann) in Jammu, J & K

D.N. Sahi and Shubhra Sharma

Department of Zoology, University of Jammu, Jammu Tawi-180006

Introduction

India is known as one of the primate rich area of the world both in species diversity and population abundance. About 14 species of non-human primates are known from India which include 6 sps. of macaques, 5 sps. of langur, 2 sps. of loris and 1 sp. of gibbon (Southwick and Lindburg, 1986). Among extant species of all non-human primates, *Macaca mulatta* (Zimmermann) is the most widely distributed. It inhabits area from Afghanistan in the west to Vietnam, Hong Kong and eastern China in the east. In India, it is widely distributed outside peninsular India in Sub-Himalayan forests from northeast to northwest India. Its distribution has been reported in south upto Godavari river and in west coast near Bombay (Hill, 1974).

Field studies on *M. mulatta* have been done in different parts by Carpenter (1942), Chance (1956), Southwick *et al.* (1961), Hinde and Rowell (1962), Altman (1962), Neville (1968), Mukharjee (1969), Lindburg (1971, 1976), Bernstein (1972, 1993), Malik *et al.* (1985) but no such study has ever been done in the state of Jammu and Kashmir which represents the northern limit of its distribution in India save for work on populations, survey of some parts of Jammu (Malhotra and Sahi, 1984). The present work deals with the group size and structure, food and feeding, home-range, grooming and reproductive activities of free-ranging monkeys inhabiting different ecological sectors in Jammu.

Study Area

The study was conducted in four different ecological sectors within the district Jammu of J & K state (Fig. 11.1.) These are:

(1) **Bahu Fort**: It is an old fort built by Maharaja of Jammu and houses a temple within its premises where large number of devotees visit everyday. It lies on the southeast of river Tawi on a small hillock at an altitude of 325 m. The area harbours a sizable population of rhesus monkey. As per the record it inhabits largest troop of rhesus monkey (Malhotra and Sahi, 1984) in J & K State.

(2) **Mahamaya**: It is a forest area 3 km North of Bahu Fort and is presently known as City forest. On the top of the hill within the forest, there is small temple which is often visited by devotees mostly on Tuesdays and Sundays. Vegetation cover found here is fairly dense and comprises a mixture of shrubs and small sized trees which are mainly of deciduous and thorny type.

(3) **Nandani**: A small village 26 km north of Jammu on Jammu-Srinangar National Highway at an approximate elevation of 410 m. On both sides of the highway, there is thick semi-evergreen forest of *Pinus roxburgii* and some broad leafed trees such as *Mallotus phillipensis, Dalbergia sisso, Butea monosperma* etc.

(4) **Panjthirthi**: It is thickly populated urban area on the north-west bank of river Tawi in the east of Jammu city. This locality is peculiar in having the old secretariat building, built by Maharaja of Jammu and presently houses several govern-ment offices. Several troops of monkeys harbour this congested human dwelling area.

Material and Method

For each habitat excepting temple troop, two troops of *M. mulatta* were chosen for detailed study. Troop residing in Bahu Fort was designated as Troop No. I, two troops confining their activity in city were numbered as Troop No. II and III, troops living in Mahamaya were called as Troop No. IV and V while two troops of monkeys living in Nandini along National Highway were designated as Troop No. VI and VII.

Rhesus monkey is diurnal in habit hence observations were made at different parts of the day from dawn to dusk till they settle down for roosting. The movement of each troop was followed and plotted on sketch map to find out home-range. Visual observations were made however when necessary prismatic binoculars of Bushnell make (7 × 50 mm) were also used. The study was conducted during April 1997 to Dec. 1999.

Observations and Discussions

Group Size and Structure

M. mulatta lives in social, heterosexual troop comprising of all age classes. The average contribution of male in different troops studied was 15%, that of female was 35%, juvenile 23.3% and infants 25.7% Southwick *et al.* (1961) had reported from total population samples of various ecological habitats that male contributed 21.4%, female contributed 43.6%, juveniles 9.5% and infants 25.5%. Rhesus monkey doesn't form exclusive male bands in contrast to that of Hanuman langur as reported by Jay *et al.* (1965), Sugiyama (1976). The size and sex-ratio of seven troops studied during the present investigation is given in Table No. 11.1. The average number of individuals studies during present investigation in different ecological sectors was 58.8. Southwick *et al.* (1961, 1965), Jay and Lindburg (1965), Neville (1968), Lindburg (1971), Rehman and Parthasarthy (1978), Gupta and Kumar (1993) however, reported average number of individual per troop to be 50, 32, 28, 23, 41, 37 and 59 respectively. Thus the average number of individuals reported in present study came to be closer to that reported by Gupta and Kumar (1993) i.e. 59.

No strikingly consistent trend in composition of troop has been observed but one thing is very clear that females outnumbered the males. Average adult male to adult female ratio observed during the present study was 2.42 whereas other workers had reported it to be 2.4, 2.7, 3.0., 2.4, 2.7, 1.2 and 1.7 respectively.

During present survey, lowest number of individuals studied in a troop was 34 while maximum was 90. This is in contrast to that reported by Jay and Lindburg (1965), Makwana (1978), Seth and Seth (1983), Gupta and Kumar (1993) who has reported minimum size as 8, 18, 10 and 25 and maximum as 98, 90, 102 and 92 respectively. This shows that there is no consistency in troop size and various factors like type of habitat availability of food, water and cover along with attitude of people determine the troop size.

Home range

It is an area that a troop habitually frequents for various life activities such as feeding, breeding, social interactions etc. Home range of seven troops studied was observed to be 6.5, 1.25, 1.05, 1.10, 1, 1, and 0.75 sq. km respectively. Largest home range of 6.5 sq. km was recorded in temple area while smallest of 0.75 sq. km in a road side troop. This is not in accordance with Lindburg (1971), Southwick and Siddiqui (1974) who have reported that populations living in areas of low food availability tend to have larger home range than those where food was abundant. Home range of seven troops occupying different habitats was noticed to vary widely and various factors determining the home range in the study area are water, food, tendency of dominant male to monopolies more female and to feed on certain plants species not available in the particular area. However, certain workers like Chance (1956), Mukherjee (1969), Pirta *et al.* (1978) held other factors like seasonal fluctuations, orientation of male, human interference inducing fluctuation in home range.

Feeding habits

Feeding plays an important role in determining the daily activity of rhesus. Studies conducted at all the study sites showed that although monkeys are basically omnivorous but herbivorous feeding activity predominates. The main three food resources are natural vegetation, agriculture crops and food provided by human beings.

The food provided by the human beings was found to be fairly similar almost ritualistic in all the study areas but at Bahu Fort and Mahamaya temple (Fig. 11.1). Level of provisionalization was much higher in comparison to other sites. In both these habitats, it has been observed that monkeys prefer to receive food provided by humans which accounts for 63.75-76.25% of total feeding period in a day.

Natural vegetation consumed by rhesus was found to differ seasonally. Throughout the year rhesus exploited the leaves of wide variety of plants but during fruiting season, fruits of some plants formed their choicest food.

Feeding pattern not only revealed seasonal fluctuation but also shows variation during different times of the day. During summer days, monkeys were observed invading feeding sites in early morning hours (6 a.m. to 7 a.m.) but during the cold and cloudy days, their feeding activity was noticed to start as late as 9 a.m. During monsoon period, depending upon rains, they were seen to start their feedings activity at irregular times. Generally two principle feeding bouts in a day were noticed, one in the morning and other in the evening.

The monkeys used their cheek pouches as temporary storage organs mostly when food is in abundance or is of choice or when the individuals are in hurry. Food picked from dusty grounds was usually rubbed between the hands before putting in mouth (Fig. 11.2).

Significant variation was noticed in the amount of time spent in feeding by different sexes and age classes in different months of the year.

Grooming

The most frequent interaction among individuals of a troop is grooming which refers to continuous act of picking some foreign material from the fur with their hands and fingers. It can be autogrooming or allogrooming. Autogrooming is called as self-grooming (Fig. 11.3) while allogrooming is a behavioural sequence in which two or more monkeys perform grooming act (Fig. 11.4). The individual which performs the grooming is called groomer and other being groomed is called a groomee. The common pattern of the grooming involves two partners. However, more than two individuals were also involved. Occasionally, grooming involving 4-6 individuals was also noticed.

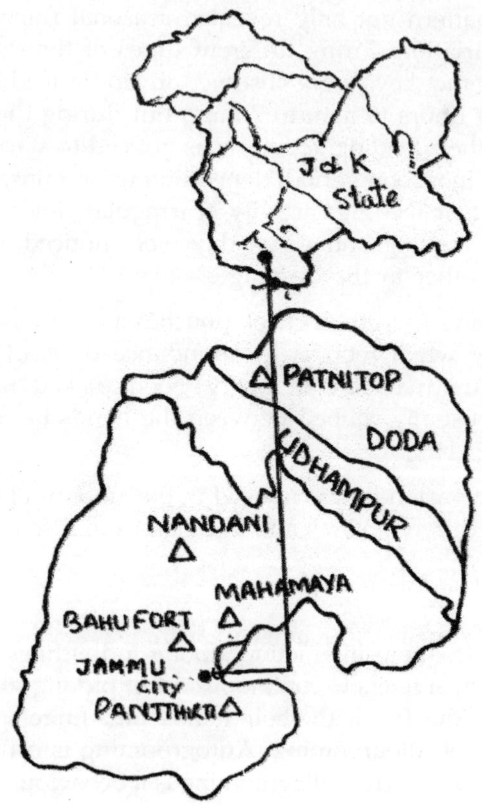

Fig. 11.1. Map of Jammu showing Study Area
(Bahu Fort, Panjthirthi, Mahamaya, Nandani and Pantnitop)

Fig. 11.2. Food provisonalization at Mahamaya Temple

The motor pattern used by rheus in grooming were noticed to be varied and intergrading. The most common pattern involved sliding movement in which whole hand with extended fingers was slided over the body of monkey, being groomed. The movement varied from fast to slow. Sliding was noticed to be complemented by scrapping in which nails and tips of forefingers

Fig. 11.3. Autogrooming by a female Rhesus Monkey

usually of middle and other fingers scratches the skin to peel of the debris. This was seen to be followed by picking in which thumbs and fore finger remove flakes of loose epidermis, dirt, ectoparasite and putting them in mouth.

During present study dominant members were noticed to groom the subordinate ones. This observation is in contrast to Carpenter (1942), Altman (1962) Southwick *et al*. (1965), Lindburg

Fig. 11.4. Allogrooming involving more than Two Individuals

(1976) who reported it to be one way process where subordinate groomed the dominant males and not vice-versa.

Survey conducted also depicted that amount of the time spent in grooming also show fluctuations with sex, females spending more time in grooming in all the seasons than other members of the troop. Grooming interaction with males as groomers was found to be less in present study. Grooming between adult males was virtually non-existent.

Grooming is daily activity which is of some prevalence and show fluctuation with sex, age, season and habitat e.g. in winters it ranks above feeding in case of female, while in case of male, it is the activity which ranked below the major activities of day like feeding, resting etc. During summer, in both male and female grooming is activity of least performance. However, juveniles didn't show any fixed pattern during both the seasons.

Reproduction

Rhesus monkey lives throughout the year in heterosexual groups having biased female sex-ratio. Females are seasonally polyestrous and mating is by and large confined to an average period of 9.66 days. Carpenter (1942) reported the average duration of estrus to below 9.2 days whereas Kaufman (1965) recorded it to be 11 days.

Reproduction behaviour in case of rhesus is marked by onset of estrus in case of females. Since the specification of estrus was less precise under natural condition so estrus in female was not detected by physical characteristics alone but also by behavioural aspects.

Physical characteristics involved predominantly changes in colour of sexual skin. Not only entire genital region but also root of tail, upper and inner parts of thighs, legs upto the ankles and sometimes face also becomes red. Most conspicuous behavioural changes during consort are described under following heads:

Approaching Behaviour

The beginning of breeding season was marked by female's approach towards male and at times similar behaviour was also reciprocated by male. During this behaviour, just to approach dominant male, female was also observed to provoke him by taking inlist aid from other members of the troop/from distant object. Approaching behaviour either by male/female towards each other, however, doesn't always end in copulation.

Affinitive Gesture

Various kinds of gesture seen in *M. mullata* during breeding season are:

1. **Rhythmic lip movement** – Male monkey was observed to make some kind of effinitive gesture to the female with which he intends to form close relation. This behaviour was reflected in rapid muscular movements of protruded muzzle and some times lead to successive momentary opening and closing of lips.

However, this behaviour was also observed in non-breeding season when subordinate monkey perform this behaviour towards dominate one to express his submission.

2. **Presentation** – Females in estrus were noticed presenting their genital region as mating invitation to male. On experiencing such type of presentation, a male has been seen to respond by simply gazing at the presenting female or by performing actual coitus.
3. **Mounting and copulation** – During mounting activity male was observed climbing on female in a series of attempts until actual coitus took place which was followed by ejaculation. Average time taken for this act was observed to be seven and a half minutes (Fig. 11.5.)
4. **Estrous call** – Male and female at the time of mating were observed to make estrous call. This sound was observed to be produced along with rhthythmic lip movement. Mating season of *M. mulatta* is between Oct. to end of Feb., reaching its peak during Nov. and Dec. But births were seen in period between March to August with maximum births occurring during May and June.

REFERENCES

Altman, S.A. (1962). A field study of sociology of Rhesus monkey (*Macaca mulatta*). *N.Y. Acad. Sci.*, 102(2): 338-435.

Bernstein, I.S. (1972). Daily activity cycles and weather influences on Pigtail monkey group. *Fol. Primatol.*, 18(5-6): 390-415.

Bernstein, I.S. (1993). Seasonal influences on Rhesus monkey (*Macaca mulatta*) behaviour. *Int. J. of Primatol.*, 14(3): 383-403.

Carpenter, C.R. (1942). Sexual behaviour of free ranging Rhesus monkey. *J. Comp. Psychol.*, 33: 113-162.

Chance, M.R. (1956). Social structure of colony of *Macaca mulatta*. *Brit. J. Anim. Behav.*, 4: 1-13.

Gupta, R.C. and Kumar, S. (1993). On the troop size and composition of Rhesus monkey (*Macaca mulatta)* in Haryana: *J. Nat. Cons.*, 5(2): 111-114.

Hill, W.C.O. (1974). *Primates: Comparative anatomy and taxonomy.* New York Wiley and Sons.

Hinda, R.A. and Rowell, T.E. (1965). Communication by postures and facial expression in Rhesus monkeys. *Proc. Zool. Soc. Lond.*, 138: 1-21.

Jay, P. and Lindburg, D.G. (1965). *Indian Primate Ecology Project.* Unpublished report to the National Centre for Primate Biology. Davis California.

Kaufman, J.H. (1965). A three-year study of mating behaviour in free ranging band of Rhesus monkeys. *Ecology.*, 46: 500-512.

Lindburg, D.G. (1971). The Rhesus monkey in North India: An Ecological and behavioural study: In: *Primate Behaviour. Rosenbleum* (ed), Academic Press, N. York, pp. 1-106.

Lindburg, D.G. (1976). Dietary habits of Rhesus monkeys (*Macaca mulatta*) in Indian Forests. *J. Bomb. Nat. Hist. Soc.*, 73(2): 261-279.

Malhotra, Y.R. and Sahi, D.N. (1984). A population survey of non-human primates in some parts of Jammu Province. In: *Current Primate Research*, Roonwal, Monhot and Rathor (eds.), University of Jodhpur, Jodhpur, pp. 75-80.

Malik, I. and Southwick, C. (1984). Feeding behaviour of free ranging Rhesus monkeys of Tughlaqabad. *J. Bomb. Nat. Hist. Soc.*, 84: 336-348.

Mukhrjee, R.P. (1969). A field study on behaviour of two road side groups of Rhesus monkey (*Macaca mulatta*) in Northern U.P. *J. Bomb. Nat. Hist. Soc.*, 66(1): 47-56.

Neville, M.K. (1968). Ecology and activity of Himalayan foot hills rhesus monkey (*Macaca mulatta*). *Ecology*, 49(1): 110-123.

Pirta, R.S. and Singh, M. (1978). Some factors affecting the size and location of home range in natural groups of Rhesus monkeys (*Macaca mulatta*) *Proc. Ind. Acad. Sci.*, (*Animal Science*).

Rahman, H and Parthasarathy, M.D. (1978). Behavioural variants of Bonnet Macaque (*Macaca radiata*) inhabiting cultivating gardens. *J. Bomb. Nat. Hist. Soc.*, 75: 406-425.

Seth, P.K. and Seth, S. (1983). Population dynamics of free ranging Rhesus monkeys in different ecological condition in India. *Am. J. Primatol.*, 5: 61-67.

Southwick, C.H., Beg, M.A. and Siddiqui, M.R. (1961a). A population survey of Rhesus monkeys in villages, towns and temples of Northern India. *Ecology,* 42(3): 538-547.

Southwick, C.H., Beg, M.A. and Siddiqui, M.R. (1965). Rhesus monkey in Northern India. In: *Primate Behaviour: Field Studies of Monkeys and Apes,* DeVore I. (ed.), New York, Holt, Rinehart and Winsten, pp. 11-159.

Southwick, C.H. and Lindburg, D.G. (1986). The *primates of India: Status, Trends and Conservation.* New York, Springer, Verlag, pp. 171-187.

Southwick, C.H. and Siddiqui, M.R. (1974). Contrasts in Primate social behaviours, Bioscience 24: 394-406.

Sugiyama, Y. (1976). Characteristics of ecology of Himalayan Langur. *J. Hum. Evol.,* 5: 249-277.

12

Control of *Musca domestica* by *Casia fistula* Seed Extract

Mangla Bhide, Sunil Kumar and Vandana Kharya

Department of Zoology, Dr. H.S. Gour University Sagar-470003 (M.P.) India

ABSTRACT

Musca domestica is pathogenic in nature and causes so many diseases in human being and their live stock, so to control the fecundity and viability by the topical application of *Casia fistula* seed extract (as antifertility agent of plant origin) has been done in this present investigation resulted as an effective ovicidal, larvicidal and pupicidal agent with high rate of morality as the time of moulting, low percentage of hatchability was also observed. Adults showed high degree of sterility which was evident by the histopathological and histo-chemical studies of gonads. Ovaries showed the presence of large number of resorbed oocytes which was the main cause of low rate of fecundity. Ovaries showed duration of treatment dependent gradual pathological changes not only in their histological architecture but also showed gradual decline in protein +ve, PAS + ve and sudanophilic materials respectively

Key words: *Musca* control by *Casia fistula*.

Introduction

Musca domestica is the domestic pathogenic flies is responsible for the transmission of so many diseases and also having high population density. The literature concerning with the research work on the reproductive physiology of insect pests to control their population by treatment with different chemical fractions extracted from plants is very scanty and has been done by Satio *et al.* (1989) in *Musca domestica,* Khanam *et al.* (1990) in *Tribolium confusum,* Zhang *et al.* (1992) in field application, Facknath and

Kawol (1993) in cabbage webworm *Crocidolomia binotalis*, Jalees *et al.* (1993) in *Anopheles staphensi*, Roychoudhary (1993) in *Sitophilus oryzae*. Srimannarayana (1993) reported the use of extracts of indigenous insecticidal plants in integrated pest management. Lee and Chiang (1994) in mosquito, Regault and Hamraoui (1995) in *Acanthoscelides obtectus*, Castillo *et al.* (1998) in *Tecia solanivora*, Prates *et al.* (1999) investigated the control of *Spodoptera frugiparda*, Desai and Desai (2000) in *Spodoptera litura*, Potenza *et al.* (2000) in *Ceratitis capitata* and Bhide (2002) in *Bagrada cruciferarum* respectively and have evaluated the insecticidal properties of the tested plant extracts.

The literature regarding the effect of plant extract on the anatomy and histological aspects of ovarian tissue has been done by a few workers as investigated by Chang (1998) in house flies, Mallini *et al.* (1994) in *Exorista larvarum*, Kumar and Bhide (2001) in *Periplaneta americana* and Bhide (2002) in *Bagrada cruciferarum* while Ait *et al.* (1991) in *Sitona discoideus* and Anwar *et al.* (1992) in *Pyrilla parpuallis* observed the morality in nymphs and adults respectively but Ait *et al.* (1991) also reported the malformation in nymphs of *Sitona discoideus carthumii*, Kumar *et al.* (1999) reported the control of safflower aphids *Parigea capensis* by the application of indigenous plant extract and by Bhide (2002) in *Bagrada cruciferarum* after *Cerbera thevetia* seed kernel extract but no work has been done on the control of *Musca domestica* after topical application *Casia fistula* seed extract treatment so this attempt has been done to know more on the various aspects concerning with control of fecundity and viability of *Musca domestica*.

Materials and Methods

Procurement and rearing of Musca domestica:

Adult flies of *Musca domestica* were collected easily from *Azadirachta indica* seeds in the month of April and May and were reared by the method adapted after Goel (1983). Adults after acclimatization for 4 days were kept in pairs separately (10 pairs in each group) and the eggs collected from control and experimental groups were used for the study of ovicidal activity.

Selection of plant seeds and extraction for Antifertility Agent:

Casia fistula seeds were collected from the pods of the corresponding plants which is abundant in Sagar University campus. Seeds were washed, shadow dried and powdered mechanically by using grinder and the antifertility agent was extracted with petroleum ether (BDH B. P. 40° C) by Soxhlet method after Sharma (1988).

Experiments

(1) Experiments were done to decide the dose of, Lc100, Lc50, Lc0 and sub lethal doses Data was collected and summarized in Table No. 12.1, 12.2, 12.3.

(2) **Ovicidal Activity**: Data was collected and summarized in Table No. 12.1 and Graph No. 12.1.

(3) **Larvicidal Activity:** Data was collected and summarized in Table No. 12.2 and Graph No. 12.2.

(4) **Pupicidal Activity**: Data was collected and summarized in Table No. 12.3 and Graph No. 12.3.

Note: Untreated groups were used as control while for the studies of ovicidal, larvicidal and pupicidal activity "**Topical Application**" was the procedure which was used in the present investigation.

Histopathological Studies:

Adults of 2, 4, 6 and 8 days control and experimental flies (*Musca domestica*) treated with sub-lethal concentration of the extract of *Casia fistula* were dissected out and their ovaries were fixed in *Carnoy's fluid* (6 : 3 : 1). Paraffin blocks were prepared in usual manner and sections were cut at 6 μm.

For Histological and Histochemical Studies:

Haematoxylin and Eosin (Delafield method from Pearse, 1960), Mercuric Bromophenol Blue (Davenport, 1960), Periodic Acid Shiff's Reagent method (McManus, 1948) and Sudan Black B methods from Pearse (1960) were applied for histological and histochemical studies.

Observations and Results

The observations are based on the following aspects:

Ovicidal, Larvicidal and Pupicidal Activities:

The data on these aspects have been collected and summarized in Table 12.1, 12.2, 12.3 and Graph 12.1, 12.2, 12.3.

Table 12.1: Effect of different concentration of *Casia fistula* seed extract on topically treated eggs of *Musca domestica.*

Exp. No.	Forms of treatment	Conc. of extract/type of dose	Total No. of eggs	Egg servi-vility	Mortality (%)	Hatc hability (%)	No. of 1st instar larvae hatched out
1.	Control group without any treatment	Not Applied	50	50	Nil	100%	100
2.	Topical application After oviposition and on every alternative day	0.05 ppm Lc100	50	Nil	100%	Nil	Nil
3.	- Do -	0.022 ppm Lc50	50	25	50%	50%	50
4.	- Do -	0.01 ppm Lc0	50	48	4%	96%	96
5.	- Do -	0.005 ppm Sub lethal	50	50	Nil	100%	100

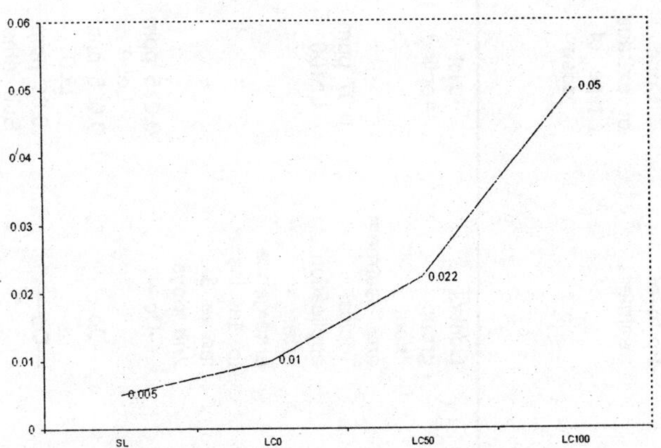

Graph 12.1: Lethal concentration value, doses and mortality rate in the Eggs of *Musca domestica* treated with *Casia fistula* seed extract

Table 12.2: Effect of different concentration of *Casia fistula* seed extract by topical application of different larval instars of *Musca domestica*.

Exp. No.	Form of treatment	Conc. of extract/ type of dose	No. of larvae	1st larvel instars hatched into 2nd instar larvae	2nd larval instars hatched into 3rd instar larvae	3rd larval instars hatched into pupae	Mortality (%)
1.	Control group ithout any treatment	Not Applied	50	50	50	50	Nil
2.	Topical application after emergence to 1st instar larvae & 2nd days	0.07 ppm Lc100	50	30	05	Nil	100%
3.	—Do—	0.035 ppm Lc50	50	45	25	25	50%
4.	—Do—	0.015 ppm Lc0	50	50	48	48	04%
5.	—Do—	0.01 ppm Sub lethal	50	50	50	50	Nil

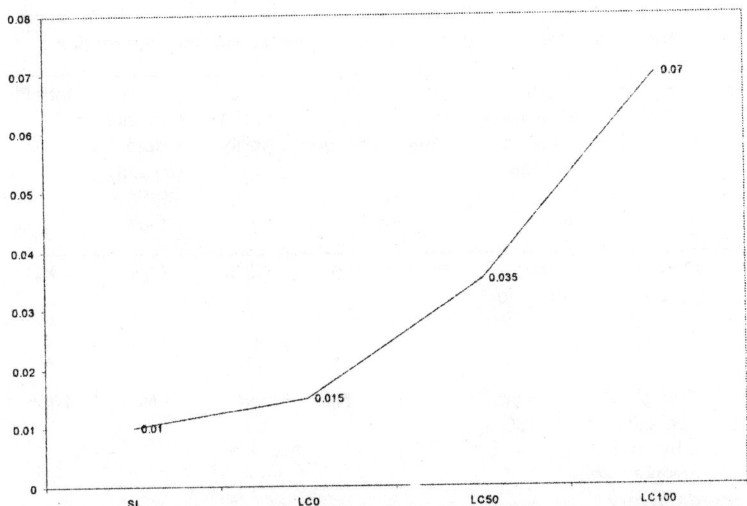

Graph 12.2. Lethal concentration value, dosages and mortality rate in the larval instars of *Musca domestica* treated topically with *Casia fistula* seed extract

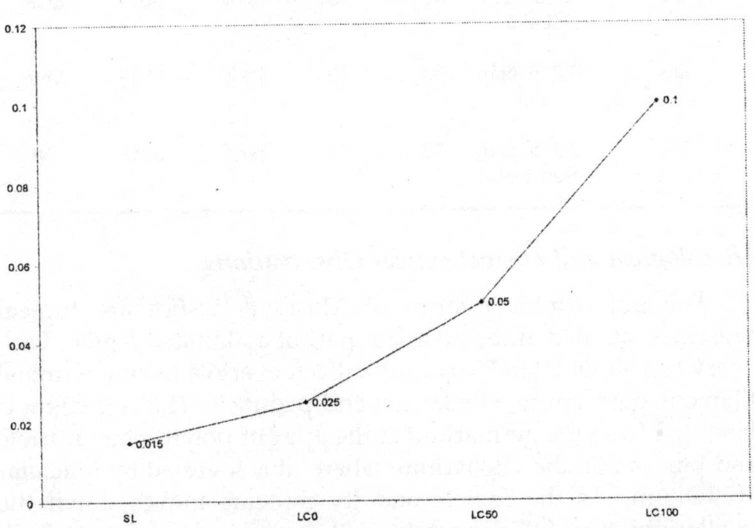

Graph 12.3: Lethal concentration value, dosages and mortality rate in the pupae in *Musca domestica* treated topically wtih *Casia fistula* seed extract

Table 12.3: Effect of different concentration of *Casia fistula* seed extract by topical application of pupae of *Musca domestica*

Exp. No.	Form of treatment	Conc. of extract type of dose	No. of puape	No. adults emerged	% of hatch ability	% of survivity upto sexually mature flies	Mortality (%)
1.	Control group without any treatment	Not Applied	50	50	50%	50%	Nil
2.	Topical application after ovi-position and on every alternate day upto emergence of Adult flies	0.1 ppm Lc100	50	Nil	Nil	Nil	100%
3.	-Do-	0.05 ppm Lc50	50	25	50%	50%	50%
4.	-Do-	0.025 ppm Lc0	50	49	98%	96%	96%
5.	-Do-	0.015 ppm Sub lethal	50	50	50%	50%	Nil

Histological and Histochemical Observations:

The polytrophic ovaries of *Musca domestica* are typical muscidae situated at the posterior part of abdominal region. Each ovary has about 50 to 75 ovarioles. Each ovariole having terminal filament, germarium, vitellarium and peduncle. The oogonium is liberated from the germarium at the apex of polytrophic ovariole and passes into the vitellarium, where it is invested by follicular epithelium. As the follicle and its contents moves down the vitellarium, new follicles are formed successively abobve it, each presumably containing a single oogonium.

In newly emerged female flies, the developing egg chamber or follicles are already clearly differentiated. Each oocyte develop in it's follicle with the aid of another type of cell (accessory). In *Musca domestica* there are two main types of accessory cells - nurse cells or trophocytes and follicle cells. The nurse cells are derived from oogonium from which the corresponding oocyte developed. An oogonium under goes four successive mitotic divisions, resulting into production of 16 cells. One of these became oocyte and remaining 15 became nurse cells. Each follicle is wrapped in highly proteinaceous follicular epithelium which is intensely PAS +ve in nature. Tunica properia which is also highly protein +ve and PAS +ve in nature found firmly attached along the entire length of the follicle. The trophocyte cytoplasm, the ooplasm and even the follicular epithelium are seen to contain dense sudanophilic materials.

It is interesting to note that in *Musca domestica* the penultimate egg follicle did not grow until the first one is fully developed and has ovulated. In *Musca domestica* the process of yolk deposition has seen in terminal oocyte only.

The development of oocyte in *Musca domestica* could be divided into three successive phases i.e.

(1) The Previtelline Phase:- In newly emerged to 2 days old flies.

(2) The Vitelline Phase with two Subphases:

(i) The Early Vitelline Phase:- in 3rd to 4th day old flies.
(ii) The Late Vitelline Phase:- in 5th to 6th day old flies.

3. The Post Vitelline Phase:- In 7th day onwards up to egg laying.

In the previtelline phase, as in newly emerged flies the egg follicle is surrounded by follicular epithelium and the germarium contains an oogonium. This follicle descends into the vitellarium and begins to grow. After 4 successive mitotic divisions the oogonium is divided into 16 cells. The oocytes are differentiated sooner than the remaining 15 trophocytes, which take an active part in oocyte growth by keeping the developing oocyte continuously supplied with contents.

The nurse cells and oocytes are indifferential from each other. Each of them has a protein +ve nucleus and the trophocytes and oocyte of the same follicle are all interconnected by cytoplasmic bridges.

The follicles of 2 days old flies are much larger and the oocyte is easily distinguishable from the trophocytes. The later have a central nucleus which is larger than the oocyte nucleus and has thick membrane and coarse chromatin network formed of large irregularly distributed chromatin granules. Each trophocyte nucleus contains 7-9 prominent and highly protein and PAS + ve nucleoli. The trophocyte cytoplasm is also intensely protein and PAS + ve in nature. The trophocyte nucleoli are rounded and are sudanophlic and stained densely with protein and PAS stain.

The entire egg follicle is surrounded by multilayered follicular epithelium, whose cells are proliferated for some way between the nurse chamber and developing oocyte. The follicular epithelial cells are less proteinaceous but their nuclei are intensely protein + ve in nature. The follicular epithelium round the nurse chamber forms only a thin cell layer, because its contents contribute continuously to the growing trophocytes. The nurse cells grow faster than the oocyte and in a short time attain their maximum size, they then receive no further nutrients from the follicular epithelium but begin to supply the developing oocyte from their own stocks of nutrients by way of cytoplasmic connections. The ooplasm is less protein +ve and PAS +ve in nature but its outer layer contains highly protein +ve granules. The protein yolk precursors (PYP bodies) can be seen in the opplasm. The lipid spheres are only L1 type. A number of lipid granules are also to be found scattered about among the PYP bodies and lipid spheres.

In experimental group treated with sub-lethal concentration of extract showed some histopathological lesions in the egg follicles of 2 days old flies and showed the sign of disintegration in the trophocytes e.g. vacuolization of trophocyte nucleus and trophocyte cypoplasm. Cells of follicular epithelium showed vacuolization and their nucleoli became pycnotic in nature.

In control group the egg follicle of 4 days old female flies is much larger as the trophocytes have stopped developing. The trophocytes immediately adjacent to oocyte continuous to grow larger than the other trophocytes and it also has a large nucleus. The trophocyte cytoplasm is highly vacuolated, the chromatin material is less dense and there is marked destruction of nuclear material. The nuclear membrane of the nurse cells subsequently breaks down and chromatin is released continuously into the cytoplasm, from which nutrients are supplied to the oocyte.

There is no partition between the nurse chamber and the oocyte and owing to the absence of follicular epithelium the trcphocytes are in direct contact with the oocyte, so the nutrients can be conveniently transported from the trophocytes to the oocyte via cytoplasmic bridges.

The trophocyte nucleus and cytoplasm are both proteins and PAS +ve while trophocyte nucleoli showed intensely protein and PAS +ve material. The ooplasm contains a large number of irregularly distributed PYP bodies. The follicular epithelium is also intensely protein and PAS +ve indicating that vitellogenin are taken up actively from the haemolymph by micropinocytosis.

The oocyte nucleus is situated close to the nurse chamber. The trophocyte nucleus is filled with sudanophlic material and can clearly seen to be slightly vacuolated. The trophocyte cytoplasm is also vacuolated, with large sized vacuoles situated near the periphery i.e. adjacent to the follicular epithelium. Numerous PYP and LYP bodies can be seen irregularly scattered about in the ooplasm. An intensely protein +v and PAS +ve vitelline membrane has been secreted below the follicular epithelium.

In experimental group treated with sub-lethal concentration of extract showed the increase in the disintegration in the follicular material. There is an increase in the vacuolization in the nurse chamber in comparison to the control group. There is a marked reduction in the oocyte of the corresponding follicle in comparison to control. Marked degenerative changes can be seen in the follicular epithelium of the corresponding egg follicle. It became highly vacuolated showing the marked depletion in the protein +ve, PAS +ve and sudanophilic materials.

The rate of oocyte growth in the 6 days old female flies is high because the oocyte now attains the full size. The follicular epithelium is more active showing that vitellogeinin passes actively from the haemolymph to the ooplasm by micropinocytosis. The nuclei of the follicular epithelium cells are also more active and are rich in protein and PAS +ve intensity. The nurse chamber has started to shrink.

The number of trophocytes is smaller. Their cytoplasm is densely sudanophlic and their nuclei are intensely proteinaceous. The follicular epithelium is of equal length along the entire egg follicle. It is composed of a single cell layer. The nuclei of the follicular epithelium cells are highly proteinaceous. The fully grown oocyte is filled with PYP & LYP bodies which are densely packed in the centre of the ooplasm and are less dense in the peripheral ooplasm. Only type L1 lipid bodies are found in *Musca domestica* in trophocytes and in the follicular epithelial cells. Slight vacuolization of the trophocytes is discernible at the periphery and below the follicular epithelium. The vitelline membrane is also intensely protein and PAS +ve in nature. Large number of lipid granules are scattered about in the ooplasm. On the sixth day the chorion is secreted on the outer surface of the vitelline membrane and inner surface of the follicular epithelium.

The six days flies treated with *Casia fistula* seed extract showed high rate of degeneration, disintegration not only in nurse chamber but also showing the active resorption of ooplasm evident by the presence of large number of vacuoles. The size of the oocytes became reduced in comparison to the control. The follicular epithelium became irregular in outline and showed the decrease in the protein +ve, PAS +ve and sudanophlic materials. Oocytes also showed the decrease in PYP, LYP bodies and also showed the decrease in the protein +ve and PAS +ve intensities. Follicular epithelium also showed the sign of degeneration and disintegration as evident by the presence of large number of pycnotic nuclei and high rate of vacuolization in their cytoplasm. Most of the egg follicles showed the total resorption of nurse chamber.

In control group the ovaries of 8 days old flies showed that the oocytes became fully developed and flies were ready for egg laying. The oocytes attain that size which is required for egg laying and contain large number of PYP and LYP bodies which are highly protein +ve and PAS +ve in nature and also showing the presence of sudanophlic material. Oocytes also showed the detachment of chorion and follicular epithelium. Total resorption of nurse chamber has been seen.

The ovaries of 8 days old experimental flies treated with *Casia fistula* seed extract showed that most of the egg follicles became empty due to the total resorption of oocytes while a few of the follicles showed the presence of pathological or sub-pathological oocytes which are going to resorbed within short time. These experimental flies did not showing the egg laying because their oocytes did not attain such maturity which is required for egg laying. The resorbed oocytes showed low intensity of protein +ve and PAS +ve material and also showed the less amount of sudanophlic material. All follicles showed the detachment of chorion and follicular epithelium. Both these membranes are irregular in outline and also showed the deformities in their structural architecture. The ultimate follicles also showed the sign of resorption and did not showing the proper progressive development as observed in control group.

In control group after ovulation the ovaries showed the presence of large number of yellow colour bodies (corpora lutea) at the junction of the ovariole and the pedicle. The number of yellow colour bodies is always equal to the number of eggs laid down by the corresponding female flies. In the corpus luteum the tunica properia is strikingly folded. Folding and refolding of tunica properia is a sign of progressive contraction of the empty follicle, which is soon resorbed.

In control flies when the oocyte has been discharged from the egg follicle, early sign of degeneration are clearly visible in the follicular epithelium which play active role in the resorption of corpus luteum. The earlier sign of disorganization has clearly marked in follicular epithelium and obvious changes have been observed in the nuclei of their cells. In the earlier stage the chromatin material is converted into chromatin granules and afterwards the nuclei became pycnotic and are found scattered about irregularly in the cytoplasmic syncytium. In the early stage the pycnotic nuclei are larger in number and smaller in sizes but in later stages their number steadily decreases, while their size increases and they rapidly degenerate. When the successive follicle is ready to ovulate, the corpus luteum contains a few pycnotic nuclei scattered about irregularly in the cytoplasmic syncytium. The tunica properia is' excessively folded. The cytoplasmic syncytium showed the presence of hypervacuolization, indicating that active resorption has taken place in the corpus luteum.

The tunica properia is highly protein and PAS + ve in nature. The pycnotic nuclei and their chromatin contain densely sudanophilic material, while cytoplasmic syncytium is less proteinoid in nature showed less PAS +ve and sudanophilic materials.

So it could be concluded from the present investigation that extract of seed of *Casia fistula* acts as antifertility agent and having potency to control the fecundity of *Musca domestica* flies which are pathogenic in nature and are responsible to spread out so many diseases in human beings.

Discussion

The plant products are environmentally safe, biodegradable, non-persistent and less toxic with less or no side effect to higher animals. Plant extracts play very important role in the resorption of oocytes as evident by the surveyed literature till date.

In the present investigation *Casia fistula* seed extract was found to be a potent chemosterilants and growth inhibitor as evident by the low rate of fecundity and survivity with high rate of morality in eggs (Table No. 12.1) as well as in larval instars (Table No. 12.2) and in pupae (Table No. 12.3) of *Musca domestica* exhibited ovicidal, larvicidal and pupicidal activity of *Casia fistula* seed extract. High percentage of mortality was also reported by Saito *et al.* (1989) in *Musca domestica* after treatment with 240 Brazilian plant extracts and noticed that *Annona* species extract was more effective, Khanam *et al.* (1990) in *Tribolium confusum* also reported the insecticidal property after treatment with indigenous plant extract, Anwar *et al.* (1992) observed 75% mortality after *Asculus indica* flower and leaves, 60 to 90% after *Azadirachta indica* leave extract treatment in field crop pests and *Pyrilla parpuallis* while Ragnault and Hamraoui (1995) in bruchid *Acanthosecelides obtectus* after administration of *Origanum vulgare* plant extract not only showed the mortality but also exhibited the inhibition in oviposition and reduced number of adult emerged from the last nymphal instars as reported in *Bagrada cruciferarum* after administration of *Cerbera thevetia* seed kernel extract by Bhide (2002) while in the present investigation high rate of mortality was observed in eggs at the time of hatching hence showed the low rate of hatchability in *Musca domestica*.

Larvae also exhibited the high rate of morality at the time of moulting to the successive instar larval stages while pupae showed the partial or total arrest of adult emergence resulted into low percentage of adult emergence in *Musca domestica* in the present investigation while Desai and Desai (2000) reported 30% mortality in *Spodoptera litura* after treatment with 11 locally available plant extracts.

So in the present investigation *Casia fistula* seed extract showed ovicidal, larvicidal and pupicidal properties as also reported by Ait *et al.* (1991) in weevil *Sitone discoideus* after Azadirachtin administration in 4th and 5th instars and has noticed the nymphicidal activity and malformation in nymphs and adult insects, Jalees *et al.* (1993) in *Anopheles stephensi, Culex quinquefesciatus, Aedes aegypti* (100% mortality within 24 hr) after treatment with 4% *Cannabis sativa* extract (which showed larvicidal activity), Lee and Chiang (1994) reported the larvicidal activity of *Stemona tuberosa* aqueous extract administration on mosquito larvae while Patil *et al.* (1997) showed the ovicidal, larvicidal action of some indigenous plant extract in *Dactynotus carthumii* and also reported high rate of mortality in adults but Prates *et al.* (1999) in 3rd instar larvae of *Spodoptera frugiperda* and Facknath and Kawol (1993) reported that cabbage webworm *Crocidolomia binotalis* when treated with *Annona, Crassiflora* and *Argimone maxicana* showed larvicidal action of the treated agents as also reported by Bhide (2002) in *Bagrada cruciferarum* after administration of *Cerbera thevetia* seed kernel extract. Zhang *et al.* (1992) evaluated the insecticidal properties of Azadirachtin. Roychoudhary (1993) applied the *Didymocarpus podocarpa* extract on *Sitophilus oryzae* and Srimannarayan (1993) reported the use of extracts of indigenous insecticidal plants in integrated pest management and have to proved the anti-feedant properties of the treated extracts as observed by Bhide (2002) in *Bagrada cruciferarum* after application of *Cerberia thevetia* seed kernel extract as reported with *Casia fistula* seed extract which also showed the insecticidal property in the present investigation.

Kumar *et al.* (1999) evaluated the control of safflower aphid *Parigea capensis* (Lepidoptera) after administration of some indigenous plant extracts while Castillo *et al.* (1998) reported the intoxication of *Barbaris samacana* extract on *Tecia solanivora* resulted into arrest of pupae development as observed in *Musca domestica* in the present investigation after treatment with *Casia fistula* seed extract while Potenza *et al.* (2000) evaluated the effect of *Alos vara* plant extract on *Ceratitis capitata* and reported the fecundity decrease from 33 to 99.16% while in the present investigation 100% mortality with 00.00% of hatchability was achieved by administration of Lc100 doses in *Musca domestica* and the same observation were reported in *Bagrada cruciferarum* by Bhide (2002) after treatment with *Cerbera thevetia* seed kernel extract.

Chang (1988) in house fly and Mallini *et al.* (1994) in *Exorista larvarum* reported the histopathological changes caused in the ovarian tissue by the treatment with 5 fluorouracil and by a parasitoid respectively. The adverse effects were reported in germarium, in the egg chamber and in follicular cells in house fly (Chang, 1988) as reported in *Musca domestica* in the present investigation where long term topical application of *Casia fistula* *w*as able to created high rate of degeneration, disintegration and resorption in pathological egg follicles in ovarian architecture which was evident by the presence of high rate of vacuolization.

Bhide (2002) has reported in *Bagrada cruciferarum* that sub-lethal concentration of extracted glucosides significantly inhibited the ovarian development as exhibited by the decrease in the oocyte diameter and the increase in follicular atresia as reported by Kumar and Bhide (2001) in *Periplaneta americana* while in *Musca domestica* in the present investigation showed not only the deformities in oocytes but also active degeneration of trophocyte nucleus and cytoplasm has been observed in ovarian architecture after topical treatment of *Casia fistula* seed extract.

In control group vitellogenin synthesis by the fat bodies and its uptake by the developing oocytes in the form of PYP bodies in the phase of vitellogenesis is the normal phenomenon while *Casia fistula* seed extract intoxication inhibited the partial or total arrest of vitellogenin synthesis resulted in the production of large number atretic, pathological egg follicles in the ovarian architecture and formation of large number of resorbed oocytes in *Musca domestica* in the present investigation as reported by Kumar and Bhide (2001) in *Periplaneta americana* and by Bhide (2002) in *Bagrada cruciferarum* but the literature concerning with histochemical aspects due to intoxication of *Casia fistula* in the ovarian tissue of *Musca domestica* or other pest insects are totally missing so the observations are original and that of author's only. The present investigation suggests the antifertility potential of the extract of seed of *Casia fistula* and should be use to control the other pathogenic pest insects.

In the present investigation in *Musca domestica* after ovulation yellow bodies are formed at the junction of peduncle and the ovariole which are equal to the number of oocytes laid down by the corresponding flies but in the experimental group the number of yellow bodies (corpus luteum) is always reduced in comparison to control group and it is equal to the number of oocyles attain the maturity which is required for ovulation or oviposition. In the present investigation the maturity of oocytes is fully effected by the treatment of *Casia fistula* seed extract as observed by Kumar and Bhide (2001) in *Periplaneta americana* after *Casia fistula* seed extract 7 by Bhide (2002) in *Bagrada cruciferarum* after treatment with *Cerbera thevetia* seed kernel extract.

It could be concluded by the present investigation that the production of large number of resorptive bodies resulted into decrease in the rate of fecundity in *Musca domestica* as reported in *Bagrada cruciferarum* by Bhide (2002) after treatment with two glucosides extracted from *Cerbera thevetia* seed kernel. The treated pupae were emerged into very weak impaired adults because of partial or total arrest of ovarian development resulted into partial or total sterility in *Musca domestica* in the present investigation.

Acknowledgements

The authors wish to express their thanks to *Prof. D. P. Gupta, H.O.D, Department of Zoology, Dean Faculty of Life Sciences, Dr. H. S. Gour University, SAGAR 470003 (M.P.)* for providing the necessary laboratory facilities and one of the author (*Dr. Mangla Bhide)* would like to express her sense of gratitude to UGC for providing the financial assistance.

REFERENCES

Ait, El, Koch A., Herbut, C. and Dakhama A. (1991). Functional type of the ovary of weevil pest of Lucerne *Sitona discoideus (Coleoptera Curlionidae). Actes-de-Institute Agronomique-et Veterinaier-Hassan-II,* 11(3): 27-32

Anwar, T., Jabbar, A., Khalique, P., Tahir, S., and Shakeel, M.A. (1992). Plants with insecticidal activities against four major insect pests in Pakistan. *Tropical Pest Management*, 38(4): 431-437.

Bhide, M., (2002). Control of *Bagrada crucijerarum by Cerbera thevetia* seed kernel extract. 2nd Conference International Sur les Organismes nuisibles aux vegetaux, held at Lille France at Palais du Nouveau Siecle from 4-7th Mar 2002: 309-316.

Castillog, G., Luque, E. and Moreno, B. (1998). Laboratory evaluation of insecticidal activity of five plants species of *Tecia solanivora (Lepidoptera: Gelechiidae). Agronomia Colombiana,* 15(1): 34-40.

Chang, Jt., Su, Dk. (1988). Effect of 5 fluorouracil on ovarian development of the housefly and the induction of mutation. *Acta Entomologica Sinica*, 31 (III): 254-258.

Davenport H, A. (1960). Histological and Histochemical techniques. Philadelphia.

Desai, S.D. and Desai, B.D. (2000), Studies on the insecticidal properties of some plants. *Shashpa* 7(I): 85-88.

Fachnath, S. and Kawol, D. (1993). Antifeedant and insecticidal effects of some plant extract on the cabbage webworm *Crocidolomia binotalis. Insect Science and it's Application,* 14(5): 571-574.

Goel, S.C. (1983). *Laboratory Exercise in Development Biology*. Rational Printers 216, Narayan Peth Pune 30.

Jalees, S., Sharma, S.K., Rahman, S.J. and Verghese, T. (1993). Evaluation of insecticidal Properties of an indigenous plant *Cannabis sativa Linn* against mosquito larvae under laboratory conditions. *Journal of Entomological Research*, 17(2): 117-120.

Khanam L.M., Talukedar, D. and Khan, A.R. (1990); Insecticidal property of some indigenous plants against *Tribolium confusum (Coleoptera: Tenebriondae) Bangladesh Jr. Zool.* 18(2): 253-256.

Kumar, S. and Bhide, M. (2001). Effect of *Casia fistula* extract on *Periplaneta American. Jr. Exp.* Bio.: 299-311.

Kumar, S.S.R., Kulkarni, G.G., Basavana, G.K. and Mallapur C.P. (1999). Insecticidal properties of some indigenous plant extract against lepidopteran pests of safflower. *Annals of Bio. Research*, 4(1): 49-52.

Lee, H.L. and Chiang, Y.E. (1994). Insecticidal activity of the herbal plant *Stemona tuberose Lour* to mosquito larvae. *Tropical Biomedicine* 11(1): 87-89.

MC Manus, J.F.A. (1948). Histological and Histochemical uses of periodic acid. *Stain Technology*, 23(3): 101-102.

Mallini, E., Gardenghi, G. and Coulbaly, A.K. (1994). Anatomical and Histological characteristics of the female genital apparatus of *Exorista larvarum l. (Diptera: Tachinidae)* a parasitoid that lays macrotype eggs on the host. *Bollettino Dell Institute di Entomologia Guido Grandi dell Universita degali Studi-di-Bolongna,* 48: 45-49.

Patil, R.K., Rayar, S., Basappa, H., Hiremath, L.G. and Patil, B.R. (1997). Insecticidal properties of indigenos plants against *Dactynotus carthumii* H.R.L. and it's predator *Chrysopersi carn. J. of Oilseed Reasearch,* 14(1) 71-73.

Pearse, (1960). A.G.E. *Histochemistry Theoretical and Applied,* J. & A. Churchill London.

Potenza, M.R., Bergmann, E.C., Nishimori, R.F. and Benedictor, L.H. (2000). Insecticidal effect of some plant extract on *Ceratitis capitata.* Revista-de-Agricultra-Piaracicaba, 75(1): 55-65.

Prates, H.T., Viana, P.A., Pimentia, L.P.S. Boaventwra, M.A.D., Oliveira, AB. OE. and Oliveria, A.B. (1999). Assessment in the laboratory on the insecticidal activity of plant extract for the control of *Sapodoptera frugiperda. Pesquisa-Em-Milho-E-Sorgo*: 38.

Regault Roger, C. and Hamraoui, A. (1995). Comparison of the insecticidal effects of water extracted and intact aromatic plants on *Acanthosecelides obtectus,* a bruchid beetle pest of kidney beans. *Chemocology,* 5-6(1): 1-5.

Roychoudhary, N. (1993). Antifeedant and insecticidal properties of some plant extract for the adults of *Sitophilus oryzae (Linn.) (Curculinidae). Uttar Pradesh Jr. of Zoology,* 13(2): 97-100.

Satio, M.L., Oliveira, P., Fell O., Takematsu, AP., Jocys, T. and Oliveira, LJ. (1989). Evaluation of the insecticidal activity of some Brazilian plants. *Biologia,* 56(I-II): 53-59.

Sharma, J.D. (1988). *Forenisc Science and Toxicology,* Suvidha Law House, 28 Malviya Nagar (Roshanpura) Bhopal. M.P., India.

Srimannarayana, G. (1993). Chemical investigation of indigenous insecticidal plants including neem and evaluation of insecticidal, antifeedant activity of the constituents: *Botanical Pesticides in Integrated Pest Management,* pp. 62-73.

Zhang, Y.G., Zhang, Z. and Zhao, S.H. (1992). Preliminary studies on azadirachtin content and insecticidal activity of the seeds from the introduced neem tress *Azadirachta indica.* A Juss planted in China. *J. of South China Agricultural University,* 13(1): 14-19.

Key to Abbreviations:

C.	=	Chorion
C.C.	=	Cytoplasmic Connections
C.L.	=	Corpus Luteum
C.S.	=	Cytoplasmic Syncytium
E.F.	=	Egg Follicle
F.E.	=	Follicular Epithelium
L.C.	=	Lecitholytic Cells
N.C.	=	Nurse Chamber
O.	=	Ooplasm
O.N.	=	Oocyte Nucleus
PAS +ve	=	PAS +ve material
P.N.	=	Pycnotic Nuclei
Pr. +ve	=	Protein +ve Material
PYP	=	Protein Yolk Precosers
R.B.	=	Resorptive Bodies
R.O.	=	Resorbed Oocyte
S. +ve	=	Sudan Black B +ve Material
T.	=	Trophocyte
T.N.	=	Trophocyte Nucleus
T.NU.	=	Trophocyte Nucleolus
T.P.	=	Tunica Proreria
V.	=	Vacuoles

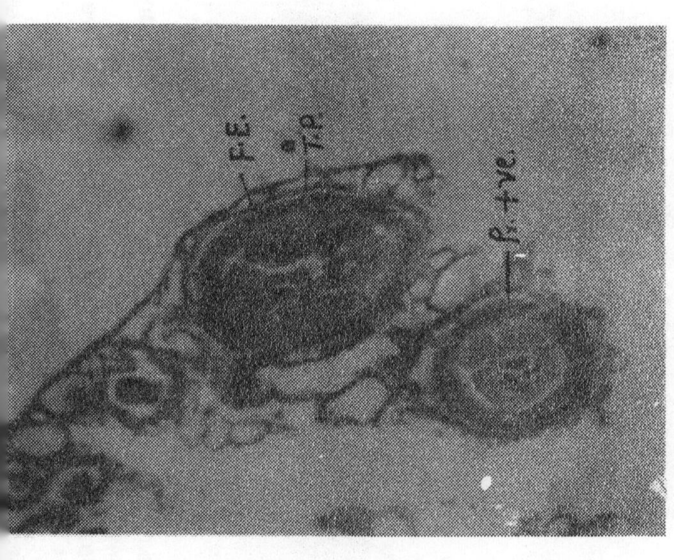

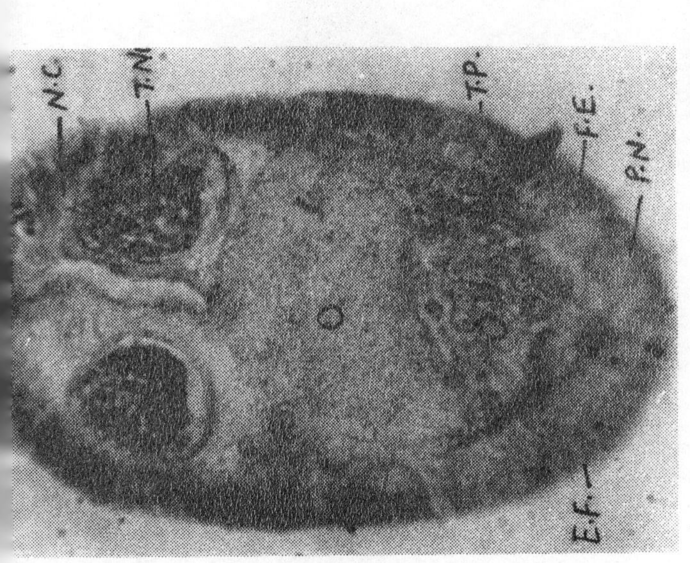

Marcuric Bromophennol Blue Stain (Hg. B.B.)

Fig. 12.1: T.S. of ovary of 2 days old *Musca domestica* showing highly protein +ve intesity of egg follicles (× 100).

Fig. 12.2: Section of egg follicel of 4 days old *Musca domestica* showing highly protein + ve trophocytes nuclei and nuclie of follicular epithelim (× 450).

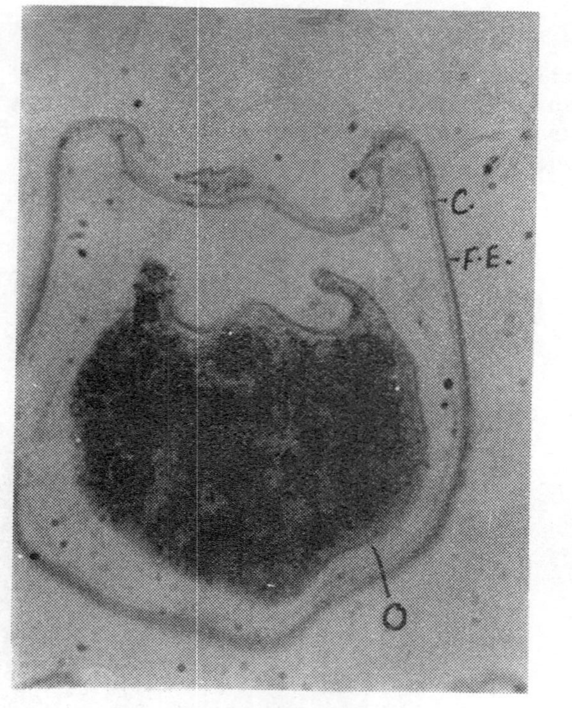

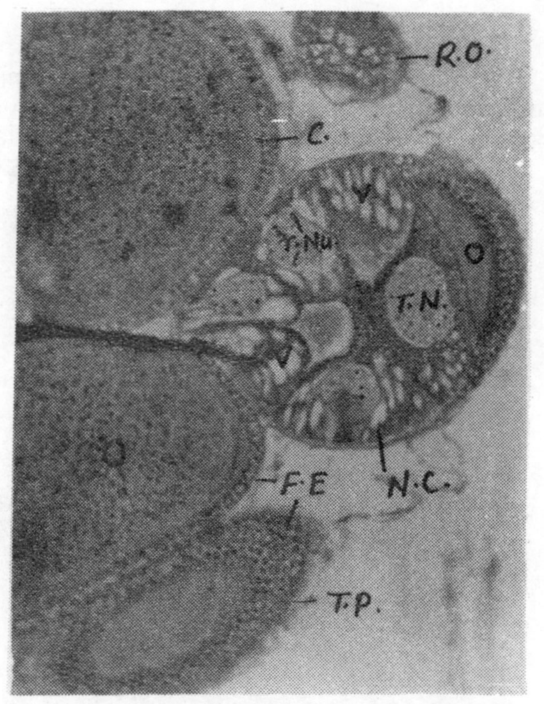

Fig. 12.3: Section of ovary of 6 days old *Musca domestica* showing highly protein +ve tunica properia of developing follicle and decreased protein +ve intesity of resorbed oocytes (× 100).

Fig. 12.4: Section of mature oocyte of 8 days old *Musca domestica* showing the detached highly protein +ve follicular epithelium and chorion (× 450).

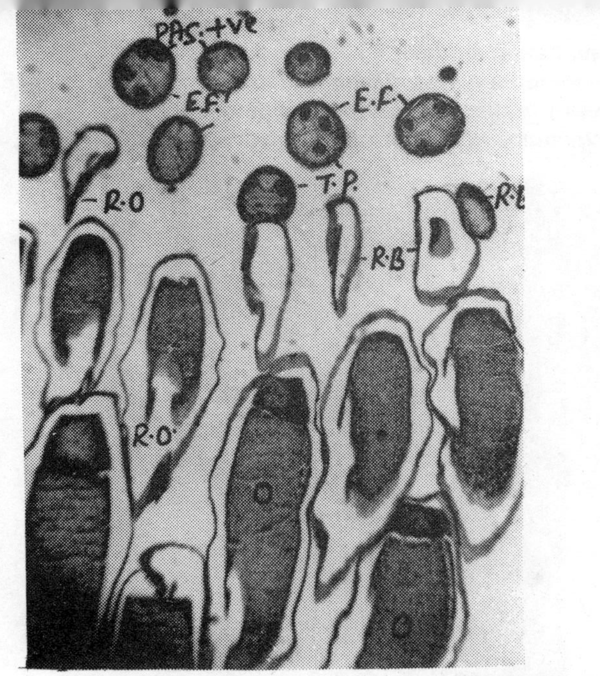

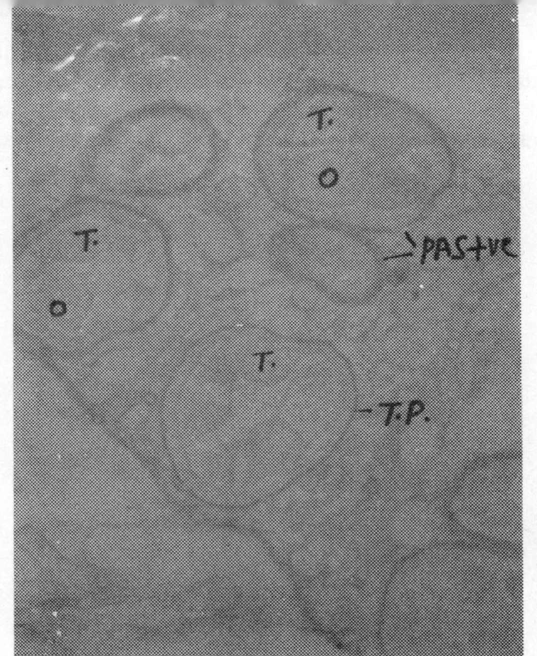

Periodic Acid Schiff's Reaction (PAS STAIN)

Fig. 12.5: Section of ovary of 2 days old *Musca domestica* showing less PAS +ve intensity of while tunica properia is highly PAS +ve in nature (× 100).

Fig. 12.6: Section of egg follicle of 4 days old *Musca domestica* showing highly PAS +ve intensity of trophocyte nuclei and nuclei of PAS +ve follicular epithelium in comparison to less PAS +ve trophocyte cytoplasm and ooplasm (× 450).

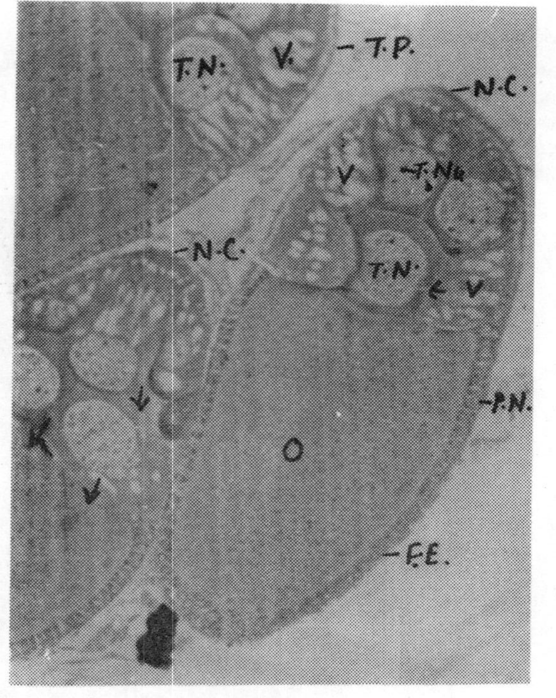

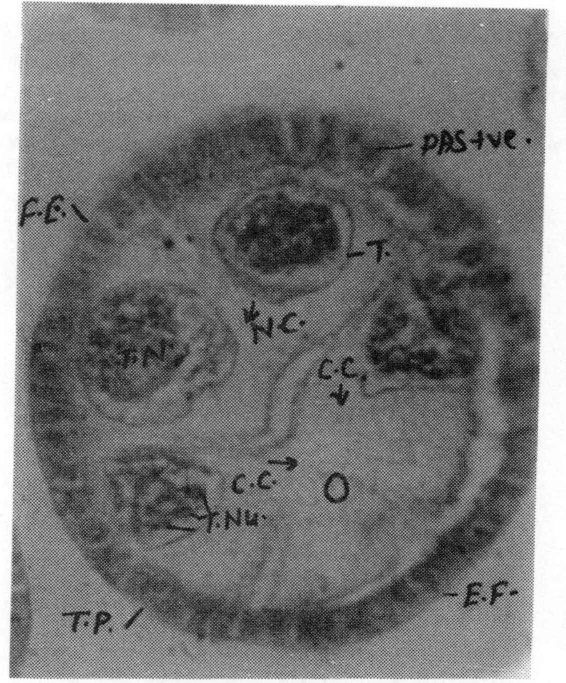

Fig. 12.7: Section of ovary of 6 days old *Musca domestica* showing less PAS +ve intensity of nurse chamber of egg follicle with high rate of vacuolization and vacuoles are totally PAS +ve in nature (× 100).

Fig. 12.8: Section of ovary of 8 days old *Musca domestica* showing the resorption of PAS + ve material from egg follicle, some follicles are empty due to the total resorption of PAS +ve material (× 100).

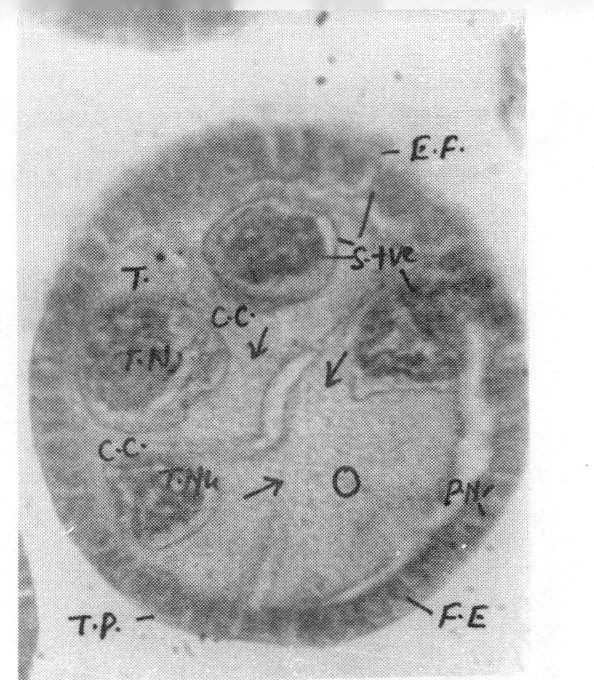

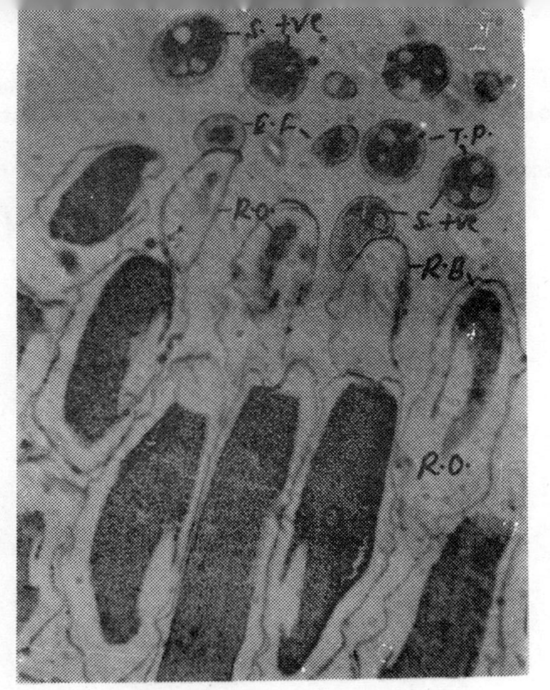

Sudan Black Stain 'B'

Fig. 12.9: Section of egg follicle of 2 days old *Musca domestica* showing sudanphilic material in trophocytes nuclei and follicular epithelial nuclei (× 450).

Fig. 12.10: Section of egg follicle of 4 days old *Musca domestica* showing the resorption of sundanophilic material in trophocytes (× 450).

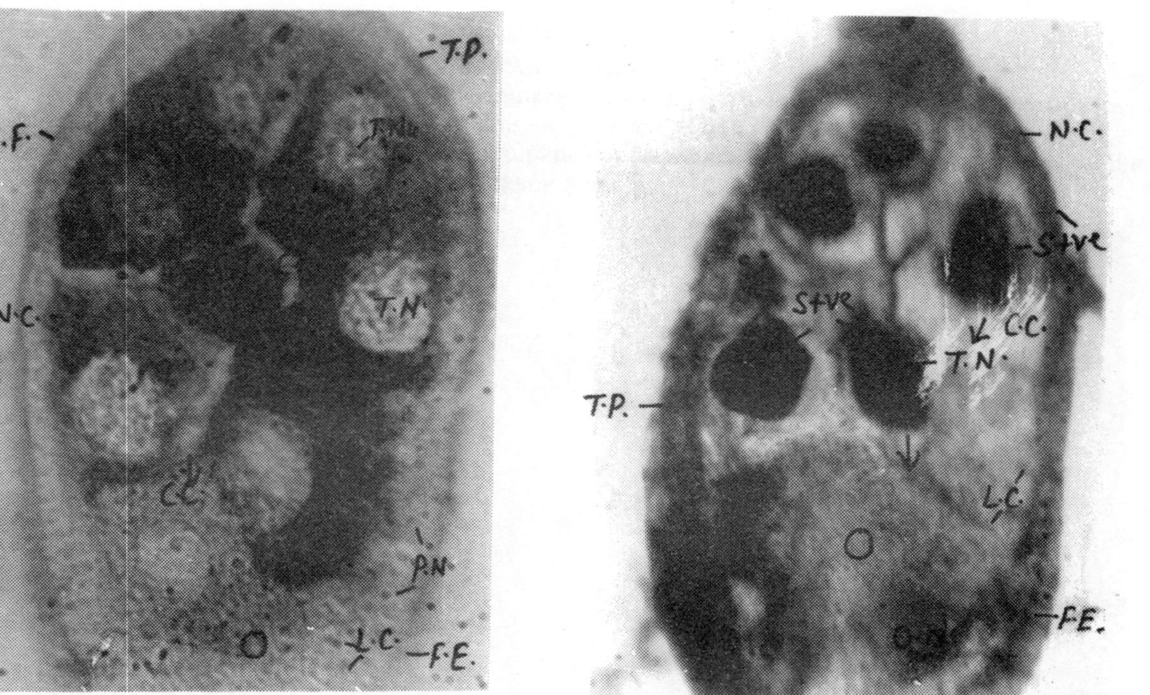

Fig. 12.11: Section of egg follicle of 6 days old *Musca domestica* showing sudanophilic tunia properia and less sudanophilic trophocyte and oocyte nuclei (× 450).

Fig. 12.12: Section of ovary of 8 days old *Musca domestica* showing the resorption of sudanophilic material from the pathological egg follicles (× 100).

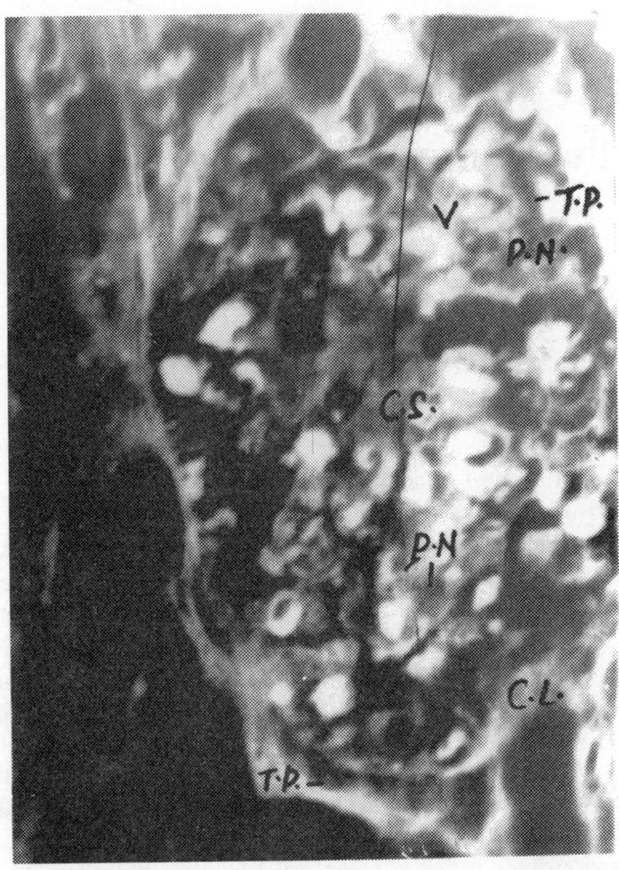

Fig. 12.13: Section of corpus luteum of 4 days old *Musca domestica* showing high rate of vacuolization of cytoplasmic syncytium and pycnotic nuclei (× 450).

13

Environmental Toxicants and their Impact on Reproduction

Charanjit Kaur Dhanju

Department of Zoology, Punjab Agricultural University, Ludhiana-141004, India

Introduction

Environmental toxicants are agents in our surroundings that are harmful to human health. Environmental toxicology is concerned primarily with the harmful effects of chemicals that are encountered by man either incidently because they are in the atmosphere, or by contact during occupational or recreational activities, by ingestion of food substance that contain naturally occurring toxic substances or chemical residues or by ingestion of water containing chemical or biological contaminants. Generally air and water pollutants are recognized for their toxicity, but other agents equally harmful to human health (viz. food additives and contaminants, bacteriotoxins, fungitoxins, phytotoxins, household products and industrial chemicals) often go unrecognized for their serious toxicity. Although naturally occurring substances are a larger concern to human health, but man-made products e.g. industrial chemicals and pesticides are thought to exhibit greater toxicity. Out of the man-made products, pesticides are high-volume, widely used, environmental chemicals which have possible role in many chronic health problems.

Since few efforts are being made to moderate either national or world population growth, demands for food and fibre must increase. Thus in a developing country like India, the overgrowing demands to meet the growing needs of rapidly increasing population has severely affected the production system which involved an extensive use of fertilizers and pesticides in order to increase the crop yield. Pesticides are being used extensively to destroy the pests of agricultural, veterinary and medical importance. Consequently, our environment is highly polluted due to fertilizer and pesticide use in agriculture. The indiscriminate or misuse of pesticides poses serious threat to human health as pesticidal residues in food chain produce ill effects when their concentration exceeds safe tolerance level (Kalra, 1989). India accounts for one third pesticide poisoning cases and spray men are most affected. According to WHO reports, the level of chlorinated pesticide intake is high among Indians than among people of other countries (Kulshrestha, 1992). Cases of blindness, cancer, deformities, diseases of liver, kidney and nervous and immune system from pesticide poisoning have been reported (Kulshrestha, 1992). Pesticide toxicity in India has caused dwarfism, loss of memory, effects on reproduction, skin and allergic reactions, behavioural changes; miscarriages and epilepsy outbreak (Dhar, 1996). In general, pesticide poisoning causes dizziness, nausea and fatigue.

Environmental pollutants are known to cause reproductive toxicity and have resulted in a decreased in the fertility level of human population, but the identification of environmental agents that have adverse effects on reproductive health is still a challenge. About 37% of all unexplained infertility in the United States is thought to be caused by environmental factors (Lasley and Overstreet, 1998) but few human reproductive toxicants have been clearly identified and the mechanisms of chemical toxicity to reproductive tract cells are not well defined. Unlike other aspects of human health, disturbances of reproduction generally go unrecognized and unreported. For example, only limited assessments of reproductive health are possible in over 50% of women in United States because of elected or age-related infertility.

The exposure of humans and animals to pesticides may be direct or indirect, but they adversely influence growth, respiration, nervous system, development, immune response and reproduction in experimental animals (Chopra *et al.*, 1979; Gopal *et al.*, 1979, 1990, Banerjee and Hussano, 1987; Randhawa, 1991; Kelce *et al.*, 1995; Dhar, 1996; Kaur, 1997). The evidence from animal experiments, wildlife studies and reports of occupational and accidental human exposures indicate that chemical contaminants can adversely affect reproduction, but a link between exposure and adverse effects has not yet been identified (Foster, 1995). Although environmentally induced reproductive disease has been documented in a human population and one or more toxicants are suspected of being responsible, it is still challenging task to identify the cellular target(s) of toxicity and the mechanism(s) of toxic action. In this article, various aspects of the impacts of environmental toxicants on reproduction and development are reviewed and concluded.

Environmental toxicants and reproduction

Reproductive hazards are considered to be any hazard that interferes with production of normal gametes, conception and in utero development. Identification of environmental hazards to reproductive health and characterization of the adverse outcomes necessitate a multi-disciplinary approach. Many aspects of reproductive biology are so specific to primate species that only a non-human primate model may be suitable as a surrogate for humans in experiments that identify the target of a reproductive toxicant. But the non-human primates are very expensive and their supply is limited. Still the results from whole animal studies are usually required to select and design *in vitro* experiments for mechanistic studies that use more widely available cells from human or other animals.

A number of pesticides have the potential to cause reproductive toxicity in animals and several of these (e.g. dibromochloropropane [DBCP], ethylene dibromide [EDB], chlordecone [Kepone] and carbaryl) are known to affect human reproduction (Gordon and Shy, 1981; Gray, 1992; Gray *et. al.*, 1989, 1994; Mattison *et al.*, 1994; Hileman, 1994). Some are direct acting agents, either through their chemical reactivity (e.g. germ cell destruction by ionizing radiation) or by structural similarity to endogenous molecules (e.g. hormone agonists or antagonists). Other xenobiotics can interrupt reproduction indirectly either by metabolism to a direct acting toxicant or by altering the endocrine system (e.g. increased steroid clearance). One important mechanism of action of reproductive toxicants results from structural similarity of the parent compound or a metabolite to an important biological molecule such as a hormone. Xenobiotics in this category then act as agonists or antagonists of endogenous hormones.

Effect of Environmental Toxicants on Male Reproduction

There is increasing concern about the gonadal effects of environmental chemicals including drugs and occupational chemicals on human reproduction but the toxic effects of the vast majority of industrial and environmental chemicals on human reproduction are not known (Foster, 1995). Nevertheless environmental chemicals are hazardous to human reproduction and embryo development. In addition to other chemicals, polychlorobiphenyls (PCBS), dibromochloropropane and glycol ethers are known to impair male fertility. Recent studies (Kelce *et al.,* 1995) demonstrate that the primary DDT metabolite p, p'DDE interferes with the action of male sex hormones, or androgens suggesting that the feminizing action of DDT resulted from the antiandrogenic action of the DDT metabolite. Bancroft (1993) suspected that organophosphorus pesticides are compared to chlorinated hydrocarbons produced temporary and reversible infertility in males. Vermeulen (1993) reported that parathion causes infertility in farm animals, inhibits testosterone metabolism in rodents but has no apparent affects on Leydig cells function in man. Dibromo chloropropane, a nematicide and 2, 3-Dibromopropanol caused partially reversibly infertility in factory employees and when the exposure ceased, FSH levels remained normal. Giwercman *et al.,* (1993) reported 3-4 fold increased incidence of malignant germ cell neoplasia in Danes during the years 1940-1980. A similar trend has been reported from Scotland, USA and Finland where the frequency of testicular cancer is very low as compared to Denmark. These authors attribute the testicular cancer to environmental rather than genetic factors.

Currently, there is scientific and public concern about whether exposure to ambient levels of so called endocrine disrupters such as p, p'-DDE [1, 1-dichloro-2, 3-bis (p-chlorophenyl) ethylene], a metabolite of DDT [1, 1-Dichloro 2, 2-bis (p-chlorophenyl)-1, 1, 1 [trichloroethane] and polychlorinated biphenyls (PCBs), alter semen quality (Sharpe and Skappebeak, 1993; Irvine *et al.,* (1996). This concern stems from studies showing that PCBs and p, p'-'DDE are found in a large proportion of general population

(Murphy and Harney, 1985; Stehr-Green 1989; Longnecker *et al.*, 1997). Moreover animal and human studies suggest possible associations of exposure to PCBs and p, p′-DDE with semen abnormalities and altered testis function. (Bush *et al.*, 1986; Kelce *et al.*, 1995 and Raychoudhury *et al.* 2000). It has also been estimated that > 99% of individuals have detectable blood levels of p, p′-DDE (Stehr-Green, 1989). PCBs and p, p′-DDE, the most stable metabolite of DDT, are persistent lipophillic chemicals that are suspected endocrine disrupters. DDT was widely used as an insecticide, whereas PCBs were used in cutting oils, lubricants and as electrical insulators. Although their use and manufacture were banned nearly 30 years ago, still they are ubiquitous and persist in the environment PCBs are distributed worldwide as environmental pollutants and have been measured in air, water, aquatic and marine sediments and in fish and wildlife (De Voogt and Brinkman, 1999). Furthermore, they are biologically concentrated and stored in human adipose tissue. The general population continues to be exposed to PCBs and p, p′-DDE through ingestion of contaminated food (fish, meat, milk and their by-products) and water, dermal contact (soil and house dust) and inhalation (indoor air in buildings that have various sources).

Many recent studies have indicated an increasing prevalence of various abnormalities of the reproductive system in human males. There is growing concern about the considerable decrease in sperm density over the last 50 years in general populations worldwide, especially in the United States and in Europe (Carlsen *et al.*, 1992; Giwercman *et al.*, 1993; Anger *et al.*, 1995; Irvine *et al.*, 1996 and Swan *et al.*, 1997). Possible explanations for this phenomenon include increased stress, life-style factors and a variety of endocrine-altering chemicals in the environment that can be linked to decrease male reproductive capacity, as indicated mainly by the results of experimental animal studies. However, the population studies published to date lack information at the individual level on life-style factors, such as cigarette smoking, which may adversely affect semen quality as well as information on other potentially important environmental exposures (Vine *et al.*, 1994; Sun *et al.*, 1998). Of the possible casual factors considered, including the most explored hypothesis of the role of various xenoestrogens and synthetic estrogens (Safe, 1995; Toppari *et al.*, 1996), none have yet been reliably identified as the cause of a generally decreasing reproductive capacity in human males, at least in some areas.

The human male is of relatively low fertility as compared to other mammals. Thus the human male may be at greater risk for reproductive toxicants. For example, the number of sperm per human ejaculate is only 2- to 4- fold higher than the number at which fertility is significantly reduced whereas the number of sperm in rat, rabbit or bull ejaculate is many times (up to 1,400 fold) the number that will produce maximum fertility (Working, 1988). Human males have markedly smaller relative testis size and the lowest rate of daily sperm production per gram of testes (by a factor of more than 3) as compared to the mouse, rat or monkey. The percentage of progressively motile sperm and morphologically normal sperm in human semen are also lower than in any of the animal models (Working, 1988). Di-n-butyl phthalate (DBP), a widely used industrial solvent and plasticizer in plastic industry has been recognized as a ubiquitous contaminant of the environment (Peakall, 1975) and has been found in tissues of plants and animals (Schneider, 1987). Shrivastava *et al.* (1990) indicated testicular atrophy caused by DBP and suggested that it is associated with alteration in the activities of enzymes related with specific events of spermatogenesis.

Pb and Cd are highly toxic metals for human and other mammals. Both are pervasive in the human environment and accumulate in the human body over a lifetime including prenatal life (especially Pb). Apart from numerous sources of occupational exposure to each metal, the most important non-occupational sources are food (especially seafood from metal-polluted areas), water (Pb, mostly from Pb pipes in contact with soft and acidic water), air (especially Pb from gasoline in dense traffic areas), Pb-based paints of housing, smoking habits [Cd (and to lesser extent Pb) from tobacco] and alcohol consumption (Pb-contaminated alcoholic beverages). There is also some evidence of the possible interaction of Pb and alcohol in humans, that is an ethanol induced increase in the biologically active or mobile fraction of Pb accumulated in the body (Prpic-Majic *et al.*, 1984; Telisman *et al.*, 1984 and Newton *et al.*, 1992).

The involvement of heavy metals in aetiology of reproductive pathology including prostatic cancer has been debated for almost four decades. Specific metals, including lead and cadmium have at different times been implicated in the aetiology of male infertility and many experimental studies also show that Pb and Cd can adversely affect the mammalian male reproductive system. On the other hand, relatively few data are available regarding the possible reproductive effects of Pb and/or Cd in men, and generally conflicting results have been reported in reviews on this subject (Clarkson *et al.*, 1985; Tas *et al.*, 1996). Nevertheless, the results of several studies suggest that relatively high occupational exposure to Pb, as indicated by blood Pb (BPb) levels, can reduce human semen quality (decreased number, motility and altered morphology of sperm) whereas reproductive endocrine function is either not affected or is only marginally affected (Alexander *et al.*, 1996; Assennato *et al.*, 1987; Braunstein *et al.*, 1978; Chia *et al.*, 1992; Cullen *et al.*, 1984; Fisher-Fischbein, 1987; Gennart *et al.*, 1992; Gustafson *et al.*, 1989; Lancranjan *et al.*, 1975; Lerda, 1992; McGregar & Mason, 1991; Ng *et al.*, 1991; Rodamilans *et al., 1988;* Saaranen *et al.*, 1987; Wildt *et al.*, 1983, Xu *et al.*, 1993; Viscum *et al.*, 1999). Some data suggest that the reproductive effects of Pb in men are reversible, that is, a trend towards normalization of reproductive parameters was found in subjects treated with Pb-chelating agent (Cullen *et al.*, 1984, Fisher-Fischbein *et al.*, 1987) or after cessation of occupational Pb exposure (Viscum *et al.*, 1999). However, recent epidemiologic studies have indicated that even moderate occupational exposure to Pb can reduce the fertility of male workers (Gennart *et al.*, 1992; Lin *et al.*, 1996) and increase spontaneous abortions in workers' wives (Lindbohm *et al.*, 1991).

Evidence for possible reproductive effects of Cd in men is scanty and less conclusive as compared to that for Pb (Clarkson *et al.*, 1985; Tas *et al.*, 1996). In groups of male subjects suspected of infertility (including non-smokers and smokers), a significant positive correlation was found between abnormal sperm morphology and blood Cd (BCd) levels (Chia *et al.*, 1992, 1994), significant inverse correlations were found between sperm density and BCd (Xu *et al.*, 1993; Chia *et al.*, 1994) and between semen volume and either BCd (Xu *et al.*, 1993) or seminal fluid Cd (SfCd) levels (Xu *et al.*, 1993; Chia *et al.*, 1994). On the other hand, significant positive correlation was found between sperm motility and linear and curvilinear velocity with respect to semen Cd

(Noack-Fuller *et al.*, 1993) or no significant correlation was found between conventional parameters of semen quality and either semen Cd (Oldereid *et al.*, 1994) or SfCd levels (Noack-Fuller *et al.*, 1993; Keck *et al.*, 1995; Saaranen *et al.*, 1989) in the same kind of subjects. No significant influence of occupational Cd exposure on testicular endocrine function (McGregor and Mason, 1991) and fertility in male workers (Gennart *et al.*, 1992) was found. Telisman *et al.* (2000) measured blood Pb, Cd, serum zinc, seminal fluid Zn, serum copper, parameters of semen quality and of reproductive endocrine function in 149 health male industrial workers 20-43 years of age and indicated that even moderate exposure to Pb (BPb 400 ug/L) and Cd (BCd < 10 ug/L) can significantly reduce human semen quality without conclusive evidence of impairment of male reproductive endocrine functions. There are indications that ingestion of excess selenium affects the testicular morphology resulting in interruptions of spermatogenesis. It severely affects the concentrations, motility and morphology of sperm (Nabbia *et al.*, 1987; Kaur and Parshad, 1994). Kaur *et al.* (1999) from their biochemical investigations hypothesized that selemum perhaps affects acido-reducase activity of glutathione resulting in testicular damage. Ghosh *et al.* (1989, 1990) showed that lithium (which has important therapeutic properties in certain kinds of mental illness) administration resulted in an inhibition of testicular steroidogenesis and spermatogenesis both in toad and in rat and they (Ghosh *et al.*, 1998) further suggested that the effect of lithium on male reproductive organs mainly depends on the amount of drug being injected and the duration of treatment to it. Very recently Hauser *et al.* (2002) have presented the results from a pilot study on the relationship between serum PCBs and p, p'-DDE and semen quality in 29 subjects recruited from the Massachusetts General Hospital Andrology Laboratory. Their data showed general trends that were suggestive of an association between PCBs and p, p'-DDE and abnormal motility, as well as with sperm concentration and morphology and their full-scale study is currently in progress.

Thus the review reveals that, there are some conclusive reports which clearly give correlative (case studies in humans) and experimental evidences (animal studies) of the toxicity caused by various types of environmental pollutants to male reproduction and there is need for further strengthening of work in this area so that preventive and curative measures may be developed to prevent human and animal population from these hazards.

Effect of environmental toxicants on female reproduction

Women's health is an area that is gaining increasing attention with the realization that men's and women's bodies do not just look different, they also react differently to environmental agents. Ovarian toxicity refers to the ways that different chemicals affect the ovaries. Because the ovaries are a clean house of hormonal activity in the female body, disruption of ovarian function greatly affects women's reproductive and endocrine health. If a chemical exposure affects ovarian function, it may eventually lead to any of several hormone and ageing related dysfunctions, including infertility, birth defects, cancer and osteoporosis. The basic requirements for comprehensive evaluation of an environmental hazard to human female reproduction are similar, regardless of the source of toxicity. Most important is to estimate or measure exposure to reproductive toxicant because without exposure estimation, a predicted dose cannot be calculated and the risk to a specific reproductive outcome cannot be characterized. There are a number of reports (Akiyama *et al.*, 1975; Holdrinet *et al.*, 1977; Szymczynski *et al.*, 1981; Trapp *et al.*, 1984; Slonach and Vaz, 1985; Chia *et al.*, 1992; Jarrell *et al.*, 1993) from different regions of the world describing trace levels of numerous contaminants found in human serum, breast milk, fat, ovarian follicular fluid and seminal plasma.

Regarding the reproductive toxicity of contaminants present in human tissues, data from animal experiments reveal that hexachlorobenzene (HCB) is both a developmental and reproductive toxicant. Data relating to the treatment of adult female animals with HCB suggests that this compound is primarily an ovarian toxicant. This conclusion is supported by the demonstration of altered steroidogenesis of rhesus monkey in the absence of changes in circulating levels of pituitary gonadotropins (Muller *et al.*, 1986). Altered ovarian steroidogenesis has also been shown in both the cynomolgus monkey (Foster *et al.*, 1992a) and the rat (Foster *et al.*, 1992b) and HCB treatment (10 mg/kg/day for 90 days) resulted in depressed ovulatory levels of oestradiol and response to ovulation induction in the cynomolgus monkey (Foster *et al.*, 1992a). Furthermore, degenerative changes in primordial follicles and ovarian germinal epithelium have been demonstrated in cynomolgus monkey (Jarrell *et al.*, 1993; Babineau *et al.*, 1991) with HCB treatments at levels as low as 0.1 mg/kg/day, which suggests that ovarian function may also be altered at levels lower than previously reported. Ultrastructural changes in the ovary appear to be the most sensitive end point and these occur with HCB treatment at a dose of 0.1 mg/kg/day for 90 days (Babineau *et al.*, 1991). Morphometric studies in which ovarian follicle numbers have been shown to be significantly reduced following exposure to reproductive toxin such as 7, 12-dimethylbenz (α) anthracene, benzo (α) pyrene, 2, 5-hexanedione, and hexachlorobenzene (Siracusa *et al.*, 1992; Miller *et al.*, 1992; Weitzman *et al.*, 1992; Jarrell *et al.*, 1993; Faddy *et al.*, 1992) suggest that HCB and these agents may induce premature ovarian failure. This hypothesis is supported by biexponential decline in ovarian follicle numbers for the human ovary (Brand and Lehert, 1978). Hexachloroethane (HCE) is a compound that is used as an inhibitor in explosives and human exposure is most likely to occur through inhalation or as a result of contaminated drinking water. However, there are virtually no reports of reproductive effects of HCE treatment in adult animals of either sex, but in the absence of any data, it is not possible to conclude what risk if any, this compound poses to human reproduction because the identification of HCE in ovarian follicular fluid in 48% women from three fertility clinics in Ontario poses a great concern (Foster, 1995).

There are no studies that establish an association between adverse reproductive outcomes in humans with circulating levels of 1, 1, 1-trichloro-2, 2-bis (p-chlorophenyl) ethane (DDT), 1, 1-dichloro-2, 2-bis (p-chlorophenyl) ethane (DDD) or 1, 1-dichloro-2, 2-bis (p-chlorophenyl) ethylene (DDE). In animal experiments, these compounds have been shown to be embryotoxic and fetotoxic in the rodent (Fabro *et al.*, 1984). Adverse effects of DDT treatment on fertility have been shown in the rat and in the beagle dog treated with DDT (Deichmann *et al.*, 1971; Jonsson *et al.*, 1976). A number of lines of evidence suggest that the adverse effects of DDT and related compounds may be mediated through their structural similarity to estrogens and thus interaction with the estrogen receptor. In particular, exposed rats have been induced to enter precocious puberty (Clemant and Okey, 1974) and in adult uterine weight has been increased by treatment with DDT and RNA, DNA and carbohydrate content was also increased (Bitman and Cecil, 1970; Clemant and Okey, 1972). Although, these contaminants are weak estrogens compared to estradiol, they represent a serious concern to human reproduction because they persist in the body at physiologically relevant concentrations, they can interact with estrogen receptor and induce estrogen response in sensitive tissues and unlike endogenous estrogens, they are not bound to sex hormone binding globulin. Thus, the potential for these compounds to adversely affect human reproduction needs to be carefully investigated (Foster, 1995).

The reproductive and developmental effects of PCBS - the ubiquitous environmental contaminants have been well documented and reviewed (Gabo *et al.*, 1991; Battersil, 1994; Safe, 1994). In adult animals (mice, rats and monkeys) treated with PCBs (Chlophen A-60 or Arocort 254), there is evidence of increased estrous/menstrual cycle lengths (Muller *et al.*, 1986; Orberg and Kihlstrom, 1973; Brezner *et al.*, 1984) and evidence of anovulation in monkeys (Muller *et al.*, 1986). Changes in circulating levels of ovarian steroids have been documented (Jonsson *et al.*, 1976; Muller *et al.*, 1986) at PCB treatment levels as low as 7.5 mg/kg/day for 36 weeks. Reduction in the number of fertilized ova that implant has also been reported (Orberg and Kihlstorm, 1973). Moreover, PCB treatment has also been shown to induce reduction in the number of ovarian follicles (Orberg and Kihlstrom, 1973) thus suggesting that prolonged exposure could induce premature ovarian failure in sensitive individuals.

There are very few data describing the effect of 2, 3, 7, 8-tetrachlorodibenzo-p-dioxin (TCDD) (occurring through the combustion of industrial and municipal waste as well as through burning automotive fuels containing chlorinated additives) exposure in the adult females. Effects on female reproduction included reduced fertility, decreased litter size, gonadal effects and changes in estrous/menstrual cycle characteristics. In the rodents chronic exposure to TCDD (0.1 μg/kg/day) resulted in altered fertility and was significant at a lower dose (0.1 μg/kg/day) in the F_1 generation than in the F_0 (Murray *et al.*, 1979). Anovulation and ovarian dysfunction were shown in rats treated with 1 to 2 μg/kg/day TCDD for 13 weeks (Kociba *et al.*, 1976). In contrast to these findings, Kociba and co-workers (1976) were unable to demonstrate any effects on the female reproductive system in a 2-year study using dioxin doses of 0.001 and 0.01 μg/kg/day. In rhesus monkeys, effects on reproductive function have been associated with chronic exposure to 50 to 500 ppt TCDD in the diet. However, chronic treatment with lower levels of TCDD (5 and 25 ppt) have not produced any adverse effects on reproduction (Bowman *et al.*, 1989; Shanz *et al.*, 1989). In a later report, it has been demonstrated that chronic exposure of the rhesus monkeys to 5 and 25 ppt TCDD in the diet for 4 years induced a dose-dependent increase in the prevalence and severity of endometriosis when evaluated 10 years after the initiation of the study.

Davis (2000) has reported her efforts on studying ovarian toxicity and its relationship to the environment. Davis's work on adult ovarian functions has concentrated around the effects of two classes of chemicals: phthalates and glycol ethers. Phthalates are used to make plastic flexible and soft. They are contained in cosmetic products such as nail polish and in paints, varnishes, or any plastic that needs to be flexible. Ethylene based glycol ethers are primarily used as solvents and can be found in paints, varnishes and deicing chemicals. The route of exposure of both these chemicals may be dermal or inhalation. People are also exposed to phthalates orally as they leach from plastic products into food. According to Davis, there is little research to date that strongly implicates phthalates as a reproductive toxicant in women, a few phthalates clearly affect the ovaries in rodents. Di (2-ethylhexyl) phthalate (DEHP), for instance, has been shown in rodents to suppress oestradiol production which ultimately leads to infertility. Several epidemiological studies have shown ethylene-based glycol ethers to have reproductive effects. Glycol ethers have been shown to induce spontaneous abortions and possibly to reduce female workers' ability to deliver a live birth, although fertility appears to be unaffected. Davis and her colleagues studied the effects of ethylene glycol monomethyl ether and its metabolite methoxy acetic acid on ovarian luteal cells of rats and humans respectively. They found that the chemical elevated progesterone production in both rat and human cells. In the rats, the chemicals also inhibited ovulation. Davis (2000) has also shown that the exposure to pesticide methoxychlor causes ovarian dysfunction resulting in reduced estrous levels of follicle stimulating hormone at all dosages and reduced estrous progesterone levels at some dosages. Davis (2000) has also studied the effects of 2, 3, 7, 8-tetrachlorodibenzo-p-dioxin (TCDD or dioxin) which has been shown to promote ovarian tumours in female rats and nitrofurantoin, an antibiotic used in urinary tract infections that has been associated with ovarian cancer in rodents, although not in humans.

Since pesticides are major agricultural chemicals invading the aquatic resources, most of the stressful events are documented in fish, where reproduction is a cyclical phenomenon displaying a synchronized rhythm with the environment. Suble environmental cues influence fish reproduction and any deviation may be highly stressful for the fish. The events of stress are expressed in terms of gross changes in both ovary (Sukumar and Karpagaganapathy, 1992) and testes (Pundir and Saxena, 1990). Xenobiotic stressed oocytes frequently contain less yolky granules (Sukumar and Karpagaganapathy, 1992), ruptured oocyte walls, damaged yolky vesicles (Kulshetra and Arora, 1984), nucleolar and cytoplasmic changes (Kirubagaran and Joy, 1988; Murugesan and Haniffa, 1992). Even safe concentrations of carbofuran and fenitrothion over a long term exposure can trigger atresia in *Channa punctatus* (Mani and Saxena, 1985). Long term treatment with phenthoate revealed destruction of stage II and III oocytes in *Channa punctatus* (Dey and Bhattacharya, 1989). The same profile was equally significant in gamma-BHC-treated *Heteropneustes fossilis* (Singh and Singh, 1990) or metacid 50 treated *Channa punctatus* and *Anabas testudineus* (Choudhary, 1991). In a recent study, adult prespawning fish *Labeo rohita*, were sublethally (1/5 96h LDL_{50}) exposed to mercuric chloride and metacid (methylparathion). Accumulation of both was studied and found that prespawning ovary appears to be a potent organ for deposition of both the pollutants (Aditya *et al.*, 2002). Using pig as a toxicological model, Campagna *et al.* (2001) reported that exposing porcine cumulus - oocyte complexes to an environmentally pertinent organochlorine mixture during *in vitro* fertilization disturbs oocyte development, thus supporting recent concerns that such pollutants harm reproductive health in humans and other mammalian species.

It can be concluded that survivability of any species depends on the integrity of its reproductive system. Since the germ cells carry genetic information from the previous generation, any interference in the production of healthy germ cells spells havoc for the evolutionary existence of a species. Stress response in biological systems in an innate mechanism of survival in the oddest circumstances. Xenobiotics or the toxic chemicals are very often labelled as endocrine disrupters and inducers of stress proteins such as, metallolthioneins, heat shock proteins, peroxisome - proliferator - activated receptors and C-reactive proteins. Signal transduction mechanisms which elicit such responses will be of major interest in the future. In the new millennium, questions are likely to be addressed in the directions of *de novo* synthesis of proteins, deciphering the mechanisms of action of endocrine disrupters, elucidation of the immunotoxic potential of xenobiotics, analysis of DNA microarray and proteomics under xenobiotic onslaught (Battacharya, 2001).

Effect of environmental toxicants on development :

Mammalian developmental toxicity refers to adverse effects initiated or evident during *in utero* development. More broadly, however, developmental toxicity also includes adverse effects on the developing organism that may have resulted from exposure of either parent before conception, of mother during parental development or postnatally to the time of sexual maturation. Development is an exceedignly complex process and the sensitivity of the conceptus to chemicals as well as the type of abnormality expressed varies with the developmental stage at which exposure occurs. The embryo is most vulnerable to the initiation of major birth defects between 3 weeks and 2 months of gestation, the critical period of organogenesis. Exposure to toxic chemicals during the first 2 weeks typically leads to fetal death, while exposure after organogenesis is more likely to cause growth retardation and functional deficits.

It has been known since the early 1970' s that fetal exposure to DDT causes male animals to develop abnormally small penises and undescended testices and these effects were thought to result from the estrogenic nature of DDT. Data currently available are inadequate to make a meaningful estimate of the degree to which pesticide exposure may be involved in human developmental problems. Animal data clearly indicate that the potential is there, provided exposure occurs during sensitive time points and at sufficient concentrations. In humans and rodents, exposure to hormonally active chemicals during the period of sex differentiation can produce a wide range of abnormal sexual phenotypes including masculinized and defeminized females and feminized and demasculinized males. Xenobiotics known to act in this capacity have been referred to as environmental estrogens or environmental androgens. Compounds structurally similar to hormones may act either as receptor agonists or antagonists. Pesticides possessing hormonal activity and having the ability to interfere with mammalian sex differentiation include methoxychlor (Gray *et. al.*, 1989 and Gray, 1992), DDT (kelce *et. al.*, 1995) and vinciozin (Gray *et al.*, 1994).

HCB has been shown to cross the placenta to the developing conceptus in mice, rats and rabbits (Villenenue *et al.*, 1974, 1975; Courtney *et al.*, 1985). Exposure of pregnant rats to doses upto 120 mg/kg/day failed to induce any adverse effects on organogenesis (Khera, 1974). In multigenerational studies, HCB exposure at doses as low as 2 mg/kg/day induced increased liver weights and decreased body weights of pups of both sexes (Grant *et al.*, 1977; Arnold *et al.*, 1985). Pup viability (Arnold *et al.*, 1985) also decreased at higher doses (8 mg/kg/day). Developmental effects of the skin, bone and nervous system have been described in children accidently exposed to HCB before attaining puberty (Peters *et al.*, 1982; Cripps *et al.*, 1984). Unfortunately, there is no exposure data to aid in the dose-response characterization for these effects, and no animal studies have been undertaken to explore the mechanism for these effects. In another report on hexachlorthane (Weeks *et al.*, 1979), Sprague - Dawley rats were treated from day to 6 day 16 of gestation either by inhalation 15, 48 or 260 ppm for 6 hr/day) or gavage (50, 100 or 500 mg/kg/day). At the highest dose, the number of fetal resorptions were increased and the number of live fetuses per litre were decreased but there were no structural effects. At higher doses of TCB; 1, 2, 4- Trichlorbenzene (360 mg/kg/day during days 9-13 of gestation) in the rat, embryonic growth was reduced (Kitchen and Ebron, 1983). In neonatal rats treated with a PCB mixture containing 21% chlorination, precocious puberty, persistent vaginal estrus and premature reproductive senescence was induced (Gillert, 1978). Postnatal effects of TCDD in the male have been shown to include altered androgenic status of the male rat (Mably *et al.*, 1992) which may be responsible for decreases in testis and epididymal weights and decreased number of sperm in the cauda epididymis.

Many organophosphate compounds are lipophilic and readily cross the placenta (Richardson, 1995). Experimental evidence has linked organophosphate exposure during gestation or early postnatal period to adverse neuro-developmental sequelae in offspring (Landrigan *et al.*, 1999; Eskenazi *et al.*, 1999). Exposuses during the spurt in brain growth (which in humans beings during the third trimester) may be particularly deleterious (Eriksson *et al.*, 1999; Ahlbom *et al.*, 1995; Whitney *et al.*, 1995; Dam *et al.*, 1998; Slotkin, 1999; Johnson *et al.*, 1998). However, epidemiologic research on this relationship is limited and has been hampered partly because of uncertainties in exposure estimates. Although biologic markers can be useful in understanding the role of environmental contaminants during fetal development (Whyatt and Perera, 1995; Perera *et al.*, 1998; Whyatt *et al.*, 1998), research on the effects of parental organophosphate exposure has been limited by the lack of biomarkers reflecting cumulative exposures. Whyatt and Barr (2001) have reported that their results on the measurement of organophosphate metabolites in postpartum meconium, indicate that this parameter has a promise as biormaker of prenatal exposure but still further research is needed to determine the time-frame of exposure represented by pesticide levels in meconium and to evaluate dose-response relationship. Recently, we have observed a significant increase in the neonatal mortality of pups of the females exposed to long-term residual level (1/8 LD_{50-}1/5 Ld_{50}) of organophosphates including monocrotophos, dimethoate and methyl parathion (unpublished observations).

Summary and Conclusion

Environmental toxicants include naturally occurring substances (viz. food additives and contaminants, bacteriotoxins, fungitoxins, phytotoxins) and man-made products (viz. industrial chemicals and pesticides) which are present in our environment and are harmful to human health. There is increasing concern about the effects of environmental chemicals on human reproduction because there are a number of reports from different regions of the world describing trace levels of numerous contaminants found in human serum, breast milk, fat, ovarian follicular fluid and seminal plasma. The reproductive health risks related to exposure to persistent organic pollutants in the environment remain controversial because most of the studies have investigated only one or two chemicals at a time, whereas populations are exposed to large spectrum of persistent chemicals in their environment. A number of environmental pollutants affect reproduction in experimental animals but the toxic effects of a vast majority of industrial and environmental chemicals on human reproduction are not known. According to Lasley and Overstreet (1998), efforts to characterize the adverse reproductive health effects of an environmental exposure can be divided into three areas which we can conceptualize as cycles of activities. These activities include human field studies, whole animal toxicological studies, and *in vitro* mechanistic studies. At present, we can just assume that the reproductive toxicants have the potential to cause adverse effects by one of the several mechanisms i.e. these may act directly through their chemical activity, due to their structural similarity to endogenous molecules or these may act indirectly through their metabolism. However, reproductive problems like decreasing sperm density over the last 50 years in general population worldwide, increasing abortion rate in humans and domestic animals are of great concern and need intense attention. Thus to prevent the human and animal populations from the toxic effects to environmental pollutants various preventive as well as curative measures need to be developed and in view of the existing grave situation, it is a big challenge before humanity as well the scientific community to tackle with it.

REFERENCES

Aditya, A.K. Chattopathyay, S. and Mitra, S. (2002). Effect of mercury and methyl parathion on the ovaries of *Labeo rohita* (Ham.) *J. Env. Biol.* 23(1): 61-64.

Ahlbom, J. Fedricksson, A., Ericksson, P. (1995). Exposure to an organophosphate (DFP) during a defined period in neonatal life induces permanent changes in brain, muscarinic receptors and behaviour in adult mice. *Brain Res* 677: 13-19.

Akiyama, K. Ohi, G. Fijitani, K., Yagyu, H., Ogino, M., Kawana, T. (1975). Polychlorinated biphenyl residues in maternal and cord blood in Tokyo metropolitan area. *Bull Environ. Contam. Toxicol.* 14: 588-92.

Alexander, B.H., Checkoway, H., Van Netten, C., Muller, C.H., Ewers, T.G., Kaufman J.D., Mueller, B.A., Vaughan, T.L. Faustman, E.M. (1996). Semen quality of men employed at a lead smelter. *Occup. Environ. Med.* 53: 411-416.

Arnold, D.L., Moodie, C.A., Charbonneau, S.M., Grice, H.C., McGuire P.F., Bryce F.R., Collins, B.T., Zawidzka, Z.Z., Krewski, D.R., Nera, E.A., Munro, I.C. (1985). Long-term toxicity of hexachlorobenzene in the rat and the effect of dietary vitamin A. *Food Chem. Toxicol.* 23: 779-793.

Assennato, G., Paci, C., Bases, M.E., Molinini, R., Landela R.G., Altamura, B.M., Giorgino, R. (1986). Sperm count suppression without endocrine dysfunction in lead-exposed men. *Arch. Environ. Health* 41: 387-390.

Auger, J., Kunstmann, J.M., Czyglik, F., Jouannet, P. (1995). Decline in semen quality among fertile men in Paris during the past 20 years. *N. Engl. J. Med.* 332 (5): 281-285.

Babineau, K.A., Singh, A., Jarrell, J.F., Villeneuve, D.C. (1991). Surface epithelium of the ovary following oral administration of hexa chlorobenzene to the monkey. *J. Submicrosc. Cytol. Pathol.* 23: 457-464.

Bancroft, J. (1993). Impact of environment, stress, occupational and other hazards on sexuality and sexual behaviour. *Environ. Health Persp.* 101 (Suppl. 2): 101-107.

Banerjee, B.D. and Hussain, Q.Z. (1987). Effects of endosulfan on humoral and cell mediated immune responses in rats. *Environ. Contam. Toxicol. 38*: 433.

Battacharya, S. (2001). Stress response to pesticides and heavy metals in fish and other vertebrates. *Proc. Indian Nat. Sci. Acad.* B 67 (5): 215-246.

Bitman, J., Cecil, H. (1970). Estrogenic activity of DDT analogs and polychlorinated biphenyls. *J. Agric. Food Chem. 18*: 1108-1112.

Bowman, R.E., Schantz, S.L., Uleerasinghe N.C., Gross, M.L., Barsotti, D.A. (1989). Chronic dietary intake of 2, 3, 7, 8-Tetrachlorodibenzo-p-dioxin (TCDD) at 5 or 25 parts trillion in the monkey. TCDD kinetics and dose-effect estimate of reproductive toxicity. *Chemosphere* 18: 243-252.

Brand, P.C. and Lehert, P.H. (1978). A new way of looking at environmental variables that may affect the age at menopause. *Maturitas, 1*: 121-132.

Braustein, G.D., Dahlgren, J., Loriaux, D.L. (1978). Hypogonadism in chronically lead-poisoned men. *Infertility* 1: 33-51.

Brezner, E., Terkel, J., Perry, A.S. (1984). The effect of Aroclor 1242 (PCB) on the physiology of reproduction in the female rat. *J. Comp. Biochem. Physiol.* 77: 65-70.

Bush, B., Bennet, A.H., Snow, J.T. (1986). Polychlorophenyl Congeners, p, p'-DDE and sperm function in humans. *Arch. Environ. Contam. Toxicol.* 15: 333-341.

Carlsen, E., Giwercman, A., Keiding, N., Skakkeback, N.E. (1992). Evidence for decreasing quality of semen during past 50 years. *Br. Med. J.* 305: 609-613.

Chatterjee, S., Roy, A., Bagachi, P. and Deb, C. (1988). Suppression of testicular steroidogenesis in rats by organochlorine insecticide Aldrin. *Environ. Pollut.* 2: 35-40.

Check, A.O., Vonier, P.M., Oberdorster, E., Burow, B.C. and McLachlan, J.A. (1998). Environmental signalling: a biological context for endocrine disruption. *Environ. Health Persp.* 106 (Suppl. 1): 5-10.

Chia, S.E., Xu, B., Ong, C.N., Tsakok F.N.H., Lec, ST. (1994). Effect of Cadmium and cigarette smoking on human semen quality. *Int. J. Fertil*, 39: 292-298.

Chia, S.E., Ong, C.N., Lee, S.T., Tsakok, F.H.M. (1992). Blood concentrations of lead, cadmium, mercury, zinc and copper and human semen parameters. *Arch Androl* 29: 177-184.

Chopra, A, Ahuja, S.P., Gupta, P.P. and Nagpal, S.K. (1979). Histopathological and lipid changes in various organs of chicks caused by acute and chronic toxicity of Metasystox - R, Labaycid. *Zbl. Vet. Med. A* 26: 822-834.

Choudhury, C., (1991). Impact of matacid 50 and carbaryl on the ovarian maturation of *Anabas testudineus* (Bloch) and *Channa punctatus* (Bloch); Ph.D, Dissertation, Visva Bharti University, Santiniketan.

Clarkson, T.W., Nordberg G.F., Sager P.R., (1985). Reproductive and developmental toxicity of metals *Scand J. Work Environ Health,* 11: 145-154.

Clement, J., Okey, A. (1972). Estrogenic and antiestrogenic effects of DDT administered in the diet of immature female rats. *Can. J. Physiol. Pharmacol.* 50: 971-975.

Clement, J., Okey, A., (1974). Reproduction in female rats born to DDT-treated parents. *Bull Environ. Contam. Toxicol.* 12: 373-377.

Compagna, C, Sirard, M. A, Ayotte, P. and Bailey, J.L., (2001). Impaired maturation, fertilization and embryonic development of porcine oocytes following exposure to an environmentally relevant, organochlorine mixture. *Biol. Reprod.* 65: 554-560.

Courtney, K.D., Andrews, J.E., (1985). Neonatal and maternal body burdens of hexaclorobenezene (HCB) in mice: gestational exposure and lactational transfer. *Fundam. Appl Toxicol.* 5 265-277.

Cripps, D.J., Peters, H.A., Gocmen, A, Dogramici, I, (1984). *Porphyria turcica* due to exachlorobenzene: a 20 to 30 years fellow up study on 204 patients. *Br. J. Dermatol.* 111: 413-422.

Cullen, M.R., Keyne, R.D., Robins J.M. (1984). Endocrine and reproductive dysfunction in men associated with occupational inorganic lead intoxication. *Arch. Environ. Health,* 39: 431-440.

Dam, K., Seidler, F.J. Slotkin, T.A. (1998). Developmental neurotoxicity of chlorpyifos: delayed targeting of DNA synthesis after repeated administration. *Brain Res Dev Brain Res* 108: 39-45

Davis B.J. (2000). Working for women's health. *Environ Health Persp.* 108(1): 18-19.

Deichmann, W.B., Macdonald, W.E. Beasley, A.G., Cubit, D. (1971). Subnormal reproduction in beegle dogs induced by DDT and aldrin. *Ind Med. Surg.* 40: 10-20.

De Voogt, P., Brinkman, U.A.T. (1989). Production properties and usage of polychlorinated biphenyls In: *Halogenated Biphenyls, Terphenyls, Naphthalenes, Dibenzodioxins and related products.* (Kimbrough, R.D. and Jensen, A.A., eds). 2nd ed. Amsterdam Elsevier-North Holland, pp. 3-45.

Dey, S. and Bhattacharya, S. (1989). Ovarian damage to *Channa punctatus* after chronic exposure to low concentrations of elsan, mercury and ammonia. *Ecotoxical Environ. Safety.,* 17: 247-257.

Dhar, B.N., (1996). Pesticide scenario in India. *Pestology. 6* 37-38.

Eriksson, P. (1997). Developmental neurotoxicity of environmental agents in the neonate. *Neurotoxicology,* 18: 719-726.

Eskenzi, B., Bradman, A., Castornia, R. (1995). Exposures of children to organophosphate pesticides and their potential adverse health effects. *Environ Health Persp.* 107 (Suppl. 3): 409-419

Fabro, S., McLachlan. J.A., Dames, N.H., (1984). Chemical exposure of embryos during the preimplantation stages of pregnancy: mortality and intrauterine development. *Am. J. Obstet. Gynecol.* 148: 929-938.

Faddy, M.J., Gosden, R.G., Gougeon, A., Richardson, S.J., Nelson, J.F., (1992). Accelerated disappearance of ovarian follicles in mid-life-implications for forecasting menopause. *Hum. Reprod.* 7: 1342-1346.

Fisher - Fischbein, J., Fischbein, A., Helnick, H.D., Bardin, C.W., (1987). Correlation between biochemical indicators of load exposure and semen quality in a lead-poisoned firearms instructor. *JAMA* 251: 803-805.

Foster, W.G., (1995). The reproductive toxicology of great lakes contaminants. *Environ. Health Persp.* 103 (Suppl. 9): 63-69

Foster, W.G., Pentick, J.A., McMohan, A., Lecavalier, P.R., (1992). Ovarian toxicity of hexachlorobenzene (HCB) in the superovulated female rat. *J. Biochem. Toxicol.* 7: 1-4.

Foster, W.G., McMohan, A., Villeneuve, D.C., (1992). Hexachlorobenzene (HCB) suppresses circulating progesterone concentrations during the luteal phase in the cynomolgus monkey. *J. Appl. Toxicol.* 12: 13-17.

Gellert, R.J., (1978). Uterotrophic activity of polychlorinated biphenyls (PCBs) and induction of precocious reproductive aging in neonatally treated female rats. *Environ Res.* 16: 123-130.

Gennart, J.P., Bernard, A., Lauwerys, R., (1992a). Assessment of thyroid, testes, kidney and autonomic nervous system function in lead-exposed workers. *(Int. Occup. Environ. Health* 64: 49-51.

Gennart, J.P., Buchet, J.P., Roels, H., Ghyselen, P., Ceulemans, E, Lauwerys, R. (1992b). Fertility of male workers exposed to cadmium, lead or manganese. *Am. J. Epidemiol.* 135: 1208-1219.

Ghosh, D., Biswas, N.M., Chaudhuri, A., Ghosh, A.K. and Ghosh P. K. (1990). Acid and alkaline phosphatase activities in lithium-treated testis, prostate and seminal vesicles of adult albino rats: Evidence of duration and dose dependent response. *Indian J. Exp. Biol. 28:* 553-556.

Ghosh D, Chaudhuri A., Biswas, N.M. and Ghosh P.K. (1990). *Life Sci.* 46: 127.

Ghosh, P.K., Sarkar, M., Ghosh, A.K., Ghosh, D. (1989). *Andrologia,* 21: 199.

Giwercman, A., Carlsen E., Keiding N., Skakkeback N.E., (1993). Evidence for increasing incidence of abnormalities of the human testis: a review. *Environ. Health Perspect.* 101 (Suppl. 2): 65-71.

Gopal, P.K., Chopra, A. and Ahuja, S.P. (1990). Effects of Fenitrothion and Disulfoton on lipid metabolism in tissues of while leghorn chicks *(Gallus domesticus) J. Nuclear Agric. Biol.* 19: 199-204.

Gopal, P.K. Chopra, A., and Ahuja, S.P. (1979). Lipid and growth changes in organs of chicks during acute and chronic toxicity with disyton and folithion. *Indian J. Exp. Biol.* 17: 1153.

Gordon, J.E., Shy, C.M., (1981). Agricultural chemical use and congenital cleft lip and/or palate. *Arch Environ. Health* 36: 213-220.

Grant, D.L., Phillis, W.E., Hatina, G.V., (1977). Effect of hexachlorobenzence on reproduction in the rat. *Arch Environ. Contam. Toxicol.* 5: 207-216.

Gray, L.E., Jr., Ostby, J., Ferrell, J., Sigmon, R., Cooper, R., Goldman, J., Slott, V., Laskey, J. (1989). A dose-response analysis of methoxychlor-induced alterations of reproductive development and function in the rat. *Fundam. Appl. Toxicol.* 12: 92-108.

Gray, L.E., Jr., Ostby, J.S., Kelce, W.R. (1994). Developmental effects of an environmental antiandrogen: the fungicide vinclozolin alters sex differentiation of the male rat. *Toxicol. Appl. Pharmacol.* 129: 46-52.

Gray, L.E. (1992). Chemical-induced alterations of sexual differentiation: a review of effects in humans and rodents. In Chemically-Induced Alterations in Sexual and Functional Development: The Wildlife/ Human Connection (Colborn, T., Clement, C, eds.) Princeton, NJ: Princeton Scientific Publishing: 203-230.

Gustafson, A., Hedner, P., Schute, A., Skerfring, S. (1989). Occupational lead exposure and pituitary function. *Int. Arch. Occup Environ. Health,* 61: 277-281.

Hauser, R., Altshul, L., Chen, Z., Ryan, L., Overstreet, J., Schiff, I. and Christiani, D.C. (2002). Environmental organochlorines and semen quality: Results of a pilot study. *Environ. Health Persp.* 110: 229-233.

Hileman, B. (1994). Environmental estrogens lined to reproductive abnormalities and cancer. *Chem. Eng. News* 31: 19-23.

Holdrinet, M.V., Braun, H.E., Frank R., Stopps, G.J., Smout, M.S., McWade, J.W. (1977). Organochlorine residues in human adipose tissue and milk from Ontario residents, 1969-1974. *Can. J. Public Health* 68: 74-80.

Irvine, S., Cawood, E., Richardson, D., Macdonald, E., Aitken, J. (1996). Evidence of deteriorating semen quality in the United Kingdom: birth cohort study in 577 men in Scotland over 11 years. *Br. Med. J.* 312 : 467-471.

Jarrell, J.F., McMachon, A., Villeneuve D., Franklin, A., Singh, A., Valli, V.E., Bartlett, S. (1993a). Hexachlorobenzene toxicity in the monkey primordial gum cell without induced porphyria. *Reprod. Toxicol.* 7: 41-47.

Jarrell, J.F., Villeneuve, D., Franklin, C., Bartlett, S., Wrinon, W., Kohut J., Zouves, C.G. (1993b). Contamination of human ovarian follicular fluid and serum by chlorinated organic compounds in three Canadian cities. *Can. Med. Assoc. J.* 148: 1321-1327.

Jonsson, H., Keil, J., Gaddy, R.G., Loadholt, C.B., Hennigar, G.R., Walker E.M. (1976). Prolonged ingestion of commercial DDT and PCB; effects on progesterone levels and reproduction in the mature female rat. *Arch. Environ. Contam. Toxicol. 3:* 479-490.

Johnson, D.E., Seidler, F.J., Scotkin, T.A. (1998). Early biochemical detection of delayed neurotoxicity resulting from developmental exposure to chloropyrifos. *Brain Res Bull 45*: 143-147.

Kalra, R.L. (1989). *Indian J. Pl. Prot.* 3: 186-189.

Kaur, H. (1997). An evaluation of organophosphorus pesticides for their effects on immunomodulation, and carcinogenicity and digestive enzymes in animals. Ph.D. Dissertation, Punjab Agricultural University, Ludhiana.

Kaur, R., Dhanju, C.K. and Kaur, K. (1999). Effect of dietary selenium on biochemical composition of rat testis. *Indian J. Exp. Biol. 37:* 509-511.

Kaur, R. and Parshad, V.R., (1994). *Mutat. Res.* 309: 29.

Keck C., Bramkamp, G., Behre, H.M., Muller C., Jockenhovel, F., Neischlag, E. (1995). Lack of correlation between cadmium in seminal plasma and fertility status of non-exposed individuals and two cadmium-exposed patients. *Reprod. Toxicol.* 9: 35-40.

Kelce, W.R., Stone, C.R., Laws, S.C., Gray, L.E., Kemppainen, J.A., Wilson E.M. (1995). Persistent DDT metabolite p, p'-DDE is a potent androgen receptor antagonist. *Nature,* 375: 581-582.

Khera, K.S. (1974). Teratogenicity and dominant lethal studies on hexachlorobenzene in rats. *Food Cosmet. Toxicol.* 12: 471-477.

Kirubagaran, R., and Joy, K.P. (nd). Toxic effects of mercuric chloride, methyl mercuric chloride and emisan 6 (an organic mercurial fungicide on ovarian recrudescence in the catfish *Claris batrachus* (L). *Bull. Environ. Contam. Toxicol.* 41: 902-909.

Kociba, R.J., Keeler, P.A., Park, G.N., Gehring, P.J. (1976). 2, 3, 7, 8-Tetrachlorodibenzo-p-dioxin (TCDD): results of a 13 week oral toxicity study in rats. *Toxicol. Appl. Pharmacol.* 35: 553-574.

Kulshereshtha, L.L.M., (1992). Pesticides and environmental pollution. Legislative control in India. *Pestology,* 10: 5-11.

Kulshetra, S.K. and Arora N. (1984). Impairment induced by sublethal doses of two pesticides in the ovaries of a freshwater teleost *Channa striatus* Bloch *Toxicol. Letters,* 20: 93-98.

Lancranjan, I., Popescer, H.I., Gavenescu, O., Klepsch, I., Serbanescu, M. (1975). Reproductive ability of workmen occupationally exposed to lead. *Arch. Environ. Health,* 30: 396-401.

Landrigan, P.J., Clandio, L., Markowitz, S.B., Berkowitz, G.S., Brenner B.L., Romero, H., Wetmur, J.G., Matte, T.D., Gore, A.C., Goldbold, J.H., Wolff, M.S. (1999). Pesticides and inner-city children: exposures, risks and prevention. *Environ. Health Perspe.* 107 (Suppl. 3): 431-437.

Lasley, B.L., Gold, E.B., Nakajima, S.T., Stewart, D.R., Overstreet, J.W., (1933). Classification of adverse reproductive effects can be improved by measurements of multiple biomarkers for ovarian toxicity and early fetal loss. *J. Toxicol. Environ.* 40: 423-433.

Lasley, B.L. and Overstreet, J.W. (1998). Biomarkers for assessing human female reproductive health, an interdisciplinary approach. *Environ. Health Persp.* 106(4): 955-960.

Lerda, M. (1992). Study of sperm characteristics in person occupationality exposed to lead. *Am. J. Ind. Med.* 22: 567-571.

Lin, S., Hwang, S.A., Marshall, E.G., Stone, R., Chen, J.H. (1996). Fertility rates among lead workers and professional bus drivers a comparative study. *Ann Epidemiol.* 6: 201-208.

Lindbohm, M.L., Sallmen, M., Anttila, A., Taskinen, H., Hemminki, K. (1991). Paternal occupational lead exposure and spontaneous abortion. *Scand. J. Work. Environ. Health.* 17: 95-103.

Longnecker, M.P., Rogan, W.J., Lucier, G. (1997). The human health effects of DDT and PCBs and an overview of organochlorines in public health. *Annu. Rev. Public Health,* 18: 211-244.

Mably, T.A., Moore, R.W., Goy, R.W., Peterson, R.E. (1992). In utero and lactational exposure of male rats to 2, 3, 7, 8-tetrachlorobenzo-p-dioxin. 2: Effect on sexual behaviour and the regulation of luteinizing hormone secretion in adulthood. *Toxicol. Appl. Pharmacol.* 114: 108-117.

Mani, K. and Saxena, P.K. (1985). Effect of safe concentration of some pesticides on ovarian recrudescence in the freshwater murrel, *Channa punctatus* (Bloch): a quantitative study. *Exotoxicol. Environ. Safety.* 9: 241-249.

Mattison, D.R., Bognmil, R.J., Chapin, R., Hatch, M., Hendrickx, A., Jarrell, J., LaBarbera, A.L., Schrader, S.M., Seleval, S. (1990). Reproductive effects of pesticides. In: *The Effects of Pesticides on Human health,* (Baker, S.R., Wilkinson, C.F., eds.) Princeton, NJ: Princeton Scientific Publishing, pp 2977-389.

McGregor, A.J, Mason, H.J. (1991). The effects of occupational exposure to cadmium, lead and mercury vapour on male reproductive endocrine function. In: *Proceedings of the International Conference on heavy Metals in the Environment,* Vol. 1, 16-20, September, Edinburgh, UK (Farmer, J.G., (ed.) Edinburgh, U.K.: CEP Consultants pp. 375-378.

Miller, M.M., Plowchalk, D.R., Weitzman, G.A., London, S.N., Mattison, D.R. (1992). The effect of benzo (a) pyrene on mutine ovarian and corpora lutea volumes. *Am. J. Obset. Gynecol.* 166: 1535-1541.

Muller, W.F., Hobson, W., Fuller, G.G., Knauf, W., Coulston, F., Korte, F. (1986). Endocrine effects of chlorinated hydrocarbons in rhesus monkeys. *Ecotoxicol. Environ. Safety,* 2: 161-172.

Murphy, R., Harvey, C. (1985). Residues and metabolism of selected persistent halogenated hydrocarbons in blood specimens from a general population survey. *Environ Health Persp.* 60: 115-120.

Murugesan A.G. and Haniffa M.A. (nd.). Histopathological and histochemical changes in the oocytes of the air-breathing of fish *Hateropneustes fissilis* (Bloch) exposed to textile-mill effluent. *Bull. Environ. Contam. Toxicol.* 48: 929-936.

Newton D, Pickford C.J., Chamberlain A.C., Sherlock J.C., Hislop J.S. (1992). Elevation of lead in human blood from its controlled ingestion in beer. *Hum. Exp. Toxicol.* 11: 3-9.

Ng T.P., Goh H.H., N.g. Y.L., Ong, H.Y., Ong C.N., Chia K.S., China S.E., Jeyaratha J. (1991). Male endocrine functions in workers with moderate exposure to lead. *Br. J. Ind. Med.* 48: 485-491.

Noack-Fuller G., De Beer C., Seibert H., (1993). Cadmium, lead selenium and zinc in semen of occupationally unexposed men. *Andrologia* 25: 7-12.

Olderied N.B., Thomassen Y., Purvis K. (1993). Seminal plasma lead, cadmium and zinc in relation to tobacco consumption. *Int. J. Androl,* 17: 24-28.

Orbeg J., Kihlstrom J.E. (1973). Effects of long-term feeding of polychlorinated biphenyls (PCB, Clophen A-60) on the length of the estrous cycle and on the frequency of implanted ova in the mouse. *Environ Res* 6: 176-179.

Peakall D.B. (1975). *Res. Rev.* 54: 1.

Perera F.B., Whyatt R.M., Jedry Chowski W. Rauh V., Manchester D., Santella RM, Ottman R. (1998). Recent developments in molecular epidemiology: a study of the effects of environmental polycyclic aromatic hydrocarbons on birth outcomes in Poland. *Am. J. Epidemiol* 147: 309-314.

Prpic-Majic D, Telisman S, Kezic S. (1984). *In virto* study on lead and alcohol interaction and the inhibition of erythrocyte delta-amino levulinic acid dehydratase in man. *Scand. J. Worl Environ. Health* 10: 235-238.

Pundir R. and Sexena A.B. (1990). Seasonal changes in the testis of fish *Punctius ticto,* and their relation to heavy metal toxicity. *Bull Environ. Contam. Toxicol.* 45: 288-293.

Randhawa, G.K. (1997). Immunomodulation in farm animals due to chronic toxicity with organophosphorus pesticides. Ph.D. Dissertation, Punjab Agricultural University, Ludhiana.

Raychoudhury S.S., Flowers A.F., Millette C.F., Finlay M.F. (2000). Toxic effects of polychlorinated biphenyls on cultured rat Sertoli cells. *J. Androl.* 21: 964-973.

Richardson R.J. (1995). Assessment of the neurotoxic potential of chlorphyrifos relative to other organophosphorus compounds: a critical review of literature. *J. Toxicol Environ. Health* 44: 135-165.

Rodamilans M., Osaba M.J.M., To-Figueras J., Rivera Fillat F., Marques J.M,. Perez P., Corbella J. (1988). Lead toxicity on endocrine testicular function in an occupationally exposed population. *Hum Toxicol* 7: 125-128.

Saaranen M., Kantola M., Saarikoski S., Vanha-Perttula T. (1989). Human seminal plasma cadmium: comparison with fertility and smoking habits. *Andrologia,* 21: 140-145.

Saaranen M, Suistoma V, Kantola M, Saarikoski S, Vanha Pertulla T. (1987). Lead, magnesium, selenium and zinc in human seminal fluid: Comparison with semen parameters and fertility. *Hum. Reprod.* 2: 475-479.

Safe, S.H. (1995). Environmental and dietary estrogens and human health: is there a problem? *Environ Health Perspect.* 103: 346-351.

Schneider B.F. (1987). *Res. Toxicol Ind. Heal,* 3: 1.

Shanz S.L., Bourman RE. (1989). Learning in monkeys exposed perinatally to 2, 3, 7, 8–tetrachlorodibenzo-p-doxin (TCDD). *Neurotoxicol Teratol.* 11: 13-19.

Sharpe R.M., Skakkeback N.E. (1993). Are estrogens involved in falling sperm counts and disorders of the male reproductive tract? *Lancet,* 341: 1392-1395.

Singh H and Singh T.P. (1980). Throid activity and TSH potency of the pituitary gland and blood serum in response to cythion and Hexadrin treatment in the freshwater catfish, *Heteropneustes fossilis* (Bloch). Env. Res. 22: 184-189.

Siracusa G., Bastone A., Sbraccia M., Settimi L. Mallozzi C., Monaco E., Frontali N. Effects of 2, 5-hexanedione on the ovary and fertility. An experimental study in mice. *Toxicology*, 75: 39-50.

Slotkin S.M. (1999). Developmental cholinotoxicants: nicotine and chlorpyriofos. *Environ. Health Persp.*, 107 (suppl 1): 71-80.

Solarch S.A., Vaz R. (1985). PCB levels in breast milk: data from the UNEP/WHO pilot project on biological monitoring and some other recent studies. *Environ. Health Persp.* 60: 115-120.

Srivastava S., Singh G.B., Shrivastava S.P. and Seth P.K. (1990, Testicular toxicity of di-n-butyl phthalate in adult rats. Effect on marker enzymes of spermatogenesis. *India J. Exp. Biol.* 28: 67-70.

Stehr - Green P.A. (1989). Demographic and seasonal influences on human serum pesticide residue levels. *J. Toxicol. Environ. Health.* 27: 405-421.

Sukumar A. and Karpagaganapathy P.R. (n.d.), Pesticide-induced atresia in ovary of a freshwater fish, *Colisa ialia* (Hamilton - Buchanan). *Bull. Environ. Contam. Toxicol.* 48: 457-462.

Sun J.G., Jurisicova A., Casper R.F. (1996). Detection of deoxyriobonucleic acid fragmentation in human sperm: correlation with fertilization *in virto. biol Reprod.* 56: 602-607.

Swan S.H., Elkin E.P., Fenster L. (1997). Have sperm densities decline? A reanalysis of global trend data. *Environ Health Perspect.* 105: 1228-1232.

Tas, S., Lauwerys R., Lison D. (1996). Occupational hazards of the male reproductive system. *Crit. Rev. Toxicol.* 26: 261-307.

Telisman S., Cvitkovic P. Jurasovic J., Pizent A., Gavella M. and Rocic B., (2000). Semen quality and reproductive endocrine function in relation to biomakers of lead, cadinium, zinc and copper in men. *Environ. Health Persp.* 108: 45-53.

Telisman S, Prpic-Magic D., Kezic S. (1984). *In vivo* study on lead and alcohol interaction and the inhibition of erythrocyte delta-aminolevulinic acid dehydratase in man. *Scand. J. Work. Environ. Health,* 10: 239-244.

Toppari J., Larsen J.C., Christiansen P., Giwercman A, Grandjean P., Guillette L.J. Jr, Jegou B., Jensen T.K., Jouannet P., Keiding N. *et. al. (1996).* Male reproductive health and environmental xenoestrogens. *Environ. Health Persp.* 104. (suppl 4): 742-803.

Trapp M., Baukloh, V., Bohnet, H.G., Heeschen W. (1984). Pollutants in human follicular fluid. *Fertil Steril* 42: 146-148.

Vermeulen, A. (1993). Environment, human reproduction, menopause and andropause, *Environ. Health Persp.* Suppl. 101 (Suppl 2): 91-100.

Villeneuve D.C., Hierlihy S.L., (1975). Placentel transfer of the hexa chlorobenzene in the rat. *Bull. Environ. Contam. Toxicol.* 13: 489-491.

Villeneuve D.C., Panopio L.G., Grant DL. (1974). Placentel transfer of hexa chlorobenzene in the rabbit. *Environ. Physiol. Biochem.* 4: 112-115.

Vine M.F., Margolin B.H., Morrison H.I., Hulka BS. (1994). Cigarette smoking and sperm density: a meta analysis. *Fertil Steril* 61(1): 35-43.

Viskum S., Rebjerg L., Jorgensen P.J., Grandjean P. (1999). Improvement in semen quality associated with decreasing occupational lead exposure. *Am. J. Ind. Med. 35*: 257-263.

Weeks M.H., Angerhofer R.A., Bishop R., Thomasino J. Pope C.R. (1979). The toxicity of hexachloroethane in laboratory animals. *Am. Ind. Hyg. Assoc. J.* 40: 187-189.

Weitzman G.A., Miller M.M. London S.N., Mattison D.R. (1992). Morphomedtric assessment of the murine ovarian toxicity of 7, 12 - dimethly ben (a) anthracene. *Reprod. Toxicol.* 6: 137-142.

Whitney K.D., Sielder F.J., Slotkin T.A. (1995). Developmental neurotoxicity of chlorphrifos: Cellular mechanism. *Toxicol Appl Pharmacol* 134: 53-62.

Whyatt, R.M. and Barr, D.B. (2001). Measurement of organophosphate metabolites in postpartum Meconium as a potential biomaker of prenatal exposure. A validation study. *Environ. Health Persp.* 109 (4): 417-420.

Whyatt, R.M., Perera F.P., (1995). Application of biologic markers to studies of environmental risks in children and the developing fetus. *Environ. Health Persp.* 103 (Suppl 6): 105-110.

Whyatt R.M., Santella R.M., Jerdychowski W., Garte S.J., Bell Da, Ottman R., Gladek-Yarborough A., Cosma G., Young T.L., Cooper T.B. *et al.* (1998). Relationship between ambient air pollution and DNA damage in Polish mothers and newborns. *Environ. Health Persp. 106* (Supp 3): 821-826.

Wildt K., Eliasson R., Berlin M. (1983). Effects of occupational exposure to lead on sperm and semen. In: *Reproductive and Developmental Toxicity of Metals* (Clarkson TW, Nordberg GF, Sager PR, eds. New York: Plenum Press, pp. 279-300.

Working P.K. (1988). Male reproductive toxicology: Comparison of the human to animal models. *Environ. Health Persp. 77*: 37-44.

Xu B., Chia, S.E, Tsakok, M., Ong, CH. (1993). Trace elements in blood and seminal plasma and their relationship to sperm quality. *Reprod. Toxicol.* 7: 613-618.

14

Effect of Sumaach Treatment on the Early Oocytes of *Channa punctatus*

Kadambri Gupta and Girija Suri

Department of Zoology, University of Jammu, Jammu – 180001, India.

ABSTRACT

In immature *Channa punctatus* collected from natural waters of Jammu and exposed to sumaach treatment (10 mg/kg, at monthly intervals), vitellogenesis comparable to that in sexually mature fish, was induced after a duration of 150 days. Controls did not reveal any vitellogenesis in the oocytes. An identical dose of sumaach but injected at shorter time intervals (once per 2 weeks) resulted in much faster maturation. The effects of sumaach treatment at short and long time intervals have been discussed.

Key words: *Channa punctatus*: immature; sumaach; monthly treatment; vitellogenesis.

Introduction

Environmental factors play an important role in regulating reproduction in teleost fishes (Billard and Breton (1979), Crim *et al.*, 1983; Malhotra and Gupta, 1985; Zaki *et al.*, 1995 and Roderick *et al.*, 1996). Environmental cues mediate secretion of hormones by the hypothalamus and pituitary which synchronize the various reproductive activities. Under artificial conditions, the required optional environmental stimuli may be lacking and thus may act as physiological barriers to block commencement of the reproductive processes. Under such conditions, hormonal therapies as biotechnological tools, need to be standardized and exploited (Lam, 1982).

The reproductive cycles of most teleosts can be divided into three periods of activity including juvenile, pubertal and adult phases (Crim *et al.*, 1982). Most of the work to-date, on endocrine control of fish reproduction deals with seasonal sexual development during the adult period (Billard and Breton, 1979), but much still remains to be known about the application of knowledge of endocrionology to a sexually immature teleost. This needs to be established by the use of exogenous hormones to promote early reproductive maturity.

Our earlier studies have shown that the fortnightly sumaach injections in *Channa punctatus* produce a significant acceleration of vitellogenesis after 90 days (Sharma, 1988; Kouril *et al* ., 1997 and Suri, 2001). The present study is a continuation of the previous work to find the effect of sumaach at longer intervals and deals with the histological changes produced in the ovaries of the immature *C. punctatus.*

Materials and methods

An initial experiment was conducted using sexually immature *C. punctatus* weighing 8 ± 2 g collected from the fresh water habitats around Jammu. The fish thus collected were maintained in hapas (14 × 10 × 10 inches) held in 2001 aquarium (3 × 1 × 1 ft) filled with tap water, under laboratory conditions.

Water was renewed on alternate days in the aquarium and fishes were fed *ad libitum*. Each experiment was conducted for a period of 150 days, during which the fish were injected with sumaach (dose, 10 mg/kg) at monthly intervals. Controls were also subjected to identical experimental conditions except the hormone administration.

Live fishes were than sacrificed and their ovaries fixed in Bouin's fixative for 24 hr 7 – 10/μ sections of paraffin embedded ovaries were stained using Mallory's triple stain.

Results and Discussion

At the start of the experiment, in both the controls and experimental fish, the ovaries were packed with stage I oocytes (Table 14.1; Fig. 14.1). By day, 150, the ovaries of treated *C. punctatus* responded by induction of vitellogenesis (Yolk vesicle

Table 14.1: Effects of sumaach on size and percentage occurrence of cocytes of immature *Channa punctatus*

Experimental Series	Observations	Oocyte groups						Oocyte stages			Egg membrane
		GI		GII		GIII					
		Size (mm)	Percentage	Size (mm)	Percentage	Size (mm)	Percentage	SI	SII	SIII	
Control	At the start of the experiment	0.016–0.049–	40	0.066–0.099	60	–	–	+++	–	–	F.E.
	At the end of the experiment	0.033–0.049	36	0.066–0.149	64	–	–	+++	–	–	F.E.
Sumaach treatment	At the start of the experiment	0.016–0.049	39	0.066–0.099	61	–	–	+++	–	–	F.E.
	At the end of the experiment	0.033–0.049	16	0.066–0.166	34	0.182–0.332	50	++	++	+++	F.E., V.M., Th.

Abbreviations: – = Absent; ++ = Medium; +++ = Abundant; F.E = Follicular epithelium; V.M. = Vitelline membrane; Th. = Theca; S = Stage; G = Group.

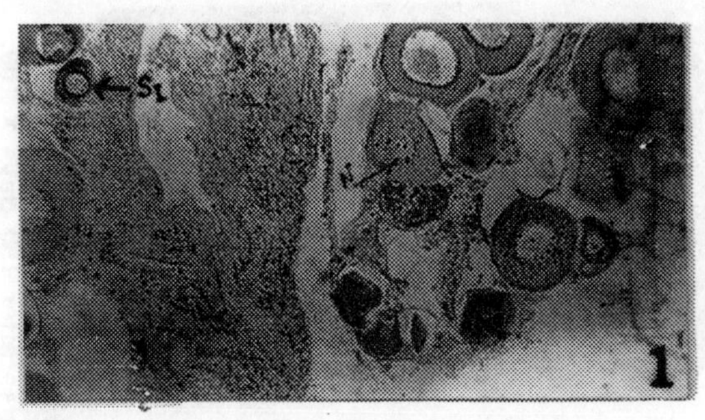

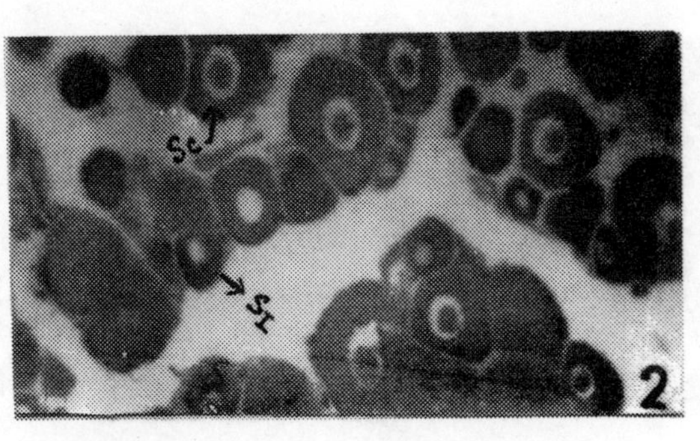

Fig. 14.1: Section of ovary of initial control × 100.

Fig. 14.2: Section of ovary of control series after 150 days × 100.

(SI - Stage I, Cocytes; S-II - Stage II eocytes; S-III - Stage III cocytes; SC - Smooth cytoplasm; ENu - Extruded nucleoli; YV = Yolk vacuoles; YG - Yolk globules; F - Follicular epithelium; VM - Vitelline membrane.

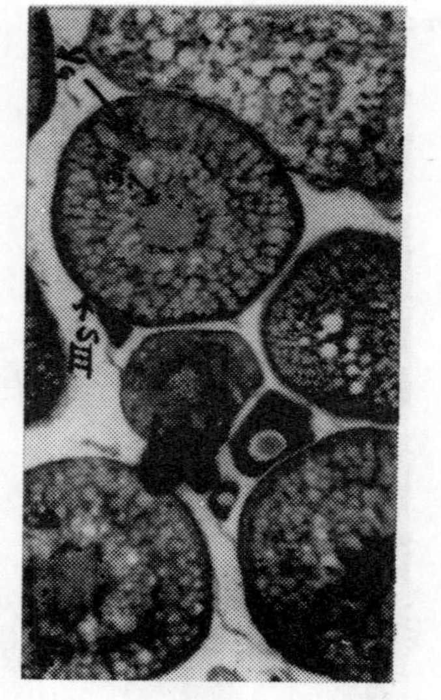

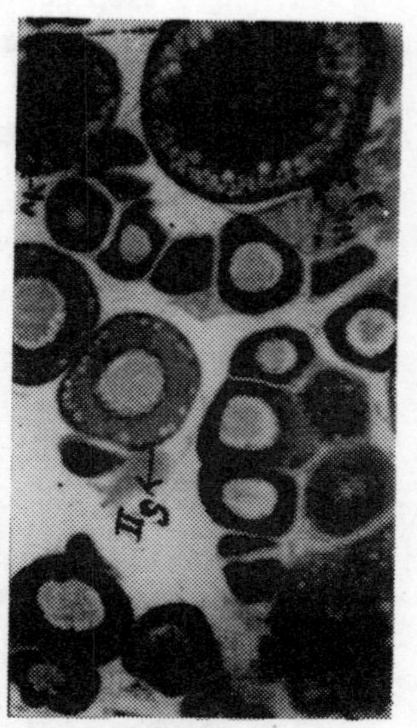

Fig. 14.3: Section of ovary of sumaach treatment series after 150 days × 100.

Fig. 14.4: Section of ovary of sumaach treatment series after 150 days × 100.

Plate: Figs 1–4: Cross-section of the ovary of immature *Channa punctatus* showing the condition of the ovary during 150 days of experimental duration.

and yolk globule) in their oocytes (Figs. 14.3, 14.4). The controls, however, did not show any such changes (Fig. 14.2). The results of the the present study and in conformity with that of earlier workers (Sunk and Donaldson, 1972; Mackinon and Donaldson, 1978; Upadyay *et al.*, 1978; Crim and Evans, 1983; Gielen *et al.* 1982 a and b, Hurk, 1982; Sharma, 1988; Raina 1999 and Suri 2001). These results indicate that the gonads of *C. punctatus* respond to exogenous gonadopropins (From H.C.G) by further secreting their own steroid hormones as already documented by Ng and Idler 1980, Hurk, 1982, Gielen *et al.* 1982 a, b, Magri *et al.* 1982 and Lee *et al.* 1986). Therefore, with respect to the entogeny of the brain-pituitary-gonadal axis, gonads do exhibit the capacity to synthesize hormones at very early stage (Hurk *et al.*, 1982).

During the present studies, monthly injections of sumaach to immature *C. punctatus* induced vitellogenesis in overies (Fig. 14.3) after 5 months duration. Comparable earlier results were achieved only after 3 months of duration with fortnightly injections of sumaach (Khajuria, 1988; Malhotra, *et al*, 1991; Malhotra, 1993 and Suri, 2001). These results indicate that the time of vitellogenesis of immature *Channa* could be manipulated by controlling the time interval of sumaach treatment.

There thus appears to be a relationship between the time interval, lasting between two injections and the rate of gonadal development in immature *Channa*. In previous studies (Sharma, 1988 and Suri, 2001) where sumaach was injected at shorter intervals (injection once per 2 weeks), gonadal development was faster than that observed during the present study. It may probably be infurred that when the interval between injections is longer, metabolic clearance may reduce the gonadotropin to an ineffective level (between treatments) thus retarding maturation of gonads (Mackinnon and Donaldson, 1978).

Further, it has been observed that there is a difference between the effect of sumaach on the size of oocytes of fortnightly and monthly treated fishes (Table 14.1). The size of oocytes, which ranged between 0.016 and 0.099 mm at the start of the experiment increased to as much as 0.332 mm (all being vitellogenic, 72%) after 150 days, of monthly sumaach treatment. Against this, the controls did not reveal any change in the oocytes except a small increment in their size (0.166 mm). Present results when compared with earlier results of fortnightly treatment of sumaach, increment in size of oocytes was to 0.398 mm after 90 days. These results again speak in favour of fortnightly treatment which proves to be more effective as compared to the monthly treatment.

The common practice of exogenous hormone injections at short intervals on a prolonged basis, is known to be stressful (Lam, 1982). Reduction of the injection frequency to once a month, though would reduce stress, but concomitantly it also reduce the efficacy of hormone and hence does not appear to find much usage in aquaculture practice.

REFERENCES

Billard, R. and Breton, B., (1979), Rhythms of reproduction in teleost fish. In: *Rhythms of Activity in Fish*, (Thrope, J.E, ed.), Academic Press, New York, pp. 31-53.

Crim, L.W., Evans, D.M. and Vickery, D.H. (1983), Manipulation of the seasonal reproductive cycle of the landlocked atlantic salmon (*Selmo salar*) by LHRH analogues administered at various stages of gonadal development, *Can. J. Fish. Aquat. Sci.*, 40: 61-67.

Crim, L.W. and Evans, D.M. (1983), Stimulation of pituitary gonadotropin by testosterone in juvenile rainbow trout (*Salmo gairdneri*), *Gen. Comp. Endocrinol.*, 37: 192-196.

Crim, L.W. and Evans, D.M. (1983), Influence of testosterone and/or luteinizing hormone releasing hormone analogue on precocious sexual development in the juvenile rainbow trout, *Biology of Reproduction*, 29: 137-142.

Crim, L.W., Evans, D.M. and Vickery, D.H. (1983), Manipulation of the seasonal reproductive cycle of the landlocked atlantic salmon (*Salmo salar*) by LHRH analogues administered at various stages of gonadal development, *Can. J. Fish. Aquat*. Sci., 40: 61-67.

Crim, L.W., Billard, R., Genge, P.D. and Idler, D.R. (1982), The influence of immature gonads on onset of gonadotropic hormone accumulation in the juvenile rainbow trout pituitary gland. *Gen. Comp. Endocrinol.*, 48: 161-166.

Funk, J.D and Donaldson, E.M. (1972), Induction of precocious sexual maturity in male pink salmon (*Oncorhynchus gorbuscha*), *Can. J. Zool.*, 50: 1413-1419.

Gielen, J.Th., Bosch, R.A. Van der, Goos, H.J. Th., Peute, J., Hurk, R. Vanden, Breton, B. and Cordt, P.G.W.J. Van. (1982a), Maturational effects of gonadal hormones on GtH-cells in the trout, *Salmo gairdneri*, Gen. Comp. *Enddcrinol.*,.

Gielen, J.Th., Goos, H.J. Th., Peute, J., Bosch, R.A., Vander and Cordt, P.G.W.J. Van. (1982b), The brain-pituitary-gonadal axis in the rainbow trout, *Salmo gairdneri*: Gonadal hormones and the maturation of gonadotropic cells, *Cell Tissue, Res.*, 225: 45-56.

Hurk, R. Van den. (1982), Effects of steroids on gonadotropic (GtH) cells in the pituitary of rainbow trout, *Salmo gairdneri* shortly after hatching, *Cell Tissue Res.*, 224: 361-368.

Hurk, R. Van der, Lambert, J.G.D. and Peute, J. (1982), Steroidogenesis in the gonads of rainbow trout fry (*Salmo gairdhneri*) before and after the onset of gonadal sex differentiation, *Repord. Nutr. Dev.*, 22: 413-425.

Kouril, J., Linhart, O. and P. Relot, (1997), Induced spawning of pench by means of a GuRH and Aquaenltive International, 5: 375-377.

Lam, T.J. (1982), Applications of endocrinology to fish culture, *Can. J. Fish. Aquat, Sci.*, 39: 111-137.

Malhotra, Y.R. and Gupta, K. (1985), Histophysiology of hypothalamo-hypophysial system in relation with reproductive cycle in *Puntius sophere* (Ham.), *Zool. Orient.*, 2: 59-65.

Mackinnon, C.N., and Donaldson, E.M. (1978), Comparison of the effect of Salmon gonadotropin administered by pellet implantation or injection on sexual development on juvenile male pink salmon (*Oncorhynchus gorbuscha*), *Can. J. Zool.*, 56: 86-89.

Magri, M.H., Billard, R., Reinand, P., and Fostier, A. (1982), Induction of gametogenesis in the juvenile rainbow trout (*Salmo gairdneri*), *Gen. Comp. Endocrinol.*, 46: 394-395.

Malhotra, N. (1990), Effect of sumaach and carp pituitary extract on ovarian maturation in *Puntuis tico* (Ham)., M.Phil. Dissertation Deptt. of Zoology, University of Jammu, Jammu.

NG., T.B., and Idler, D.R. (1980), Gonadotropic regulation of androgen production in flounder and salmonids, *Gen. Comp. Endocrinol.* 42: 25-38.

Roderick, E.E., Santiago, L.P Garcia, M.A and Mair, G.C. (1996), Induced spawning in 0. Niloticks. International Centre for living Aquatic Resources Management Manil in, 3rd International Symposium on Tilafia in Aquaculture, p. 548.

Sharma, A. (1988), Effect of sumaach (a crude form of HCG) on immature ovaries of *Channa punctatus* (Bloch), M.Phil dissertation, University of Jammu, Jammu.

Suri. G., (2001), Influence of supplementary Diets and Harmones on growth and maturation of *Puntius sophore* (Ham.), from Jammu. Ph.D. Thesis. Deptt. of Zoology, University of Jammu, Jammu.

Zaki, M.I, E.L Glasabawy, M.M., E.l, Greisy, Z.A. (1995), Baghdadi, H. Effect of some environmental and physiological factors on some reproductive parameters of *Mugil Capito* during the breeding season. *Bull. Natl. Inst Oceanogr. Fish (Egypt)* 21: 585-603.

15

Ovarian Maturation Cycle of *Puntius ticto* (Ham.) from a Lotic Water Body of Jammu

A. Khajuria[1] and Kadmbri Gupta[2]

1. *Department of Zoology, Govt. Degree College, Bemina, Srinagar*
2. *Department of Zoology, University of Jammu, Jammu-180006*

ABSTRACT

Reproductive cycle, maturation mode and body size were examined by analysing the seasonal trends in gonadosomatic index (GSI) and gonadal maturity as revealed by histological observations in the fish *Puntius ticto* (Ham.) collected from a Iotic water body of Jammu, J & K State. Fish spawns once in a year from May to August. Gonads were assigned to the following six stages of maturity viz. (S I) – chromatin nucleolar stages, (S II) – Perinucleolar stage, (S III) – Yolk vesicle stage, (S IV) – Vitellogenic stage, (S V) – Maturation stage and (S VI) – Ripe egg stage, based on conditions of the most of developed oocytes, presence of empty follicles or corpus luteum.

Key words: Oocyte maturation, Gonado-somatic index, spawning, *Puntius ticto*.

Introduction

A knowledge of oocyte maturation is prerequisite for evaluation of reproductive conditions and to develop an appropriate technology for culture practices in given ecological set up. Several stage system to express maturation of oocytes have been used for different teleosts for understanding the reproductive patterns (Belsare, 1962; Lal, 1963; Breakevelt and McMillan, 1967; Srivastava and Rathi, 1970; Guraya *et al*., 1975; Gupta, 1980; Agarwal, 1982; Qadri, *et al*., 1983; Howell, 1983; Malhotra *et al*. 1989; Garay and Luna, 1994 and Raina, 1999).

In the present communication a sequential account of oocyte growth and development of *Puntius ticto* from Jammu waters has been discussed.

Material and Methods

Collection of *P. ticto* were made from Ghomanhasan. Ovary was removed, fixed (Bouin's fixative), sectioned at 5-8 μ and stained with Mallory's triple stain. The methodological details employed during this investigations has been given elsewhere (Khajuria, 1988). Weight of fish and ovary were also determined and Gonadosomatic Index (GSI) calculated by the formula as:

$$\frac{\text{Ovarian wt.}}{\text{Fish wt.}} \times 100$$

Observations

Ovaries are ensheathed by tough covering the tunica albuginea which contains muscles, connective tissue and blood vessels. The tunica is internally lined with germinal epithelium which at places obliterates the ovocoel by protuberant ovigerous lamellae (Fig. 15.1). Embedded in the ovigerous lamellae are observed the oogonia and oocytes at various stages of development (Fig. 15.1).

For convenience of description of oocyte maturation can be divided into following stages (Table 15.1) and the characteristic feature of each stage are given in Table 15.1:

Observations on Gonadosmatic Index (GSI) over a period of one year is graphically represented in Fig. 15.10. A look at the Fig., reveals that GSI has a low value of 2.57 during the month of September which rises gradually till January. After January there is rapid increase in GSI reaching as high as 8.4 in July. Hence after it declines being low again in the month of September. The sequential rise and fall in GSI value synchronizes with the maturity stages from oogonium to ripened egg. Average Oocyte diameter followed the same pattern (Fig. 15.10).

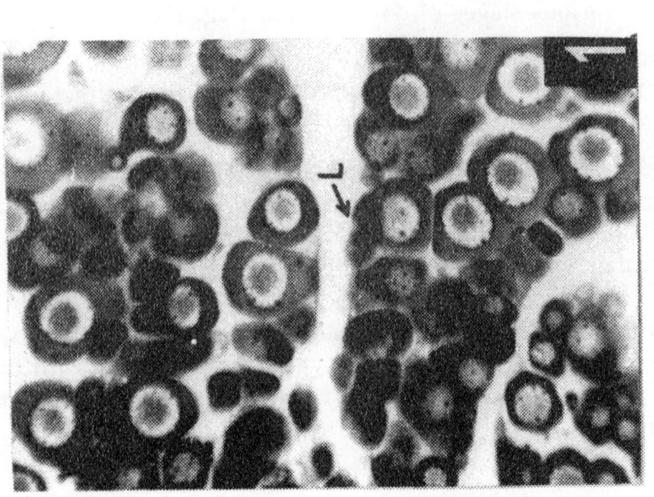

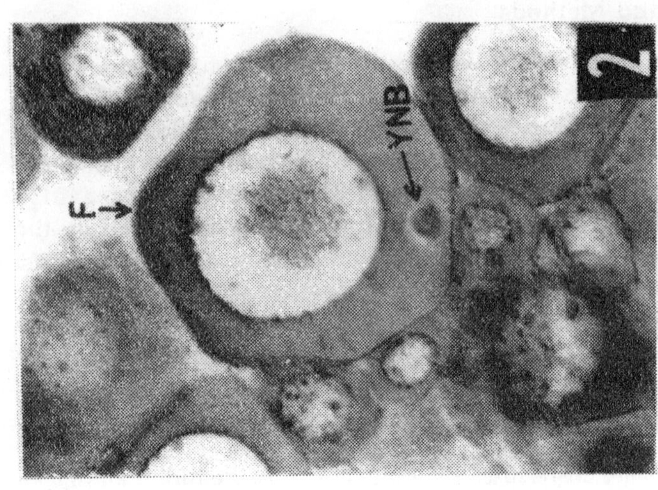

Fig. 151: Represent the lamellar arrangement of Stage 1 and 2 oocytes (L).

Fig. 15.2: Stage 2 oocyte showing enlarged view of an oocyte with darkly stained yolk nucleus of Balbian (YNB), Follicular epithelium and extruded nucleoli (en).

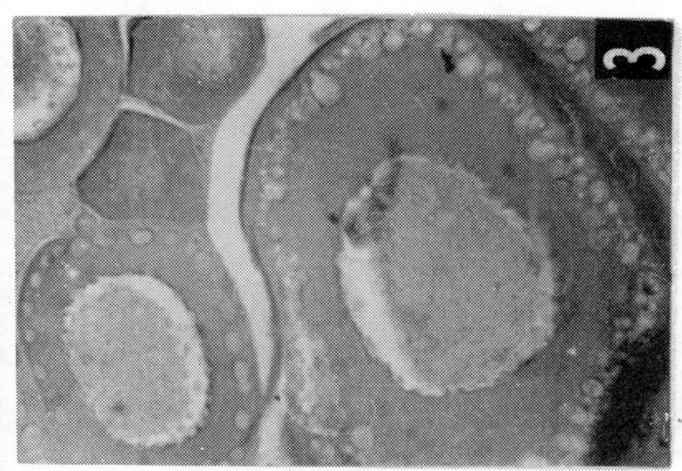

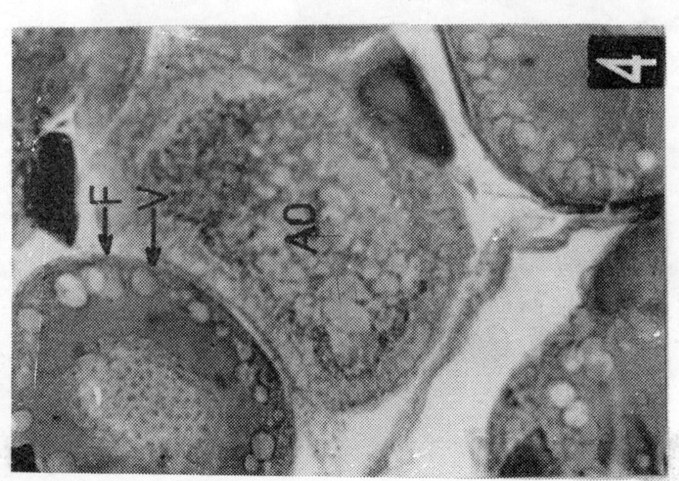

Fig. 15.3: Peripheral arrangement of yolk vacuoles in stage 3 oocytes.

Fig. 15.4: Oocyte is surrounded by follicular epithelium (F) as well as faint vitelline membrane. As atritic oocyte (AO) too can clearly seen.

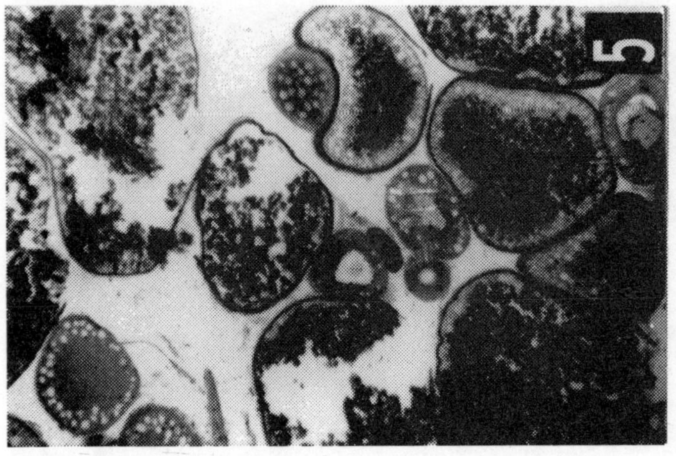

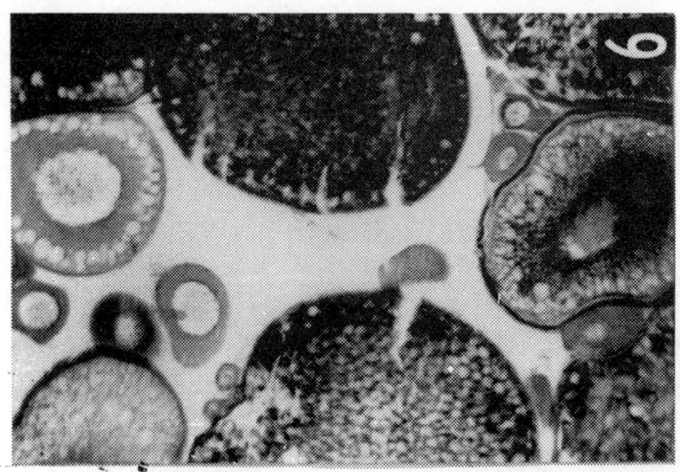

Figs. 15.5 and 15.6: The initiation of yolk deposition in most of the oocytes.

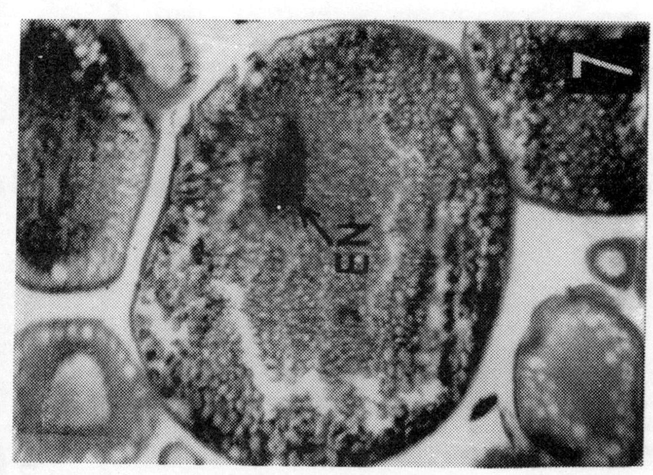

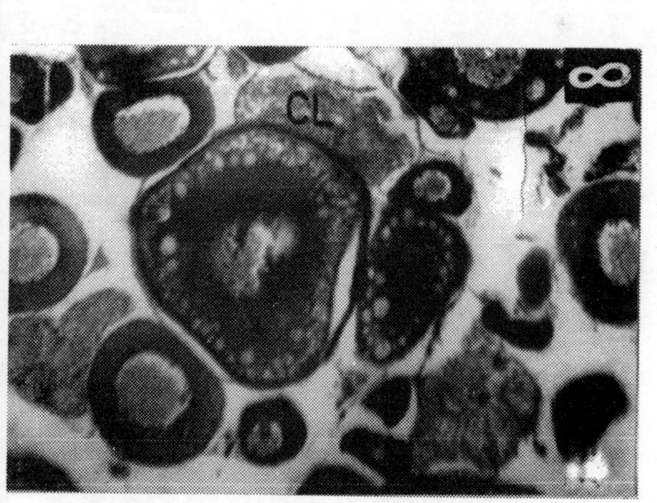

Fig. 15.7: Yolky oocyte with eccentric nucleus (EN).

Fig. 15.8: Occurrence of corpus leuteum (CL) as a consequence of ovulation.

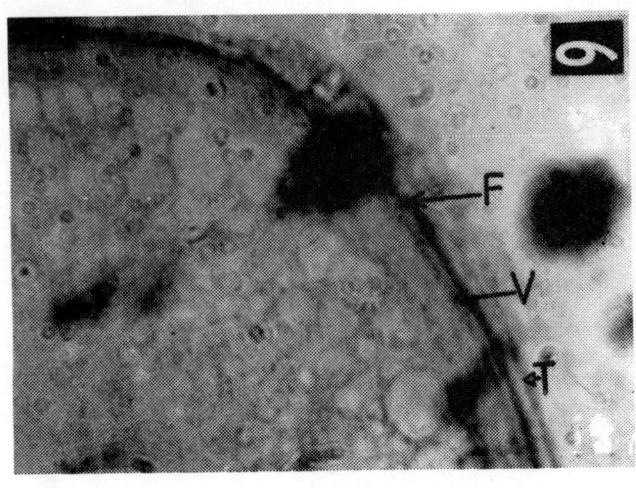

Fig. 15.9: The fully mature oocyte with all the three egg membranse i.e. Follicular epithelium (F), Vitelline membrane (V) and Theca (T).

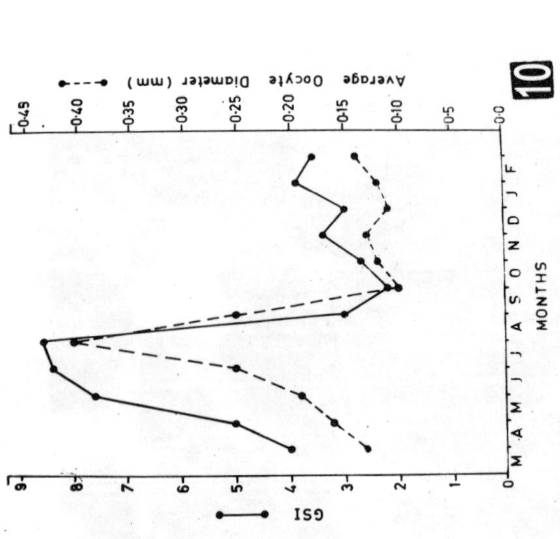

Fig. 15.10: The monthly variations in GSI and average oocyte diameter.

Table 15.1. Description of oocyte maturation

Sl. No.	Sequential stages through which oocyte passes	Oocyte diameter (in mm)	Nuclear diameter (in mm)	Characteristic features
1.	Stage 1 (S I) Chromatin nucleolar stage (Fig. 1)	0.01-0.09	0.009-0.03	Thin anastomosing chromatin threads fills the nucleus which entangles within. A prominent nucleus is present which varies in diameter from 0.008 to 0.03d mm.
2.	Stage 2 (S II) Perinucleolar stage (Fig. 2)	0.016-0.11	0.02-0.11	Nucleus contains large number of nucleoli which peripherally arranged along the inner surface of nuclear membrane. Phenomenon of nucleolar extrusion has also been observed. Yolk Nucleus of Balbiani (YNB) is also present.
3.	Stage 3 (S III) Yolk vesicle stage (Fig. 3 & 4)	0.11-0.24	0.03-0.24	Cytoplasmic volume increases yolk vesicles appears in the cytoplasm.
4.	Stage 4 (S IV) Vitellogenic stage (Fig. 5)	0.22-034	0.03-0.20	Yolk globules start depositing in the yolk vesicles. A faint vitelline membrane also appears.
5.	Stage 5 (S V)	0.36-0.42	0.01-0.20	Oocyte observed heavy accumulation of yolk both in and out of the vesicles.
6.	Stage 6 Ripe egg stage (Fig. 7, 8 & 9)	0.44-0.49	0.01-0.9	Theca the third egg membrane appears, nucleus is seen shifted towards periphery corpus lutea make appearance.

Discussion

The developmental events observed in ovaries of *P. ticto* are almost similar to those described for most of the teleosts. In *P. ticto* the development of oocytes have been divided into two phases viz. the previtellogenic phase in which growth is slow and comparatively few cytoplasmic changes occur and vitellogenic phase when occyte observed rapid growth and deposition of large amount of yolk in their cytoplasm.

Oogonia, the first differentiable product of ovary, representing the initial stage of oogenesis occur through out the year in variable numbers. They have been regarded as the reserve from where all the oocyte ultimately develop (Khajuria, 1988).

The yolk nucleus of Balbiani (YNB) was observed in stage 2 oocytes in the vicinity of nucleus (Fig. 15.2). Similar structure has been reported in fish ovaries by number of workers (Guraya 1979, 1986 and Malhotra *et al.*, 1978, 1989). As far as the origin of this structure is concerned there is lot of controversy. While some authors regards its origin from nucleus while the others assign it a cytoplasmic origin. On the basis of the appearance of YNB in the vicinity of nucleus in *P. ticto* present author also believes in nuclear origin of YNB. Speaking about its functional aspects, Malhotra *et al.* (1978) regarded it as probably substance needed for subsequent oocyte growth for it may have some connection with initiation of vitellogenesis. Guraya (1979) regarded yolk nucleus to act as a centre for formation and multiplication and accumulation of organelles and materials needed for yolk deposition. The yolk nucleus of Balbiani in *P. ticto* appears in the vicinity of nucleus and then travels towards periphery where it disintegrates at the time oocyte enters in the maturing phase. On the basis of this observation the present authors are in agreement with workers cited above who supports its role in oocyte growth, promoting organelles preparing the oocyte to enter the vitellogenic phase.

The phenomenon of nucleolar extrusion has also been observed during stage 2 of oocyte maturation in *P. ticto* of present investigation. Similar phenomenon has been observed by many workers (Malhotra, 1963; Malhotra *et al.*, 1978, 1990, Gupta, 1980; Malhotra *et al.*, 1989) but its role during oocyte maturation has remained obscure. As this process is speeded up, the process of yolk deposition is seen to be initiated. It may thus be suggested that the extruded nucleoli stimulates the oocyte to enter the yolk deposition phase as also held by Narain, 1937; Chaudhary, 1952 and Malhotra, 1963.

The spawning period of fishes varies from place to place (Malhotra *et al.*, 1989). Review of literature on spawming of fish reveals that *Puntius chonchonius* spawns twice in a year at Jammu once during April-June and secondly from Oct.-Dec. (Sharma, 1985). However, Agarwal (1982) while working on same fish species i.e. (*P. chonchonius*) has reported that fish spawns over prolonged period from Feb. – Sept. whereas, *P. ticto* studied for the first time in this area has been found to spawn from May to August. Presence of discharged follicles coupled with decline in GSI and oocyte diameter happens to be the authenticated initiation of spawning phase as also held by Garay and Luna 1994 and Raina (1999).

Acknowledgements

First author is highly grateful to the CSIR and Department of Environment, Government of India for financial assistance during the course of investigation, and the University of Jammu, Jammu for providing necessary facilities to carry on these investigations.

REFERENCES

Agarwal, A. (1982). Studies on the karyotype and histological changes in the gonads of some fishes of J & K. Ph.D. Thesis, Jammu University, Jammu.

Belsdre, D.K. (1962). Seasonal changes in the ovary of *Ophiocephalus punctatus* (Bloch). *Ind. J. Fish* 9A(1): 104-145.

Breakevelt, C.R. and D.B. McMillan (1967). Cyclic changes in the ovary of Brook Stickleback *Encalia incenstans*. *J. Morph*. 123(4): 373-396.

Chaudhary, H.S. (1952). The yolk nucleus of Balbiani in teleostean fishes. *Z. Zellforsch*, 37: 455-466.

Garay, E, J.F. and S. Luna, R. (1994). Gonadal development and spawning of female ocean white fish caulatilus princeps (*Pisces, Brachiostegidae*) in the Bay of Lapaz, B.C.S. Mexico. *J. Fish Biol*. 44: 583-586.

Gupta, K. (1980). Cyclic changes in Pituitary gland in relation to gonadal maturity in *Channa punctatus* and *Puntius sophar,* Ph.D. Thesis, University of Jammu, Jammu.

Guraya, S.S., R. Kaur and P.K. Saxena (1975) Morphology of ovarian changes during the reproductive period of the fish *Mystus tongares* (Ham.). *Acta Anat*. 91: 222-260.

Guraya, S.S. (1979). Recent Advances in morphology and histochemistry of steroid synthesizing cellular sites in the gonads of fishes. *Proc. Ind. Sci. Acad.* 3(44): 432-461.

Howell, W.H. (1983). Seasonal changes in the ovaries of adult yellow tail flonnder *Linanda ferruginea*. Fishery Bulletin. 81(2): 341-355.

Khajuria, A. (1988). Effect of Sumaach (a crude from HCG) on oocyte maturation of *Puntius ticto* exposed to polluted and non-polluted aquatic environments. M. Phil Dissertation, submitted to University of Jammu, Jammu.

Lal, B. (1963). Morphological and cytochemical studies on the oocytes of *Cirrhinus mrigala* (Ham.) with particular reference to lipids. *Proc. Nat. Inst. Sci. India,* 29B: 585-601.

Malhotra, Y.R. (1963). On the nucleolar extrusions in the developing oocytes of Kashmir fish, *Schizothorax niger* (Heckel.) *Icthyologica,* 11: 57-60.

Malhotra, Y.R., M.K. Jyoti and K. Gupta (1978). Ovarian cycle and spawning season of *Ophiocephalus punctatus* inhabiting Jammu waters India *Jap. J. Ichthyol.,* 25(3): 190-196.

Malhotra, Y.R., M.K. Jyoti and K. Gupta (1989). Reproduction cycles of Indian freshwater fishes. In: *Reproductive cycles of Indian vertebrates,* S.K. Saidapur (ed.), Allied Publishers Ltd., pp. 58-105.

Narain, D. (1937) Cytoplasmic inclusions in the oogenesis of *Saccobranchus fossilis, Clarias batrachus* and *Anabas scandens. Zellforsch,* 26: 625.

Qadri, M.Y., S. Mir and A.R. Yousuf (1983). Breeding biology of *Schizothorax richardsonii. J. Ind. Inst. Sci.,* 64(C): 73-86.

Raina, S. (1999). Seasonal variations in biochemical composition of ovary, liver, muscles and histology and histochemistry of ovary in *Channa gachua* (Ham.) and *Puntius chonchonius* (Ham.) Ph.D Thesis, University of Jammu, Jammu.

Sharma, K.K (1985). Larval ecology of some fishes from Jammu and determination of limits of tolerance of early development stages of *Puntius chonchonius* (Ham.) to same environmental factors. Ph.D. Thesis submitted to University of Jammu, Jammu.

Srivastava, P.N. and S.K. Rathe (1970). Effect of radiation on reproductive system in Indian catfish *Heteropneustes fossilis* (Bloch.) Annual cycle in development of ovarian eggs. *Acts. Anat.* 75: 114.

16

Histogenesis and Development of Corpus Luteum in Freshwater Turtles

V.K. Gupta

Division of Pharmocology, Regional Research Laboratory, (CSIR), Canal Road, Jammu-180001

ABSTRACT

Corpus luteum in the ovaries of all the three turtles, *Lissemys punctata punctata, Kachuga tectum tectum* and *Kachuga smithi* develop in a manner typical of other reptiles. Immediately after ovulation, the ruptured and puckered follicular cavity is filled with coagulum and a few blood corpuscles. Corpus luteum is formed by gradual hypertrophy and hyperplasia of the granulosa cells of the ovulated follicle which eventually fill out the cavity of the follicle. Luteinization of thecal layers is also seen, of which theca interna cells project into the central mass of the developing corpus luteum to impart it a septate appearance. The absence of capillaries in the glandular portion of corpus luteum during the active secretory phase appears to be peculiar to turtles. Disintegration of corpora lutea is accomplished by the hypertrophy of the thecal layers. The life of corpora lutea in these turtle species is approximately three months and seems to be functionally related to egg retention in the oviducts in turtles.

Key words: Corpus luteum, ovary, *Lissemys punctata punctata, Kachuga tectum tectum, Kachuga smithi.*

Introduction

Corpus luteum in reptiles was first seen in only a turtle (*Trionyx*) long ago by Agassiz (1857), but a detailed histological account of reptilian corpus luteum was undertaken only in 1903 by Lucien, who studied the formation of corpus luteum in the lizard *Angus fragilis* and *Chalcides chalcides.* Lucein (*op. cit*) concluded that a well developed corpus luteum is not necessarily associated with vivipary or maintenance of gestation as has also been maintained by Paingel (1956) in his studies on ovoviviparous lizard *Zootoca*. However, these works do not support the conclusions of Weekes (1934) that corpora lutea occur only in viviparous forms.

While a host of workers (Mingazini 1893; Hett, 1924; Cuningham and Smart, 1934; Weekes, 1934; Boyd, 1940; Samuel, 1944; Miller, 1948; Badier, 1968; Goldberg, 1970; Varma, 1970; Koul and Duda, 1977; Saidapur, 1982; Xavier, 1987) have described the structure and formation of corpus luteum in lizards and snakes, similar studies in turtles are very rare (Agassiz, 1857; Altland, 1951; Cyrus *et al.*, 1978).

This paper is intended to present observation on corpus luteum in three freshwater turtles, *Lissemys punctata punctata* (Trionychid), *Kachuga tectum tectum* and *Kachuga smithi* (Emydid) from a single geographic area, Jammu (J & K State, India), the group of reptiles from which similar studies are relatively poorly known.

Materials and methods

Monthly samples of adult turtles of *Lissemys p. punctata, Kachuga t. tectum and Kachuga smithi* were collected from two different ecological habitats. Specimens of *Lissemys p. punctata* were collected from a lake, Mansar, while *Kachuga t. tectum* and *Kachuga smithi* from a slow running stream, New Khul, from Jammu (J & K, State).

After preliminary weighings and measurements, the turtles were dissected for examination of their ovarian morphology. The ovaries were fixed in smith's fixative and kept away from light for 48 hr. After fixation, the material was processed for sectioning and cut at 5-8μ thickness. Heamatoxylin, Eosin and Mallory's triple stains were employed for staining the serial sections.

Observations

Immediately after ovulation, the follicles collapse and the follicular cells of the follicles undergoing hypertrophy resulting eventually in the formation of corpora lutea, measuring from 8-12 mm long and from 6-10 mm wide at its formation. It is pinkish-white in colour with the point of rupture of the follicle appearing as a small slit. The percental presence of corpora lutea in the ovaries of *Lissemys p. punctata, Kachuga t. tectum* and *Kachuga smithi* is depicted in Table 16.1.

The histomorphological changes through which an ovulated follicle passes before it is transformed into a corpus luteum in these fresh water turtles are arranged into 4 stages as described below:

Stage I : During the initial stage of development of corpus luteum the ruptured follicle from the walls of which it originates, shrinks in its size and presents a more or less collapsed appearance. There is a lumen "the antrum follicle" in the centre of corpus luteum at this stage of its development, bearing a faintly staining coagulum and a few scattered blood corpuscles and cells (Figs. 16.1, 16.8, 16.18). These cells which are 12 μ in maximum diameter have each large spherical nucleus and are identical to the follicular epithelial cells in their other morphological characteristics.

No clear distinction between theca interna and theca externa is observed in this stage in *Lissemys p. punctata*. However, a discreate theca externa and theca interna are distinguishable at this stage in the emydid turtles, *Kachuga t. tectum* and *Kachuga smithi* (Figs. 16.9. 16.19). In *Lissemys p. punctata,* theca interna is seen as a narrow layer, about 40 μ in thickness and theca externa is 390 μ in thickness. In *Kachuga t. tectum* and *Kachuga smithi,* the theca interna is approximately 51 μ thick and theca externa 279 μ. Between the two thecal layers, develops a regular space which interlaced by fibers and fibroblasts. The theca externa is composed primarily of dense connective tissue whereas the theca interna is less dense and thinner. Blood capillaries seen here and there in the theca externa (Figs. 16.1, 16.9) give it a partially loose consistency. Besides, it contains massed collagenous fibres and a few scattered fibroblasts. Theca interna contains only a few fibres but many fibroblasts. Some of the macrophages and leukocytes observed in the thecal layers also are more common in the inter-thecal space. A few of these are present in the central lumen. In *Kachuga t. tectum* the theca interna forms regular villi like projections (Fig. 16.10) into the central cavity. Whereas, in

Table 16.1: Monthly incidence of Corpora lutea in the ovaries of three turtles *Lissemys p. punctata*, *Kachuga t. tectum* and *Kachuga smithi*

Species	Percent Frequency											
	Jan.	Feb.	Mar.	Apr.	May	June	July	Aug.	Sept.	Oct.	Nov.	Dec.
Lissemys p. punctata	–	–	–	–	–	–	14.28	33.33	100	100	66	40
Kachuga t. tectum	100	80	83.33	33.3	14.28	–	–	–	–	100	100	100
Kachuga smithi	–	50	–	–	–	–	–	53.84	100	100	100	90

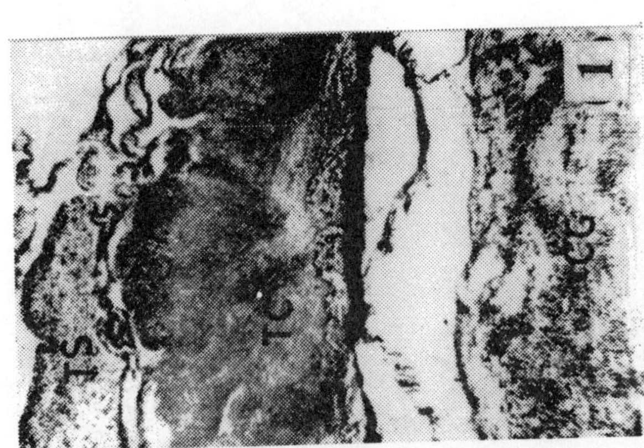

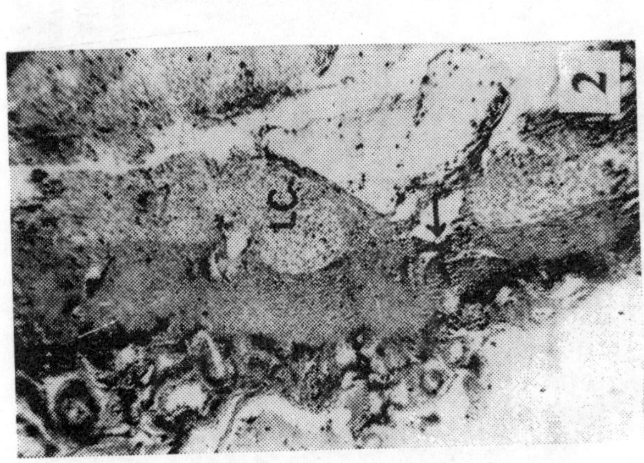

Plate I

Fig. 16.1: Portion of newly ruptured follicle of *Lissemys p. punctata* showing deeply staining Coagulum (CG) in the central cavity and theca externa having outer Theca Spongiosa (TS) and inner Theca Compacta (TC) x 360.

Fig. 16.2: Portion of corpus luteum of *Lissemys p. punctata*. Note obliteration of follicular cavity by few Luteal Cells (LC) and plugging point (Arrow) x 360.

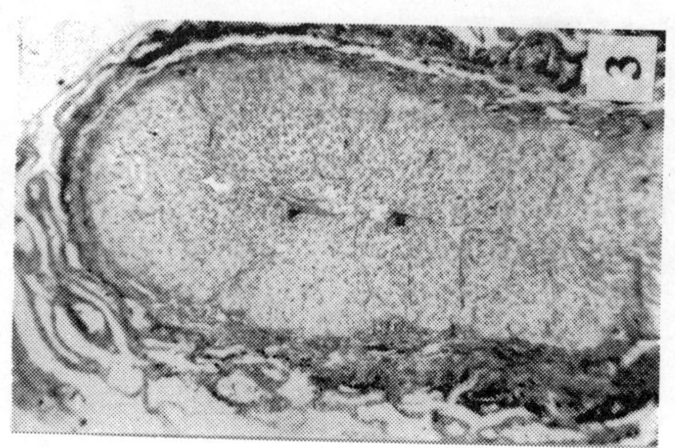

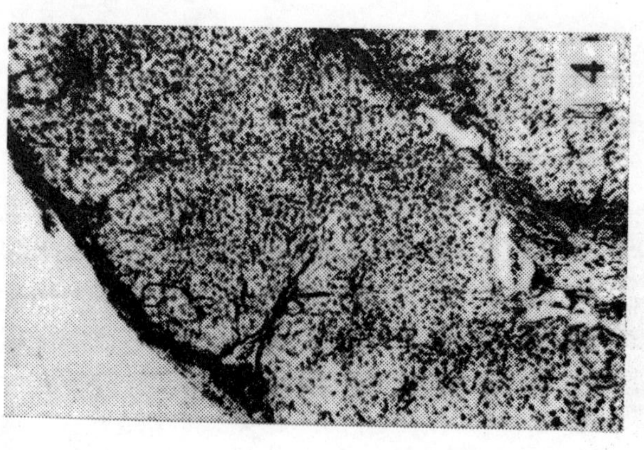

Fig. 16.3: Portion of corpus luteum of *Lissemys p. punctata* showing complete obliteratin of follicular cavity by luteal cells mass.x 80.

Fig. 16.4: Portion of corpus luteum in section of *Lissemys p. punctata* showing penetration of connective tissue fibres (Arrow) into the luteal tissue, giving a partial septate condition.x 360.

Lissemys. p. punctata and *K. smithi* no such projections were seen from theca interna.

Stage II : Externally the ovulated follicles transformed into corpus luteum during this stage are perfectly similar to the ones described in the previous stage, measures about 9 mm in length. It is more or less oval and coloured light yellow. However, histologically marked changes are noticeable. There is an increase in thickness of the follicle cells. No mitosis was observed. With continued hypertrophication, the central cavity of the corpus luteum is gradually filled up. The hypertrophied follicular cells filling up the central cavity of the corpus luteum are the luteal cells (Figs.2, 11, 19). The luteal cells vary in their shape and size, do not show consistent cell boundaries and possess almost spherical or slightly elongated nuclei. The average diameter of lutein cell nuclei when measured along the longest axis, where 8.58μ. Thecal layers of this stage and in later stages some luteal cells also take part in plugging up the point of bursting of the ripe follicle (Fig. 16.4, 16.13). Superficial septa of fibroblasts invade into the central lumen in between the hypertrophied luteal cells.

Stage III : The thecal layers become compact during the stage III of corpus luteum development. The central core of the corpus luteum is now more or less completely filled with luteum cells (Figs. 16.3, 16.12, 16.14, 16.20), which have by now become extremely hypertrophic and the corpus luteum solid. The connective tissue fibres have grown inwards among the luteal cells from all over the inner surface of the theca interna and as a result, the number of 'nests' of luteal cells at the periphery has become very numerous (Figs. 16.4, 16.15, 16.21). In some instances these fibres penetrate almost to the core of the gland. A few scattered erythrocytes and an acidophilic colloid like deposit remain in the central portion of the gland. No vascularization was traceable within granulosa lutein cells. Measurement of lutein cell nuclei along the longest axis showed little change when compared to lutein cell nuclei at the previous stage. However, the change in nuclear shape from round of stage II to an irregular somewhat shrunken shape in this stage (Figs. 16.5, 16.6, 16.16, 16.22) was markedly evident. Small vacuoles were also observed in these luteal cells.

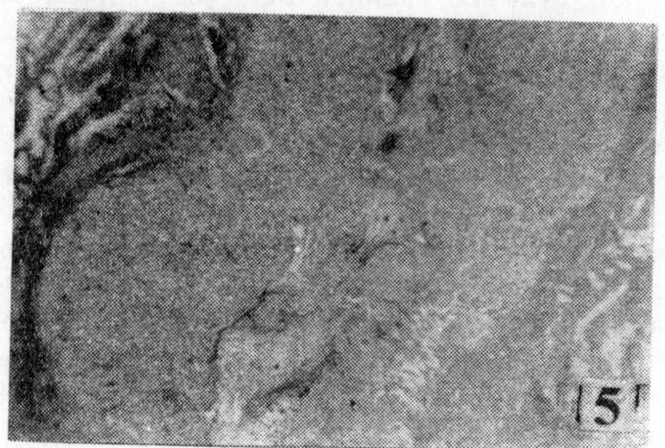

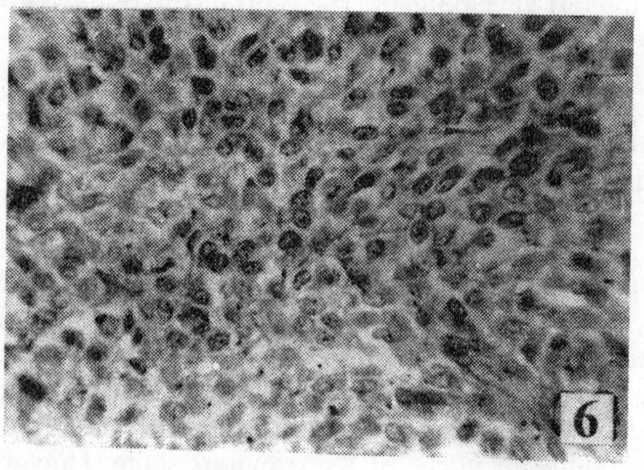

Plate II

Fig.16. 5: Portion of corpus luteum of *Lissemys p. punctata* showing colloid likes deposit in the central portion of the gland.x 120.

Fig. 16.6: Central portion of corpus luteum of *Lissemys p. punctata* showing granulosa lutein cells.x 360.

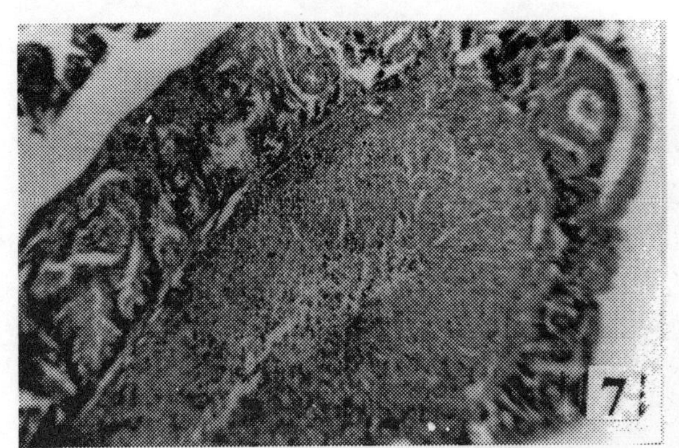

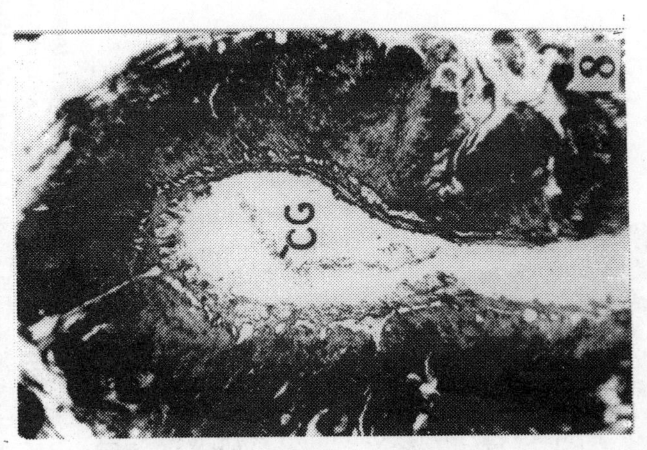

Fig. 16.7: Last stage of corpus luteum of *Lissemys p. punctata* showing luteal cells interspersed in a fibrous network. x 120.

Plate III

Fig. 16.8: Early corpus luteum of *Kachuga t. tectum* showing a central cavity having faintly staining coagulum (CG). x 80.

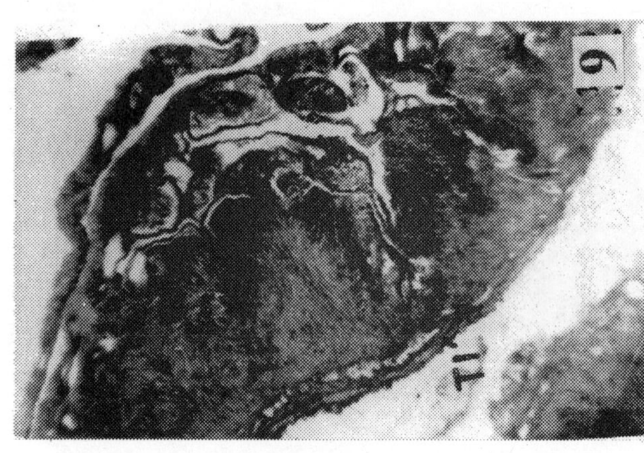

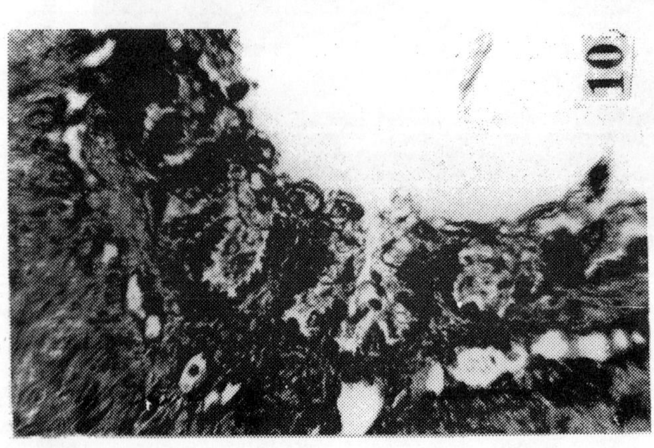

Fig. 16.9 A portion of early corpus luteum of *Kachuga t. tectum* showing Theca Interna (TI) and Theca Externa (TE). Mark the presence of blood capillaries (Arrow) in the theca externa.x 360.

Fig. 16.10: A portion of early corpus luteum of *Kachuga t. tectum* showing the villi like projections of theca interna.x 360.

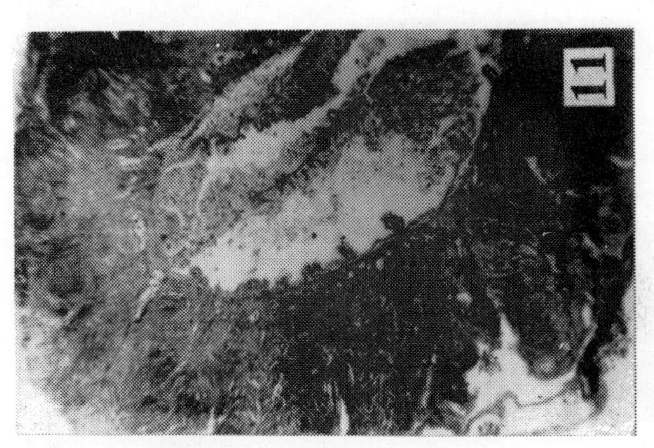

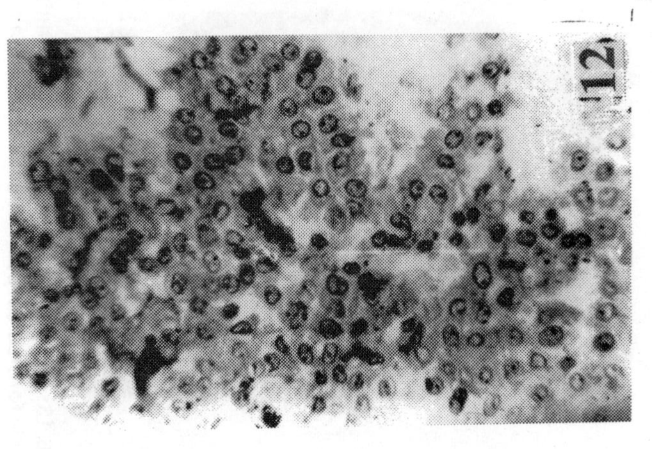

Plate IV

Fig. 16.11: Corpus luteum of *Kachuga t. tectum* showing obliteration of follicular cavity by few luteal cells.x 360.

Fig. 16.12: Central portion of corpus luteum of *Kachuga t. tectum* showing luteal cells.x 360.

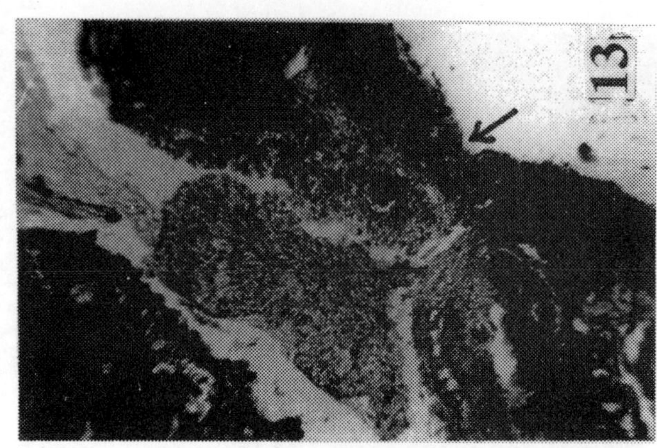

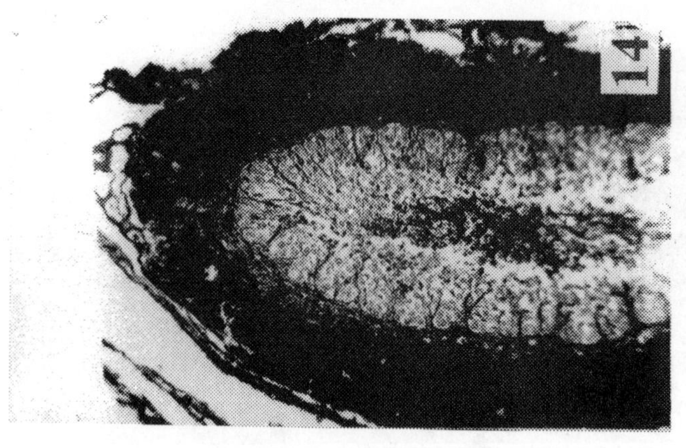

Fig. 16.13: A portion of corpus luteum of *Kachuga t. tectum* showing plugging point (Arrow).x 160.

Fig. 16.14: Fully formed corpus luteum of *Kachuga t. tectum* showing complete obliteration of follicular cavity by luteal cells.x 80.

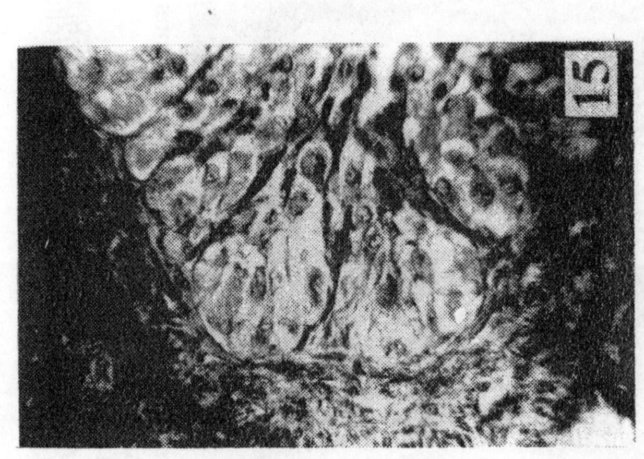

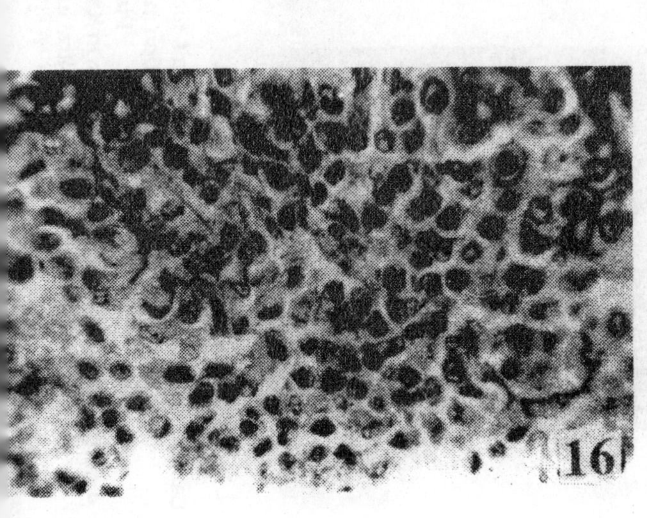

Fig. 16.15: A portion of the fully formed corpus luteum of *Kachuga t. tectum* showing the number of 'nests' of luteal cells towards the periphery.x 360.

Plate V

Fig. 16.16: Luteal cells in the advance stage of corpus luteum development in *Kachuga t. tectum* showing somewhat irregular nuclei.x 360.

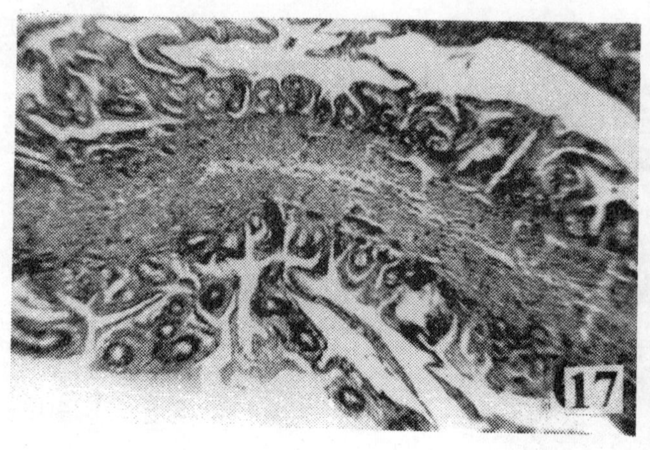

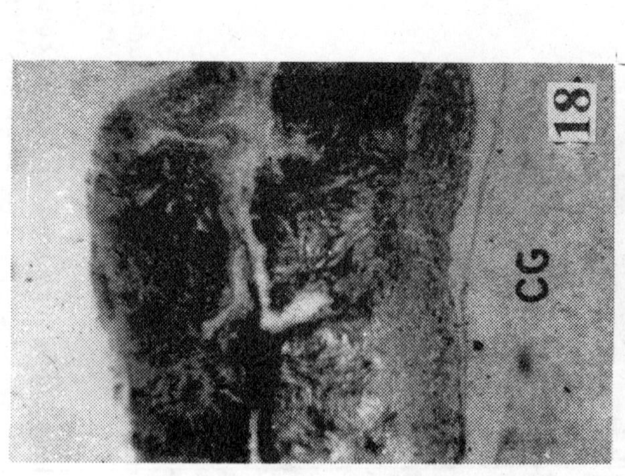

Fig. 16.17: Last stage of corpus luteum of *Kachuga t. tectum* showing no distinction into theca interna and externa and vacuolization in the central cavity.x 80.

Plate VI

Fig. 16.18: A portion of early stage of corpus luteum in *Kachuga smithi* showing lightly stained Coagulum (CG) in the central lumen.x 360.

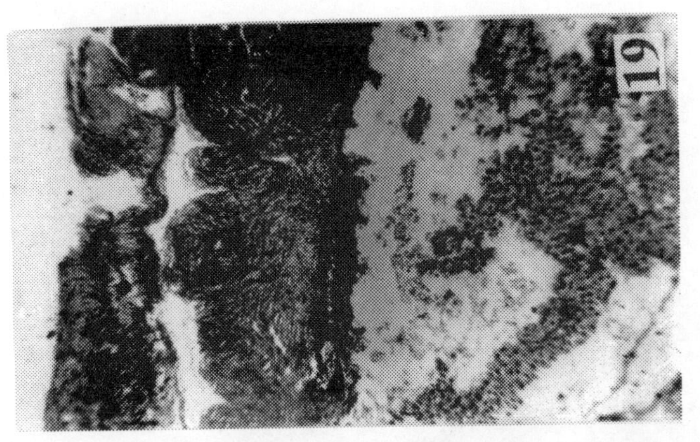

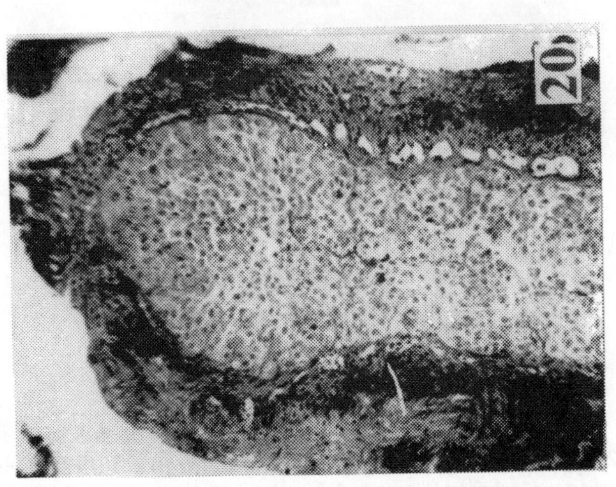

Fig. 16.19: A portion of corpus luteum in *Kachuga smithi* showing few luteal cells in the lummen.x 360.

Fig. 16.20; Fully developed corpus luteum in *Kachuga smithi* showing complete obliteration of central activity with luteal cells.x 80.

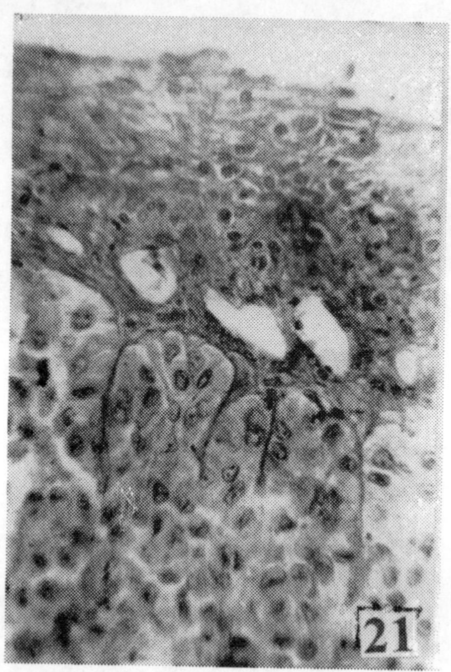

Fig. 16.21: A portion of corpus luteum in *Kachuga smithi* showing penetration of thecal elements from theca interna into the luteal cells.x 360.

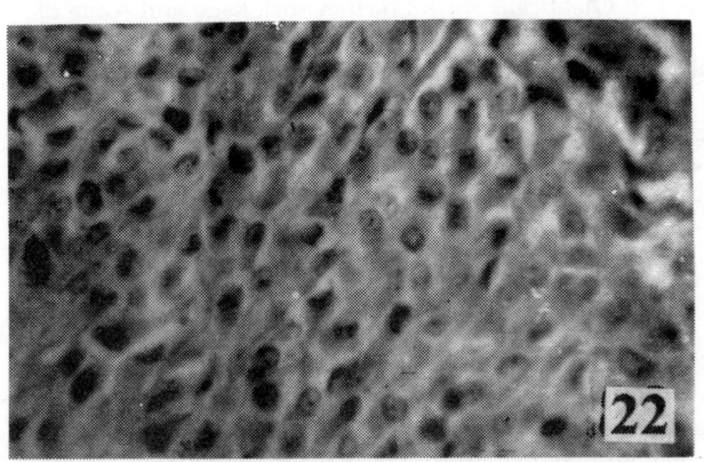

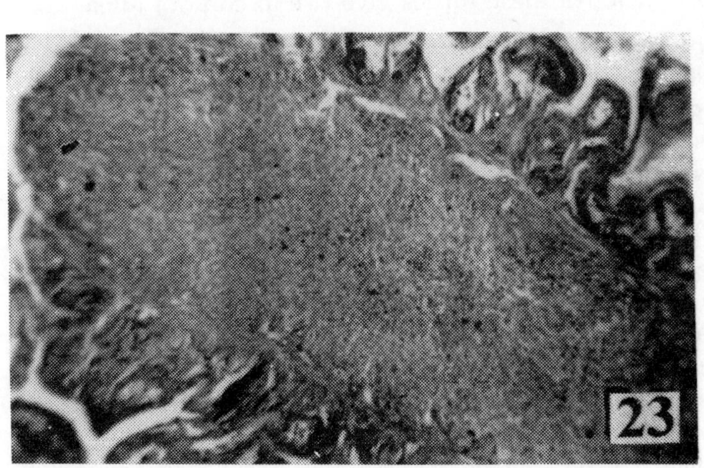

Plate VII

Fig. 16.22: Central portion of corpus luteum of *Kachuga smithi* showing luteal cell mass.x 360.

Fig. 16.23: Degenerating corpus luteum of *Kachuga smithi*.x 80.

Stage IV : Finally, in the last stage of corpus luteum history, distinction between theca interna and theca externa (Figs. 7,17, 23) is lost. The luteal cells shows further shrinkage and some of them assumes a more elongated appearance. Meanwhile the inter-epithelial connective tissue has become more finely distributed. There is an appreciable increase in the size of the vacuoles in the cytoplasm which have now become irregular. The regressive changes of corpora lutea include the resorption of the collagenous elements of the theca and dwindling of the number of luteal cells.

Discussion

The cells remaining in the ovarian follicle after ovulation reorganize with many blood vessels, to produce to corpus luteum, which is essentially a mammalian feature (Young, 1961) have been found in other tetrapods including reptiles (Agassiz, 1857; Luicen, 1903; and Asdell, 1928). Subsequent to ovulation in the months of August, September and October in *Lissemys p. punctata*, October, November, December, January and February in *Kachuga t. tectum* and September, October and November in *K. smithi*, the collapsed follicles of these turtles give rise to corpora lutea.

The present study on *Lissemys p. punctata*, *Kachuga t. tectum* and *K. smithi* shows that corpora lutea arise as a result of hypertrophy of the follicle granulose cells, which fill the interior of the discharged follicles in a way similar to the one described in other reptiles (Lucien, 1903; Loyez, 1906; Hett, 1924; Dutta, 1944; Miller, 1948; Bragdon, 1952; Annamalai, 1966; Varma and Guraya, 1973; Goldberg and Bezy, 1974; Koul and Duda, 1977). Although, frequent mitotic divisions have been recorded in the follicular epithelial cells in the early stages of the formation of the corpus luteum in the lizard *Amphibolurus muricatus* (Weekes, 1934), the present study reveals mitotic divisions, if at all, are not very extensive. The major portion of the increase in the luteal tissue must, therefore, be brought about by simple hypertrophy of the cells.

The clear distinction of connective tissue cover in the newly formed corpus luteum in *Kachuga t. tectum*, and *K. smithi* into two layers, is obviously a thecal legacy of the discharged follicle. These two turtles, thus resemble *Terrapene carolina carolina* in which a similar differentiation of thecal layers has been reported in the initial stages of corpus luteum development (Altland, 1951). However, in *Lissemys p. punctata* no such clear cut differentiation of outer cover is observed in the initial stages of development of the corpus luteum, which obviously cannot throw much light on the extent of the role played by theca of the follicle during luteinization in the trionychid turtle.

Whereas in *Hoplodactylus maculatus* (Boyd, 1940), *Xantusia vigilis* (Miller, 1948). *Mabuya carinata* (Annamalia, 1966) and *Agama tuberculate* (koul and Duda, 1977) the follicular rupture point (or the Cicatrix) is plugged by follicle cells after ovulation, theca playing no part in its closure, a condition regarded as usual for reptiles (Harrison, 1948). In all three turtles, *Lissemys p. punctata, Kachuga t. tectum,* and *K. smithi* of the present investigation, this opening is plugged with luteal mass and thecal tissues precisely in the manner described by Altland (1951) in *Terrapene c. caroline* which, therefore could be a chelonian pattern of plugging of the cicatrix.

Similar to those described in lizards (*Calotes versicolor,* Varma, 1970; *Agama tuberculata,* Koul and Duda, 1977) villi like projections were seen to develop in the initial stage of development in corpora lutea in *Kachuga t. tectum.* But, no such projections were observed in other two species of turtles, i.e., *Lissemys p. punctata* and *Kuchuga smithi* during the present investigation, nor is there are any recorded evidence of such villi like structure in any other turtle worked out so far. Presence and absence of villose corpora lutea, thus emerges as an important interspecific difference between the two emydid turtles of the present work.

Lissemys p. punctata, Kachuga t. tectum, and *K. smithi* resembles *Terrapene carolina* (Altland, 1951) and *Chelydra serpentina* (Rahn, 1938) in the absence of capillaries in the glandular portion during the active secretary phase which is not peculiar to turtles alone. Absence of vascularity within the luteal mass is, also on record in certain viviparous lizards (Hett, 1924; Weekes, 1934; Boyd, 1940) and in an ovoviviparous lizard *Lygosoma himalayanum* (Koul and Duda, 1977). Absence of vascularity in the central epithelial portion of the gland in turtle corpus lutein as a distinguishing feature of turtles as suggested by Altland (1951), thus appears does not get established. All the same this feature strongly differentiates reptiles from mammals wherein an intimate relationship between the blood vessels and the lutein cells is reported (Cowdry, 1932). Apart from difference in the pattern of drainage the direction of diffusion of glandular secretions from luteum cells in corpora lutea in the two groups also stands to differ. Whereas in mammals the secretion diffuses centripetally, in reptiles the diffusion is apparently centrifugal.

Although, not necessarily an inseparable adjuct to the histology of the corpus luteum (Dutta, 1944) the fibroblasts and the connective tissue which invade the glandular portion of the corpus luteum in reptiles has been regarded to originate mostly from theca interna (*Lacerta agilis: Hett*, 1924; *Enhydrina schistosa*: Samuel, 1944; *Terrapene carolina;* Altland, 1951; *Amphibolurus muricatus, Leiolopisma wekesee, L. entracastenanxii*: Fox, 1977) and only rarely from theca externa (*Hoplodactylus maculatus*: Boyd, 1940). The findings from the present investigation lend support to dominant view and Boyd's (*op. cit.*) observation appears, thus, an exception.

Conflicting opinions have been expressed on the para-lutein cells in the corpus luteum of reptiles. Most workers (Brambell, 1956; Betz, 1963; Varma, 1970; Guraya, 1973; Koul and Duda, 1977) have reported complete absence of these cells in some lizards and snakes. The present findings on *Kachuga t. tectum* and *K. smithi*, also indicate that the thecal layers in these turtles remains distinct (partially so in *Lissemys p. punctata*) and are not luteinized to give rise to theca lutein cells of Altland (1951), thereby not substantiating the findings of Mingazimi (1893), Weekes (1934), and Bradgon (1952) who have reported the occurrence of such cells in other reptiles. Obviously the presence or absence of para lutein cells is non-specific and not unexpected.

In only the fully formed corpora lutea in turtles of the present study, the lutein cells were found to contain many small vacuoles, similar to those described by Hett (1924) in the luteal cells of *Lacerta agilis*. Although she does not express any opinion regarding the nature of the secretory products or the cytoplasmic vacuoles in the lutein cells in her lizard corpus luteum, it would well be summarized that the vacuoles in the lutein cells of turtles of the present investigation are the blanks left out through dissolution of fatty globular material under the influence of the acid content of the fixatives used (smith's and Bouin's) in the present study and therefore, not stained by Mallory's stain. Such a view would be consistent with Boyd (1940), Altland (1951), Varma and Guraya (1973), who all have described the presence of lipoid substance in lutein tissue of *Hopladactylus, Terrapene caroline,* and *Lycodon aulicus* investigated by them respectively. Weekes (1934) mentions in the cytoplasm of luteal cells in older corpora lutea in two ovoviviparous lizards, *Amphibolurus muricatus,* and *Lygosoma quoyi,* large vacuoles are seen which become very pronounced at the time of the birth of their young and dismisses these as mere signs of regression.

It may is thus be reasonably, concluded that the development of the corpus luteum of turtles is similar to that of lizards and snakes without any serious departures from the generalized pattern as outlined above. The three turtles, *Lissemys p. punctata, Kachuga t. tectum,* and *K. smithi* resembles majority of reptiles in (i) the theca being differentiated into an outer theca externa and an inner theca interna (*Lissemys p. punctata,* excepted, where the demarcation between the two layers is not so as sharp as in the emydid turtles); (ii) hypertrophication of granulosa cells during luteal cells mass formation; and (iii) partially septate appearance of the luteal tissue, a condition that has so far been most frequently reported in lizards and snakes (Brambell, 1956; Franchi, 1962; Betz, 1963; Varma, 1970; Koul and Duda, 1977).

The presence of functional corpora lutea is commonly associated with viviparity and placental maintenance in mammals (Young, 1961). Yet, based on morphological and histological evidence, postovulatory follicles similar in structure to mammalian corpora lutea, thus appear to be ubiquitous in reptiles where viviparity is the exception rather than the rule (Altland, 1951; Brambell, 1956; Koul and Duda, 1977). Therefore, the immediate functional importance of corpus luteum in reptiles appears to be related to egg retention within the oviduct in reptiles, a function, which also is still not clearly understood. Miller (1959) suggested a definite correlation between the corpus luteum longevity and egg laying or retention in reptiles, a relationship which has subsequently come to be recognized by other workers also (Chieffi and Botte, 1970; Varma and Guraya, 1973). The findings on *Lissemys p. punctata, Kachuga t. tectum,* and *K. smithi* indicate an approximate three months life of corpus luteum in these forms, which more or less corresponds with that in *Terrapene carolina* (Altland, 1951) and *Chrysemys picta* (Ernst, 1971). In comparison to the oviparous lizards, the life of corpora lutea in turtles is much longer and could thus be functionally related to some part of reproductive cycle. In view of the time that elapses between the onset of ovulation, the termination of oviposition, and the entire spectrum of developmental changes of the corpus luteum (as evidenced by their first and last appearance in turtle ovaries), the corpus luteum in these turtles may well be associated with egg retention in oviducts. Surprisingly, Moll and Legler (1971) have pointed out that corpora lutea regress and disappear within a few weeks after ovulation which according to them is a feature typical of turtles, what is so typical about it is not immediately clear. Regression and eventual disappearance of corpus luteum in vertebrates is a known phenomenon and the duration of its life is subject to variation.

Acknowledgements

The author take this opportunity of thanking Dr. P.L. Duda, former Prof. of P.G. Deptt. of Zoology, University of Jammu, for his advice and guidance during this investigation and to CSIR, New Delhi for financial assistance.

REFERENCES

Agassiz, L. (1857). *Contribution to the Natural History of the United States of America.* Little Brown and Co.; Boston, Vol;. I, II: 1-452.

Atland, P.D. (1951). Observations on the structure of the reproductive organs of the box turtle. *J. Morphol.,* 89: 599-621.

Annamalai, P.M. (1966). The morphology and histology of the ovary, the corpus luteum and the testis of *Mabuya carinata* Boulenger *J. Zool. Soc. India,* 18: 19-28.

Asdeil, S.A. (1928). The growth and function of the corpus luteum. *Physiol. Rev.,* 8: 313-345.

Badir, N. (1968). Structure and function of corpus luteum during gestation in the viviparous lizard *Chalcides ocellatus. Anat. Anz.,* 122: 1-10.

Betz, T.W. (1963). The ovarian histolgy of the diamond-backed water snake, *Natrix rhombifera* during the reproductive cycle. *Copeia,* pp. 692-697.

Boyd, M.M. (1940). The structure of the ovary and the formation of the corpus luteum in *Hoplodactylus maculatus*, Gray, *Quart. J. Micros. Sci.*, 82: 337-376.

Bragdon, D.N. (1952). Corpus luteum formation and follicular atresia in the common garter snake, *Thammophis sirtalis. J. Morphol.*, 91: 413-443.

Brambell, F.W.R. (1956). Ovarian changes In: *Marshall's Physiology of Reproduction*, A.S. Parker, (ed.), Longmans, Green and Company, London, Third ed. Vol. I, pp. 397-542.

Chieffi, G. and V. Botte (1970). The problem of luteogensis in non-mammalian vertebrates. *Boli. Zool.*, 37: 85-102.

Cowdry, E.V. (1932). *Special Cytology III*, Hoeber, New York.

Cunningham, J.T., and W.A.M. Smart. (1934). The structure and origin of corpora lutea in some of the lower vertebrates. *Proc. roy. Soc.*, 116: 258-281.

Cyrus, R.V., I.Y., Mahmoud and J. Klicka. (1978). Fine structure of the corpus luteum of the snapping turtle, *Chelydra serpentina. Copeia* pp. 622-827.

Dutta, S.K. (1944). Studies on the sexual cycle in the lizard *Hemidactylus flaviviridis* (Ruppel). Allahabad Univ. Stud. *Zoology Sect*, pp. 59-153.

Ernst, C.H. (1971). Sexual cycle and maturity of the turtle, *Chrysemys picta. Biol. Ball.*, 140: 191-200.

Fox, H. (1971). Urinogenital System. In. *Biology of the Reptilia.*, Carl Gans and S. Parsons (eds.) Academic Press, London, and New York, Vol. 6: 505

Franchi, L.L. (1962). The structure of the ovary. Vertebrates. In: *The ovary, S. Zuckerman*, (ed.), Academic Press, London, Vol. I, pp. 121-142.

Goldberg, S.R. (1970). Seasonal ovarian histological studies of the ovoviviparous Iguanid lizard *Sceloporus Jarrovi* cope. *J. Morphol.*, 132: 265-276.

Goldberg, S.R. and R.L. Bezy. (1974). Reproduction in the Island night lizard, *Xantusia riversiana, Herpetologica*, 30: 350-360.

Harison R.J. (1948). The development and fate of the corpus luteum in the vertebrate series. *Biol. Rev.*, 23: 296-331.

Hett, J. (1924). Das corpus luteum der Zauneidechse (*Lacerta agilis*). *Z. Mikr. Anat. Forscb.*, I: 41-84.

Koul, O., and P.L., Duda (1977). Studies on the corpus luteum of some high altitude lizards from Kashmir. *Bull. Chi. Herp. Soc.*, 12(4): 110-119.

Loyez, M. (1906). Researches sur le development ovarian des oeufs ineroblastigues d vitellus mutrituf abundant. *Arch. Anat. Micro.*, 8: 69-397.

Lucien, M. (1903). Note preliminaire sur. les premieres phase de la formation des corpe jaunes chez certain reptiles. *C.R. Soc., Biol., Paris,* 55: 1116-1117.

Miller, M.R. (1948). The seasonal histological changes occurring in the ovary, corpus luteum and testis of the viviparous lizard, *Xantusia vigilis. Univ. Calif. Publ. Zool.,* 47: 197-224.

Miller, M.R. (1959). The endocrine basis for reproductive adaptations in reptiles, In: *Comparative Endocrinology,* Gorbman. (ed.), J. Willey & Sons, New York, pp. 499-516.

Mingazini, P. (1893). Corpi lutei versie falsi dei rettii Ric. *Lab. Anat. Norm. Univ. Romom.* 3: 105.

Moll, E.O., and J.M. Leger. (1971). The life history of a neoropical slider turtle, *Pseudomys scripta* (Schroepff) in Panama. *Bull. Los. Ang. Country Mus. Nat. Hist. Sci.,* 11: 1-102.

Paingel, M. (1956). Contribution a' L'etude de L'Ovoviparite chez les reptiles; Gestation et parturition chez le lizard vivipora *Zootoca vivipara. Ann. Sci. Nat.,* 18: 569-668.

Rahn, H. (1938). The corpus luteum of reptiles. *Anat. Rec.,* 72: 55 suppl.

Saidapur, S.K. (1982). Structure and function of post-ovulatory follicles (corpora lutae) in the ovaries of non-mammalian vertebrates. *Inv. Rev. Cytol.,* 75: 243-285.

Samuel, M. (1944). Studies on the corpus luteum in *Enhydrina scistosa* (Daudin) and *Hydrophis cyanocinctus* (Daudin) of the Madras coast., Proc. *Indian Acad. Sci.,* 20: 143-174.

Varma, S.K. (1970). Morphology of ovarian changes in the garden lizard, *colotes versicolor. J. Morphol.,* 131: 195-210.

Varma, S.K. and S.S. Guraya (1973). Histochemical observation on the granulosa (follicular) cells in the preovulatory and postovulatory follicles of the garden lizard ovary. *Acad Ant.,* 85: 563-579.

Weekes, H.C. (1934). The corpus luteum in certain oviparous and viviparous reptiles. *Proc. Linn. Soc. N.S.W.,* 59: 380-391.

Xavier, F. (1987). Functional morphology and regulation of the corpus luteum. In: *Hormones and Reproduction in fishes, Amphibians, and Reptiles,* O. Norries and E. Jones (eds.), Plenum Press, New York: 241-282.

Young, W.C. (1961). The mammalian ovary. In: *Sex and Internal Secretions,* W.C. Young (ed.), Balliere, Tindell and Cox, London, pp. 449-496.

17

Seasonal Variations in Follicle Number in Common House Gecko *Hemidactylus flaviviridis* Rupell (Reptilia: Geckonidae) in Jammu

Bhawana Abrol and D.N. Sahi

P.G. Department of Zoology, University of Jammu

ABSTRACT

Seasonal fluctuations in folliciale number have been studied in a population of *Hemidactylus flaviviridis* inhabiting Jammu, J & K State. Follicle number was found to be maximum in February where as minimum in June. Such a seasonal increase or decrease in follicle number in the female lizard in relation to atresia under subtropical conditions of Jammu has been discussed.

Key Words: Lizard, *H. flaviviridis,* follicle number, size, weight, atresia.

Introduction

Little precise information exists in the literature concerning the biological significance of increase and decrease in follicle number. The available literature explains the different aspects of ovarian cycle including periodicity, preovulatory and ovulatory phase in the previtellogenic and vitellogenic follicles. However, follicle number as well as diameter with respect to ovulation and ovarian cycle has not been taken in detail save for Miller (1948) in *Xantusia vigilis,* Duda (1964) in *Agama tuberculata,* Mayhew (1966) in *Uma scorporia,* Varma (1970) in *Calotes versicolor,* Rubail *et al.* (1972) in *Anolis acutus,* Koul and Duda (1977) in *Agama tuberculata,* Jones *et al. (1976) in Anolis carolinensis,* Ballinger *et al.* (1980) in *Sceloporus undulatus garmani,* Ortega and Barbault (1984) in *Sceloporous graminicus,* Ballinger and Hipp (1985) in *Crotaphytus collaris,* and Flores *et al.* (1990) in *Sceloporous muronatus.* This host of instances indicate that in lizards, the adult ovary has a stock of oocytes which populate the ovary year after year. However, Boyd (1940) in *Haplodactylus maculatus* found that yolk deposition occurs immediately after ovulation and the mature ova remained in the ovary throughout the winter and do not ovulate until the next spring, but on the other hand, Kitada and Asana (1951) have maintained that the yolk deposition in *Hemidacytylus flaviviridis* occurs gradually throughout the year. Present investigations were undertaken to describe seasonal variations in the follicle number (Previtellogenic and Vitellogenic) in females of a population of the *Hemidactylus flaviviridis* inhabiting Jammu and so as to compare this data with the results published on other populations from elsewhere.

Material and Methods

Monthly collection of 5-6 adult females *Hemidactylus flaviviridis* [>45 mm SVL] from within 2 km radius around department of Zoology in the Jammu University campus (32.67 N; 74.79E) at an elevation of 300 m above sea level were carried on during the year (1993-1996). The lizards were captured by hand net, taken to laboratory and processed with in 12 hr. The captured lizards were weighed (g) and snout-vent length (SVL in mm) recorded prior to autopsy. Lizards were dissected under the (Ether/chloroform) anaesthesia, ovaries excised washed in clear water; and soaked with filter paper. Assessment of different categories of follicles was done on the basis of diameter (in nearest mm).

Observations and Discussion

In the ovaries of *Hemidactylus flaviviridis,* 2 types of follicles i.e., the stromal follicles, (Figs 17.1 and 17.2) lying deep in the ovarian stroma and visible only in the sections and the extra stromal follicles occupying a superficial position were observed (Figs 17.3 and 17.4). These follicles are macroscopic about 0.05 mm across, and easily countable. The follicles are the

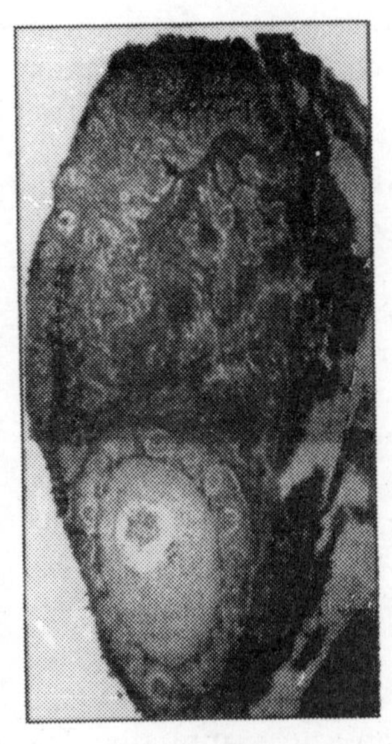

Fig. 17.1

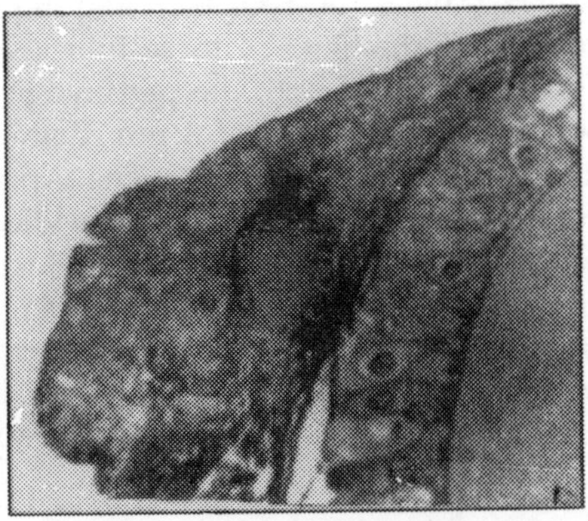

Fig. 17.2

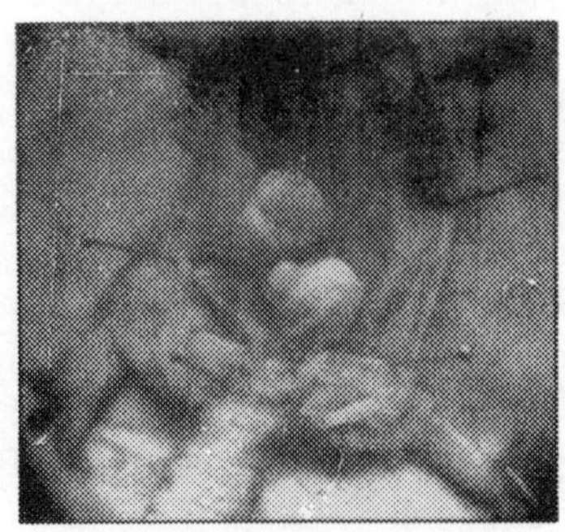

Fig. 17.3

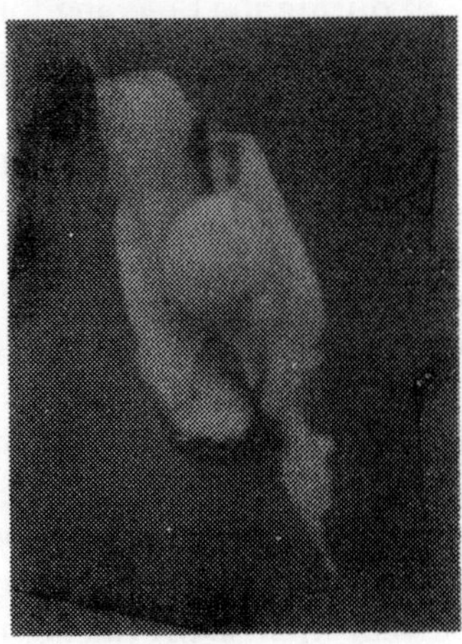

Fig. 17.4

previtellogenic and vitellogenic, varying in their number, size and weight during the different stages of ovarian development.

Maximum of as many as two such follicles were found in an ovary at a time but presence of one such follicle is a common condition in this gekkonine. During the process of yolking, the follicles turned opaque white to straw yellow and finally to yellow. The vitellogenic follicles start making their appearance in the last week of February and represented only 17.6%. The maximum diameter of these follicles in this month is 6 mm. In the month of March, the percentage increases to 58.82% along with the appearance of 8 mm diameter follicles i.e., ovulatory size (Fig. 17.7, Table 17.3) The April and May months again represented 58.06% and 52.8% while the percentage of follicles with maximum diameter i.e., 8mm is higher in the April (22.2%) but it declines in May (12.76%). There is complete disappearance of follicles in the month of June, November, December and January. While July, August, September and October months represent vitellogenic follicles respectively. (Table 17.2 Fig. 17.7).

The present observations reveal that the lizard *Hemidactylus flaviviridis* experiences a single ovarian cycle annually. The increase and dcrease in the number of follicles is however related to the atresia. There is a sudden spurt in follicular growth (Fig. 17.5) during the breeding follicles present per ovary minimum in June (Quiescent phase) and maximum in February (preovulatory phase).

The percentage of female with yolked follicles and oviducal eggs in March, April and May samples indicates that breeding occurs seasonally. This is further supported by sharp decrease in the percentage of the sample with oviducal eggs after oviposition May onwards. Since the females with yolked follicles were present in the samples of only a few months it appears that the reproduction in this gekkonine too is seasonal and breeding confined to only a short restricted period of the year. This is further supported by the presence of corpora-lutea for a brief period during restricted breeding season. During most part of the year there is prevalence of previtellogenic follicles (Figs. 17.6, 17.7) with a very few exceptions of delayed vitellogenesis but they undergo degeneration or resorption or become atretic due to unfavourable conditions for its further development.

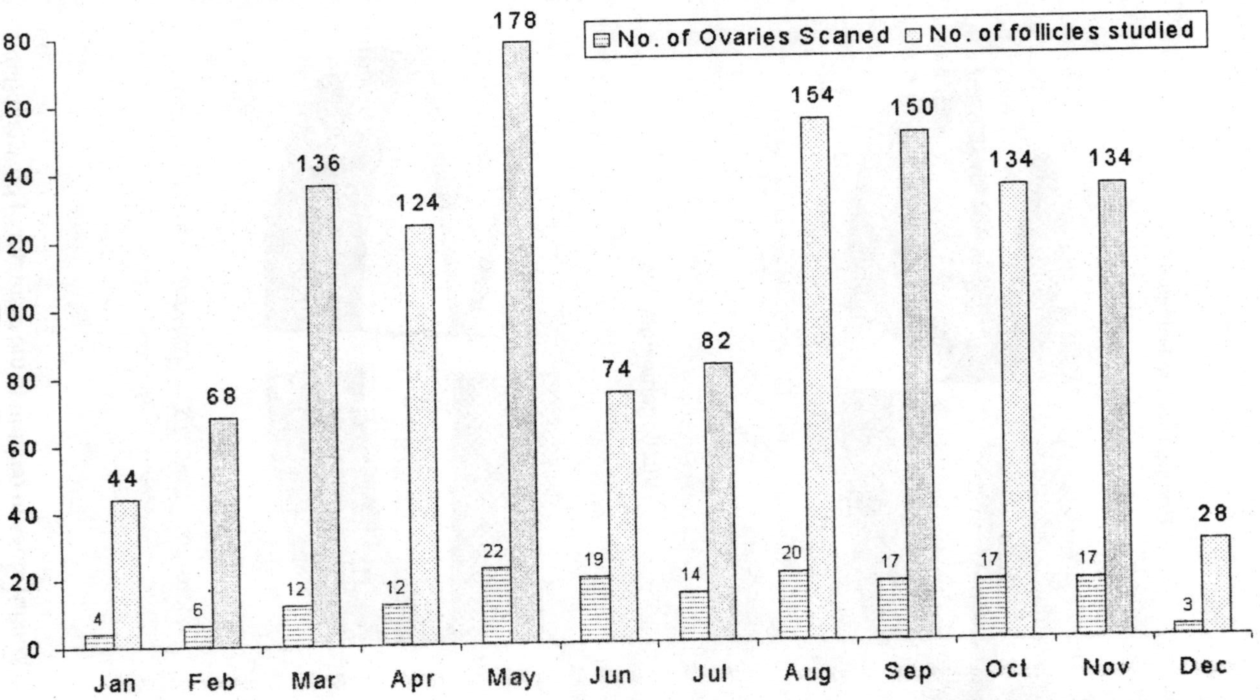

Fig 17.5: Average number of follicle per ovary per month during the period of investigation (Yr.1 993-1996).

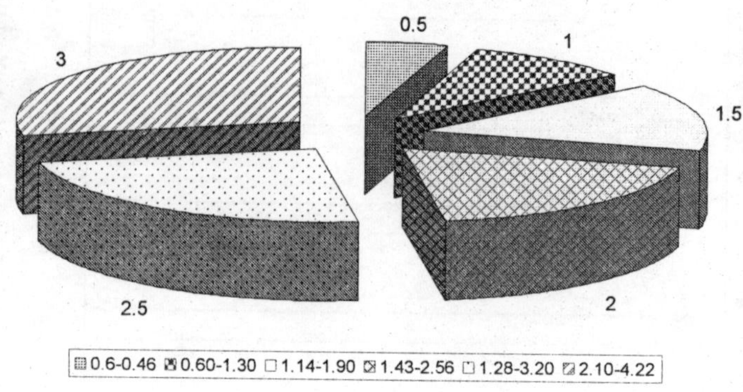

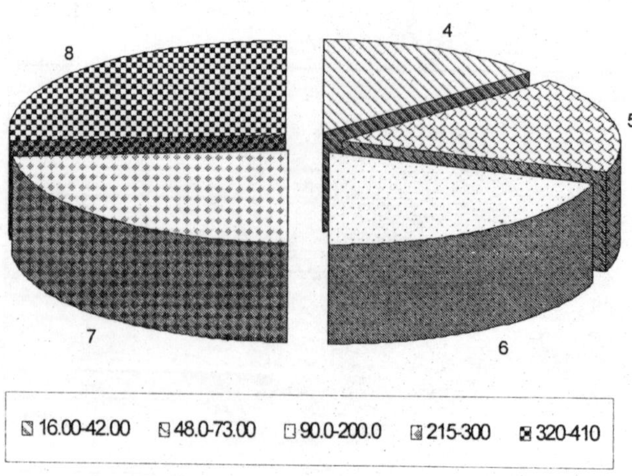

Fig 17.6: Weight index in relation to the diameter of previtellogenic follicles.

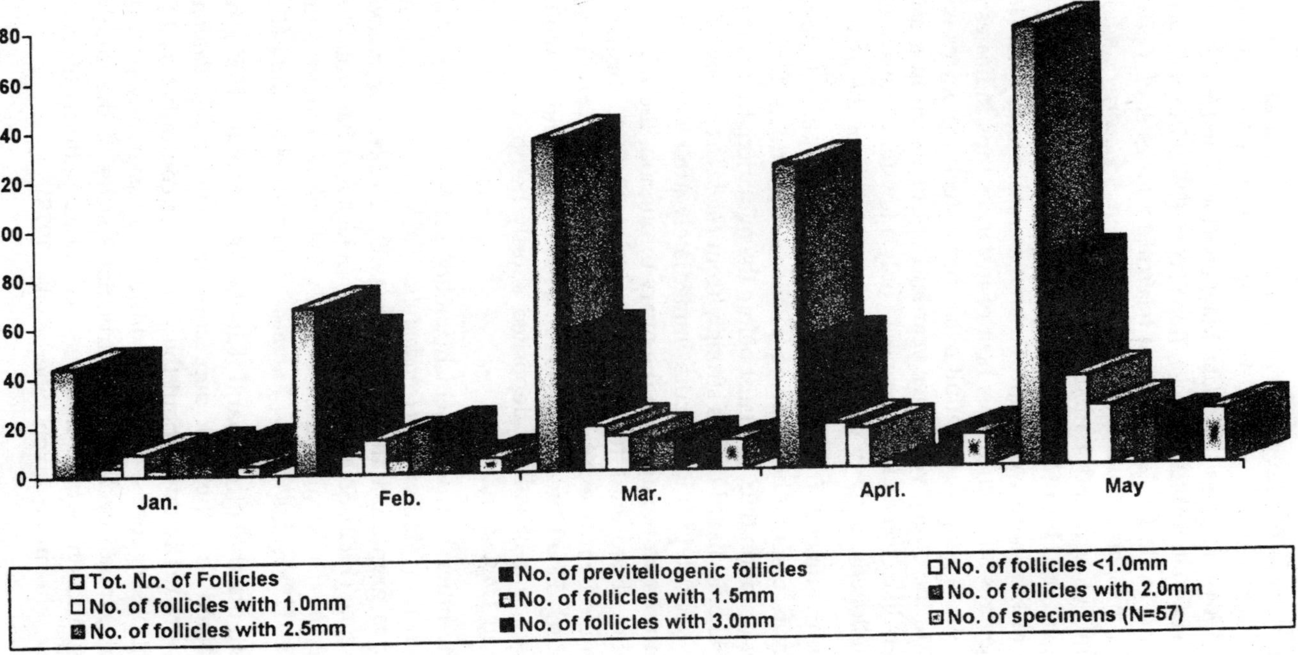

Fig 17.7: Pooled data of mean number of ppreviteIlogenic follicle with different size groups. (from Jan. to May).

Prevalence of previtellogenic follicles during most of the time of the year is also supported by Miller, (1948) in *Xantusia vigilis*, Duda (1964) in *Agama tubeculata*, Mayhew (1966a) in *Uma Scorporia*, Verma (1970) in *Calotes versicolor*, Rubail *et al*. (1972) in *Anolis acutua*, Koul and Duda (1972) in *Agama tuberculata;* Jones *et al*. (1976) in *Anolis carolinensis*, Ballinger *et al*. (1980) in *Scelaporus undulatus* garmani, Ortega and Barbault (1984) in *Sceloporus grammincus*, Ballinger and Hipp (1985) in *Crotaphytus collaris*, Kwait and Gist (1987) in European well lizard *Podarcis muralis*, Flores *et al*. (1990) in *Sceloprous muronatus*. This host of instances thus indicates that in lizards, in general, the adult ovary has a stock of oocytes which populate the ovary year after year and out of which in seasonal breeders like *Hemidactylus flaviviridis* only a few may embark upon the vitellogenic activity during the specific time of the year.

As evidenced from the present investigation vitellogenic activity is essentially a spring phenomenon and is for a very shorter duration (Fig. 17.6) just before the ovulation. In this respect this gekkonid lizard differs strongly from *Haplodactylus maculatus*, where yolk deposition occurs immediately after ovulation and the mature ova remained in the ovary throughout the winter and are not ovulated until the next spring (Boyd, 1940). In this respect, gekkonine lizards exhibit two different types of ovarian cycles, as distinguished by Miller (1948). These being one in which a large amount of yolk is deposited shortly before ovulation (as found in *Haplodactylus*).

The other lizards in which vitellogenesis is a preovulatory phenomenon similar to *Hemidactylus flaviviridis* are *Sceloporus undulatus* gramini, Ballinger *et al*. (1980), *Holobrookia muculata*, Dorge *et al*. (1982), *Sceloporus gramicus*, Ortega and Barbault (1984), *Podaric muralis*, Kwait and Gist, (1987), *Eublephoris Kuroiwae* Tanka and Gist (1987), *Eublepharis Kuroiwae*, Tanka and Nishihira (1989).

But on the other hand Kitada and Asana (1951) have maintained that the yolk deposition in *Hemidactylus flaviviridis* occurs gradually throughout the year hence contradictory to the present observations. Thus it is difficult to assign any particular type of ovarian cycle to the particular species of the family of lizards, and one would find instances of all patterns of ovarian activity with in an taxon (Sherbrooke, 1975).

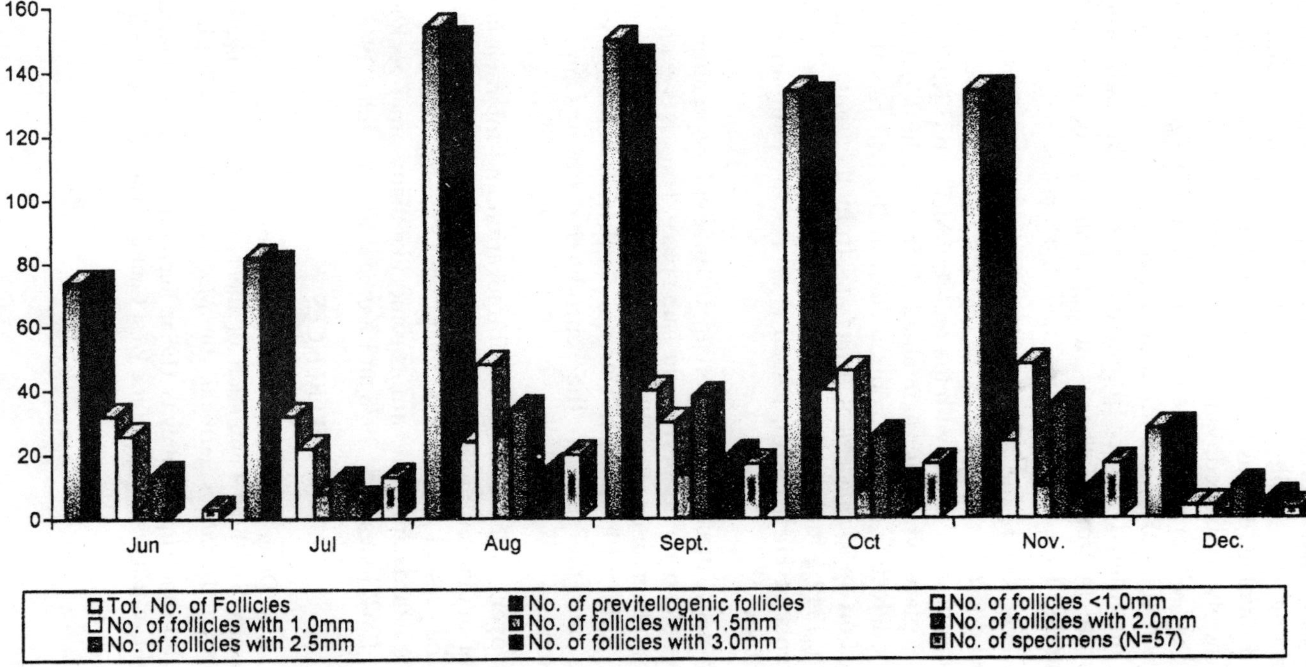

Fig. 17.8 Pooled data of mean number of previtellogenic follicle with different size groups. (from Jun. to Dec.).

This obviously indicates that inter specific differences not withstanding the vitellogenic activity and for the hormonal concert if involved in this activity of the ovaries of these lizards is extrinsically controlled; whether photoperiod, or temperature or both, as obtained at the time of their emergence from hibernation (Fig. 17.8) are implicated directly. Vitellogenesis is largely modulated by environmental factors like light and temperature in tropical lizards, if not at all has been earlier demonstrated by Licht (1973) in *Anolis Carolinensis*.

However Kitada and Asana (1951) who have maintained that the lizard doesn't exhibit any cyclic reproduction in Bombay and therefore comes in conflict with the report of McCann (1949) from the same place who has maintained definite cyclical pattern. Though interesting yet this contradiction in the two reports of reproduction in *Hemidactylus flaviviridis* from the same place (McCann, 1949; Kitada and Asana (1951) may only supports the view points of Fitch (1970), Rubail *et al.* (1972) Gorbman and Licht (1974).

Obviously, these reports pertain to different population of the house gecko, no doubt from the same place. Thus, it is evident from these observations that different variable control reproductive cycles in different localities in extreme plasticity in reproductive rhythms and therefore in the reproductive timings among populations.

Moreover, on the basis of the intraovarion and interovarian pattern of ovulation (Smith *et al.*, 1973); the data on the ovarian cycle of the present gekkonine in the present investigations shows that these lizards are mono autochronic, like many other geckos (Church, 1962; Jones *et al.*, 1978 and *Xantusia Vigilis*, Miller, 1948).

REFERENCES

Ballinger, R.E. Dorge, D.L. and Jones M. (1980). Reproduction in a Nebraska Sandhills Population of the Northern prairie lizard *Sceloporous Undulatus* garmani. *Am. Med. Nat* 106 (1): 157-164.

Ballinger, R.E. and Hipp Tommy G. (1985). Reproduction in the collared lizard *Crotaphytus collaris*, in the West Central Taxas. *Copeia*, 9: 976-980.

Boyd, M.M. (1940). The structure of ovary and formation of corpusluteum in *Haplodactylus maculatus* Gray. *Q.J. Microsc. Sci.* 82: 337-376.

Church, G. (1962). The reproductive cycle of the Javanese house gecko *Cosymbotus platurus, Peropus mutilatus. Copiea,* (2): 262-269.

Dorge, D.L., S.M. Jones and Ballinger, R.E. (1982). Reproduction of *Holbrookia maculata* in Western Nerbraska. *Copeia* pp. 356-362.

Duda, P.L. (1964). Functional morphology of the rinogenital organs of *Agama tuberculata,* Kashmir. *Scicnce.* 1/2: 53-63.

Fitch, H.S. (1970). Reproductive cycles of snakes and lizards. *Univ. Kansas Mus. Nat. Hist. Misc. Publ.,* 52: 1-247.

Flores, E.E; Santacruz, V.M., Mendez-Del-la, F.R. and Andreu, C.G. (1990). Gonadal changes throughout the reproductive cycle of the viviparous lizard *Sceloporus mucronatus* (Sauria: Lguonidac) *J. Herpetol.,* 46(1): 43-50.

Gorbman G.C. and Litch, P. (1974). Seasonality in ovarian cycles among tropical Anolis lizards. *Ecol.,* 55: 360-369.

Jones, R.E. Tokarz, R.R., La Greek, F.T., Fitzerald, K.T. (1976). Endocrine control of clutch size in reptiles VI. Patterns of FSH induces ovarian stimulation in adult *Anolis carolinensis, Gen. Comp. Endocrinol,* 50: 101-116.

Jones, R.E., Fitzgerald, K.T., and Duvale, D. (1978). Quantitive analysis of the ovarian cycle of the lizard. Lepidodactylus/ingubsis. *Gen. Comp. Endocrinol.,* 35: 70-76.

Kitada J. and Asana J.J. (1951) Notes on the seasonal spermatonenic cycle of a gecko, *Hemidactylus flaviviridis* from India. *Jap. J. Genetic,* 263: 45-147.

Koul and Duda P.L. (1977). The histo-cytological studies of atretic follicles in *Agama tuberculata* Gray and *Lygosoma himalayanum* Boulenger. *Zool. Anz.,* 1987: 115-119.

Kwait, A. and Gist H. Danniel (1987). Annual reporductive cycle of an introduced population of Europian wall lizards (*Podarcis muralis*) in Ohio. *J. Herpetol.,* 21(3): 205-209.

Licht, P. (1973). Influerce of temperature and photo period on the annual ovarian cycle in the lizard *Anolis carolinensis*. *Copeia:* pp. 165-472.

Licht, P. (1974) Mayhew, W.W. (1966). Reproduction in the psamnopheloas lizard *Uma scoporia*. *Copiea:* 114-122.

McCann, C. (1949). A reptile and amphibian miscellany Part-1. *J. Bombay Nat. Hist. Soc.,* 41: 742-764.

Miller, M.R. (1948). Seasonal histological changes occurring in the ovary corpus luteum and testis of viviparous lizard *Xantusia vigilis*. *Univ. Calif. Publ. Zool.*, 47: 197-224.

Ortega A. Barbault R. (1984). Reproductive cycles in mosquito lizard *Sceloporus grammicus*. *J. Herpetol.*, 18: 168-175.

Rubail, R. Philibosian, R. and Adeins, J.L. (1972). Reproductive cycle and growth in the lizard *Anolis acutus*, *Copeia*, 3: 509-518.

Sherbrooke, W.C. (1975). Reproductive cycle of a tropical teiid lizard, *Neustricurus ecploepus* in Peru Biotropica *Copeia*, 7: 194-207.

Smith, M.M., Sinelik, G., Fawcett, J.D. and Jones, R.E. (1973). A survey of the chronology of ovulation in Anoline lizard genera. *Trans. Kane. Acad. Sci.* 75: 107-120.

Tanka, S., and Nishihira, M. (1989). Growth and reproduction of Gekkonid lizard Eublephoris Kuroiwae, *J. Heripetol*, pp. 349-357.

Varma, S.K. (1970). Morphology and ovarian changes in the garden lizard *Calotes versicolor* *J. Morphol.*, 131: 195-209.

18

Radiation as an Environmental Agent Affecting Intrauterine Development

M.R. Saini

Radiation Biology Lab., Department of Zoology, University of Rajasthan, Jaipur-302004 (INDIA)

Presence of ionizing radiations is universal in the form of natural background. Their sources are present in the ground beneath, in the sky above, and even in the body of living organisms. Exposure to ionizing radiations is increasing and will continue to increase as man explores outer space and the depths of the ocean and sets off more atomic detonations. Exposure may be massive or minute but it can spread over a life time or a millionth of a second.

Effects of radiation on the organisms are manifested in the form of various initial pathological alterations (anorexia, nausea, diarrhoea, hypertension, vomiting, tachycardia, anaemia, lymphopenia, leucopenia, thrombocytopenia and granulocytopenia) and late effects (leukemia, sterility, shortening of life span, neoplasm, cataract, developmental anomalies and mutations). However, the most serious consequences of exposure to ionizing radiations are the effects it may have on future generations. These may be produced as a result of irradiation of a mother carrying an embryo or damage to the germ cells before or after fertilization. A definite advancement in the understanding of effects of prenatal irradiation in mammals began, when Job *et al.*,[1] demonstrated clearly the presence of well defined critical embryological periods susceptible to X-irradiation. Radiations severely interfere with embryonic or fetal development. Their embryotoxic effects have been presented in UNSCEAR[2]; BEIR[3] in the review articles of Mole[4], and Brent[5], in SSK publication[6] and in various symposia.[7-9]

The extremely high radiosensitivity of an organism in the prenatal, intrauterine period of development can easily be explained, because at this time it is a conglomerate of dividing and differentiating cells, which are exactly the ones having the highest radiosensitivity. The radiosensitivity of an embryo is determined by the most sensitive system that is in the state of active growth at the given moment. Thus, it is well known that intrauterine and neonatal animals are more susceptible than adults to injury by ionizing radiations.[10-11]

It has been shown very convincingly that the embryogenic effect of radiation is chiefly direct. Abscopal influence on impairment of fetus development may not be 5% of the total damaging action of radiation.[12] Noticeable deviations of an embryo from normal conditions can be easily produced on the seventh or eighth day of pregnancy by irradiation in a dose of 0.1-0.25 Gy, which the mother's organism does not respond to in essence, for example, can be a brilliant proof of the correctness of the above statement. When a mother is irradiated prior to conception in doses known to be damaging (up to four grays), no indications of damage to the developing fetus are detected.

A report of the UNSCEAR (The United Nations Scientific Committee on the Effects of Atomic Radiations) states that the radiosensitivity of a fetus is 10-300 times higher than that of an adult organism as regards the late effects of radiation is concerned.[13]

Classical effects of radiation on different gestation periods

Intrauterine death, malformations and growth retardation are the classical effects of radiations on the developing animal.[13-20] Each of these categories exhibits varying degree of sensitivity according to the stage of development.[19-21] The stages of prenatal development are divided into three periods: *pre-implantation* (extending from fertilization to setting of embryo into the intrauterine wall), *organogenesis* (period in which main body structures are formed) and *fetal period* (period of growth of already formed structures). This classification of stages for embryogenesis applies to both human and other mammals including mice.[19] During the pre-implantation period (cleavage, morula and blastocyst) the radiosensitivity of the conceptus is relatively high with increased mortality being the principal effect. Mortality in this stage was noted by a reduction in apparent conception rates with the embryo being resorbed and by deaths at birth or at early neonatal ages.[22]

Major organogenesis begins shortly after implantation, during which the embryo has a high degree of radiosensitivity and hence various malformations are produced.[23, 22, 24] This period is from 7 to 12 days post conception in mouse corresponding to about 14 to 50 days in humans. The induced abnormalities depend on the organs undergoing differentiation at the time of radiation, the stage of differentiation and dose of radiation.[25]

Fetal period is characterized by modification in growth and development, during which there is a reduced radiosensitivity for the occurrence of both malformations and early mortality.[20]

Teratologic Effects/Malformations

Rugh[26] has found that even diagnostic exposure levels may be teratogenic to embryo/fetus. Other workers have also reported that *in utero* exposure to ionizing radiation is associated with a variety of teratogenic and growth defects.[27, 22, 20, 28] A significant increase in the teratogenic effects was observed in fetuses after exposing them *in utero* to 0.5 and 1.0 Gy as compared to the control fetuses.[29]

Hande and Uma Devi[30] exposed pregnant Swiss mice to 9 mGy X-rays or 10 min of ultrasound on Days 6.5 and 11.5 of gestation in four combinations: X-rays on both days (X+X), ultrasound on both days (U+U), X-rays on Day 6.5 and ultrasound on day 11.5 (X+U) and ultrasound on 6.5 days and X-rays on day 11.5 (U+X). There was a significant change in the locomotion activity with a reduction in the total activity at 3 and 6 months of age in the U+U group. Latency in learning capacity was also noticed in this group. The results indicated that repeated exposures to ultrasound or its combination with X-rays could be detrimental to the embryonic development and could impair adult brain function when administered at certain stages of organogenesis.

Jacquet *et al.*,[31] reported that radiation induced a dose-dependent increase of the pre-implantation loss in the BALB/c strain and of the early postimplantation loss in the CFI strain. Embryos of the BALB/c strain were refractory to the induction of teratogenic effects after such pre-implantation irradiation. In CFI mice, the frequency of malformed fetuses increased regularly after irradiation, the difference with controls being significant for the doses of 10, 50 and 100 cGy. Irradiation did not increase the frequency of skeletal anomalies. A careful examination of the various data indicated that radiation might possibly be teratogenic in several mouse strains, when administered as early as during the one-cell stage and, to a lesser extent, during the following pre-implantation stages. However, early prenatal mortality will remain by far the greatest risk associated with an exposure to radiation during this period. Moreover, the relativity of the risk of abnormality due to such irradiation should be considered in the context of the high prevalence of developmental defects spontaneously occurring during human pregnancy. Gastroschisis (abdominal wall defects) occurred with a high frequency in the mouse inbred strain HLG as compared to C57BL/67 mice.[32] The risk of gastroschisis increased significantly after exposure to X-rays during pre-implantation development and followed a recessive mode of inheritance for the HLG susceptibility alleles. A backcross strategy and genome-wide microsatellite typing was used for chromosomal maping of this trait. A suggestive linkage for a locus responsible for radiation-induced gastroschisis (Rigs 1) was found in a region of mouse Chromosome 7.

In order to evaluate the importance of gestational age and the dose-incidence relationship by gamma radiation, pregnant ICR mice at gestational days from 2.5 to 15.5 days post-coitus (p.c) were exposed to a single dose of 2.0 Gy and also at day 11.5 p.c. (post-coitus), which was the most sensitive stage for the induction of major congenital malformations.[33] The animals were sacrificed on day 18 of gestation and the fetuses were examined for various defects. During pre-implantation period, the embryos irradiated and survived; did not show any major fetal abnormalities. A small head, cleft palate, dilatation of the cerebral ventricle, a renal pelvis, and abnormalities of the extremities and tail after exposure were prominent during the organogenesis period, especially on day 11.5 of gestation. As for the dose incidence relationship, the incidence of a small head, growth-retarded fetuses, a cleft palate, dilatation of cerebral ventricle and abnormalities of the extremities in live fetuses rose as the radiation dose increased. The result indicated that the late period of organogenesis in the development of the brain, skull and extremities of a mouse was particularly a sensitive phase. The threshold doses of radiation that induced a cleft palate and dilatation of the cerebral ventricle, and abnormal extremities were between 1.0 and 2.0 Gy, and between 0.5 and 1.0 Gy, respectively.

LD_{50} Dose

Rugh and Wohlfromm[10] reported the LD_{50} dose which ranged from 1.0 to 8.0 Gy for mouse embryo/fetus, depending primarily on the stage of development at which mothers were exposed. In most of the case, the LD_{50} dose was less than that for adult. The $LD_{50/30}$ for females exposed *in utero* was 5.80 Gy as compared to 6.25 Gy for controls and that of the males was 5.50 Gy and 6.30 Gy respectively. A dose dependent increased degeneration of cells and mitotic delay in embryos of rats, exposed to 2.0 and 4.0 Gy at 3.5 days of pregnancy was noticed.[34]

Mortality

Mortality (resorption and embryonic/fetal death) increased during pre-implantation or early post-implantation stage, when pregnant mice were exposed to 1.0 Gy of X-rays at various hours in the first day of gestation.[35] Zhou and Wang[36] found an increased mortality when mouse embryos were exposed to single or fractionated doses of gamma rays (1.0 Gy) at pre-implantation period. Roux *et al.*,[37] also reported an increased mortality of embryo and fetus after intrauterine exposure to 0.05, 1.0 and 0.25 Gy of gamma radiation.

Friedberg *et al.*,[38] observed the highest prenatal mortality after exposure to pregnant females to 0.39 Gy of neutron or 1.60 Gy of X-rays; when the time of irradiation was 12.00 hr at gestation day 7 and 11.00 h. on day 8. After exposure with X-rays, mortality was highest in-groups irradiated between 20.00 hr at gestation day 7 and 4.00 hr at day 8. A decrease in whelping rates was observed for female breeders irradiated on day 8 and 28 p.c. [39] They also reported a significant decrease in litter size from female breeders irradiated on day 8 and 28 of gestation.

Both of these findings are indicative of increased embryonic mortality. Uma Devi and Hande[40] exposed pregnant females to low dose of X-rays (29 mGy) in the range of diagnostic exposition on day 3.5, 6.5 or 11.5 of gestation and reported a significant increase in prenatal mortality. Uma Devi *et al.,*[41] also reported fetal mortality after exposure to pregnant female to 0.05 to 0.50 Gy of gamma radiation on day 11.5 p.c. The only demonstrable effect of irradiation during the pre-implantation period was an increase in prenatal mortality.[33]

Dose as low as .05 to .25 Gy caused an increased resorption in mice.[42] A great increase in resorption was found during the early stage of mesoderm formation following X-irradiation. With a dose of 1.0 Gy, the incidence showed a peak on day 8.5 p.c. [43] Number of resorbed fetuses was significantly higher in irradiated groups as compared to control groups and resorption increased with dose.[44] He also concluded that the effect of low level irradiation on early pre-implantation mouse embryo was manifested primarily by an increased number of dead fetuses. Salko[45] exposed rats to 1.35 or 2.7 Gy of[60] Co gamma rays on 11 p.c.and observed an increase in the degree of resorption even with the lower dose. Coppenger and Brown[46] noticed a very short rise in the number of resorption when mothers were irradiated at 12 p.c. Pradhān and Sreenivasan[47] reported an increase in resorption rate even with 0.5 Gy given at early pre-cleavage stage in mice. An exposure at first day of pregnancy in rats increased the rate of embryonic resorption to 11.0% as compared to 4.7 in the control.[48] It is found that the effect increased with dose and decreased with age of the embryo at the time of exposure.[13, 49,-50]

Miller and Mulvihill[51] also reported embryonic death after irradiation of mothers during pregnancy. Aldeen and Konermann[52] exposed pregnant females to a single dose of 3.0 Gy of X-rays on day 1, 2 and 3 p.c. and reported that percentage of embryonic resorption was lowest (31.8%) at 1 p.c. and highest (94%) on 2 p.c. Complete resorption of embryos was observed when females were irradiated with 2.5 Gy on 11.25 p.c.[53] They also reported that with 1.5 Gy, 11.25 days gestating females showed 76.10% resorption, while females irradiated with .5 Gy showed very less resorption. Michel and Fritz Niggli[54] also reported 9.6% resorption after exposure to 1.0 Gy of X-rays.

Litter size

Litter size is a function of embryonic resorption. Litter size was found to be reduced as the dose of X-rays closely approached to that required to kill all the conceptuses.[1] Kaven[55] reported reduced litter size (4.3 or 4.7) in rodents irradiated on day 7 or 8 gestation as compared to their controls (6.8). In rats, 2.0 Gy on 10 p.c. resulted in 80% reduction in average litter size at birth. However, the same dose when given just before birth, resulted in about 29% reduction.[56] An exposure of 1.0 or 2.0 Gy during pre- and post implantation and early organogenesis period affected litter-size, but the same dose was found ineffective during fetal period.[57] Rugh *et al.*,[58] found greatest reduction in litter-size after exposure to 1.0 Gy at day zero of pregnancy or immediately after fertilization. A exposure to 1.0 Gy at gestation day 3, 4 and 10 caused severe congenital anomalies and hence the mother killed and ate more than 10% of the live delivered fetuses. Philippe[59] gave a single exposure in the range of .5-5 Gy to pregnant Swiss albino mice on 9, 10, 11 and 12 days of gestation and stated that exposure on day 9 below 1.5 produced no significant deviation from the controls whereas higher exposures were lethal to the conceptus. Irradiation on day 10 with 1.0 Gy was threshold for placental weight reduction.

At 1.5 to 1.75 Gy, fetal mortality was over 60% and surviving fetuses were malformed. At 2.0 Gy, fetal loss approached 100%. On day 11 the feto-placental complex was reduced in weight after exposure to .75 Gy. At 2 Gy, prenatal death was over 60% and at 3.0 Gy it was close to 100%, while 5.0 Gy produced total loss of the fetuses. Ward *et al.*,[60] exposed pregnant females on day 4 or 9 of gestation and found that embryos located in the overian or in the cervical ends of the uterus had higher rates of mortality than their litter mates implanted in the middle part of the uterine horns. The influence on survival of the uterine position increased with the dose of radiation and number of implants. At higher dose of radiation and in a crowded uterus the probability of survival of embryos was less.

Sex ratio

An altered sex ratio was found in mice irradiated during day 11 to 13 p.c. with .35 to .5 Gy of X-rays (Job *et al.*, 1935). However, no significant alteration in the sex ratio of irradiated mice *in utero* was observed by Russell and Russell[57] and Rugh and Wohlfromm.[11] An alteration in sex ratio in the human offspring exposed *in utero* to diagnostic X-rays was also reported by Meyer *et al.*[61] When 3×1.1 Gy fractionated dose was given to pregnant mice from 11-13 days of gestation, a high mortality in females had distrubed the sex-ratio.[62] There was a significant decrease in the percentage of female beagles born after exposure at 28d.p.c..[39]

Post Partam Weight Loss

Mean birth weight of embryos irradiate *in utero* with 2.0 or 3.0 Gy between day 8.5-13.5 p.c. was found to be lower than that of the controls.[63] An average weight reduction of 0.22 g for every 1.0 Gy given at 11d.p.c. was noted.[64] Growth retardation in 2-11 day old embryos was observed one day after exposure *in utero* to .5 or 1.0 Gy.[56] A decrease in average weight of rat embryo exposed *in utero* at 10-14d.p.c. was also reported by Wilson *et al.*,[13] and Ershoff and Bavetta.[65] Nash and Gowen[66] showed that developing mice irradiated on day 6.5, 10.5, 14.5 and 17.5 p.c. with different doses (.2, .8, 1, 1.6 and 3.2 Gy) of X-rays had reduced body weights. There was a statistically significant reduction in average body weight of male embryo following exposure at gestation day 5 and 9 through 18, with the greatest reduction following exposure from day 12-14 p.c. For females, reduction occurred following exposure day 8 through 18, with the most drastic reduction at days 11 to 14 of gestation. The body weight response was found to be age and dose dependent. Thus, the second half days of gestation appeared to be the period of the most significant stunting by X-ray.[58, 67] Martin and Murphee[68] studied growth changes in rats exposed to 2.2 Gy of 60 Co gamma radiation on day 18 of gestation and noted a reduction in body weight upto 365 days. Retarded fetuses were also noticed after exposure *in utero* to fractionated doses of .8, 1.2 and 1.6 Gy of X-rays per day at 11 to 16 and 14 to 16 p.c. [69] Tietz *et al.*,[70] exposed beagle embryos *in utero* to 1.0 Gy X-rays on day 13, 16, 19, 22 and 25 p.c. and observed a significant reduction in body weight at the time of birth.

According to Brent,[17] exposure to radiation during early organogenetic period leads to growth retardation which is generally recoupable postnatally. However, irradiation during fetal period results in permanent growth retardation. Rats exposed on day 18 of pregnancy to a single whole body dose of fission neutron (.2 to 1.5 Gy) showed reduced body weight in their fetuses. After .2 Gy there was a small decrease in body weight which lasted from birth to weaning.[71] Sasaki *et al.*,[72] exposed mice embryos *in utero* to 2.0 Gy of X-rays on day 12 or 16 to 18 p.c. and observed a persistent growth retardation, severe congenital malformations and amyloid degeneration in those irradiated in the middle of intrauterine life. Decreased mean fetal weight was recorded after exposing pregnant mice and pre-implanation period to single or fractionated doses of gamma radiation.[36] Michel and Fritz-Niggli[54] stated that the percentage of growth retarded fetuses varied between 9.5 (0.125 Gy) and 10.9 (1.0 Gy). Jensh and Brent[73] reported a dose dependent weight reduction in term fetuses whose mothers were exposed to 0, 0.15 or C.30 Gy of X-rays/day from 14-18 days of gestation. Uma Devi and Hande[40] reported an increased incidence of retarded fetuses after *in utero* exposure on day 3.5 and 6.5% to X-ray (~9 mGy) in the range of diagnostic exposition. This dose is much lower then the occupational exposure limits (5 cGy) for pregnant women[74] and the dose to which medical termination of pregnancy is recommended.[2] Uma Devi *et al.*,[41] and Hande and Uma Devi[30] have also reported fetal growth retardation after *in utero* exposure to gamma, X-rays and ultra sound respectively. Later on, this effect was also confirmed by Jaquit *et al.*,[31] and Kim.[33]

Effects on CNS

The brain, undoubtedly is the most complex organ in the mammalian body. It is culmination of a long and interrelated sequence of molecular, cellular and tissue events. Brain function hinges on the orderly progression of these, each of which must occur correctly, temporally and spatially. Impingement on any one will give rise to a less developed system of cellular connections, and hence impaired function. Moreover, the neurons of the central nervous system are not self-renewing and thus neuronal loss cannot be repaired through repopulating. Reanalysis of the data on the prenatally exposed survivors of the atomic bombing of Hiroshima and Nagasaki suggests that severe mental retardation occur primarily, if not exclusively in the period from the 8th through the 15th week following fertilization. Within this window of vunlerability, the increase in mental retardation appears linear with dose and without threshold. More subtle functional effects also occur as reflected in diminished performance on intelligence tests and in school.[75]

Roentgen irradiation of the mammalian embryo at the beginning of neurogenesis period caused the highest percentage of gross CNS anomalies.[76-79] this period is 7.5 to 9.5 days in the rodents and 18 to 25 days in the human beings. Ionizing radiations are so penetrating and instantaneous in their impact that as many as 48% of the mouse embryos exposed on 8.5 days developed exencephaly or brain hernia.[42, 80, 81] Rugh and Wohlfromm[11] and Rugh[26] reported a variety of anomalies of central nervous system (hydrocephaly, microcephaly, exencephaly, anencephaly and cerebral blisters) in embryos than in any other system. He also stated that after completion of neurogenesis congenital anomalies of CNS are rarely produced but exencephaly is produced in 2-cell stage mouse embryo after exposure to as low as 0.1 Gy Konermann[82] exposed developing mice prenatally during organogenesis (6-13 p.c.) or fetogenesis (14-18 p.c.) to gamma rays (.1-.8 Gy/day) and found that under these conditions the capacity for compensating the brain growth deficit became lesser in later development that took place after irradiation. Heinzmann[83] describe pathogenesis of the hydro-micrencephaly in NMRI-mice after single X-irradiation with 0.9 Gy or 1.9 Gy on gestation day 12. The ultra structural alterations in the ventricular walls were sequentially investigated up to gross developmental damage. The ventricular zone is the most sensitive region in the developing brain. Its undifferentiated and proliferating cells lose their palisade like orientation and fail subsequently to differentiate into primitive neurons, glia cells, or ependymal cells. Structurally, this results in the thinning of ventricular walls by more than 50% associated with periventricular edema and a dilation of the brain ventricles by 20-60%. Damage was clearly more pronounced with the higher dose. Repair processes originated from regions with intact Zonulae adherents, which also give rise to typical globular or cylindrical heterotopic structures; these are known as rosettes and made up from undifferentiated proliferating ventricular cells. Perinatally in these rosettes cell replication persists, at a time when cell production in the ventricular zone has ceased. Histological changes were the most prominent in and around the telencephalic roof consisting in replacement of the ependymal lining by a fault of glial fibres, faulty myelinisation, and periventricular edema; postnatally these structural alterations led to hydro-micrencephaly. Results from this animal model can be translated to the human situation because the fundamental developmental processes in the brain of mammals are similar despite species related differences of the time scale.

Central nervous system morphology was disturbed acutely after embryonic exposure to X-radiation at levels in excess of 1.0 Gy.[84-89] However, Hicks and D' Amato[90] also reported CNS alterations at very low level (less than 0.5 Gy) after prenatal irradiation. Alterations in central nervous system morphology have also been shown, which persisted postnatally in some instances.[91-96]

Using neonatal rats as a model system, Yang and Tobias[97] investigated the response of the brain vascular system to ionizing radiation and found that distinct petechial hemorrhages developed in the cerebral cortex within a few hours after irradiation, reached a maximum about 13 to 24 hours, and decreased exponentially with time. No brain hemorrhage was found in neonatal rats 12 days after irradiation. Their experimental results indicated that a dose of a few hundred rad of X-rays could induce a significant number of hemorrhages in the brain, and the number of lesions increased exponentially with dose. Heavy ions induced more hemorrhages than X-rays for a given dose, and the RBE for 670 MeV/u neon particles ranges from about 2.0 for low doses to about 1.4 for high doses. A histological study of the hemorrhages indicated that a large number of red blood cells leak from the blood vessels. The radiation-induced hemorrhages may be a result of some capillary membrane damages to reproductive death of some blood vessel epithelial cells. The fast onset of hemorrhage after irradiation suggested that some membrane damage might be involved. The effect of heavy-ion radiation on the embryonic development was studied with energetic iron particles. Friedberg *et al.*,[38] irradiated pregnant mice at selected times between 10.00 hours at gestation day 7 and 16.00 hours at day 8 with 0.39 Gy of neutrons or 1.60 Gy of X-rays and found that incidence of exencephali in live embryos rose and then declined with increasing dose of neutrons. They clearly stated that embryos with neutron injury of the type that resulted into exencephalia were at a greater risk of dying *in utero,* than were similarly irradiated embryos not so injured and that this risk increased with dose.

At a dose level of 2.0 Gy 100% hydrocephalic offspring were obtained by irradiation at embryonic days 11 and 14.[98] When pregnant rats were subjected to 2.0 Gy X-irradiation at embryonic day 11, numerous ventricular cells of the telencephalic wall of the embryo became necrotic during the first 2 hours post-irradiation, but the paraventricular cell to cell interconnections made up of zonulae adherents were less affected. Mitosis took place in the surviving paraventricular surface cells throughout subsequent development. The full-term fetus exhibited little change in the cytoarchitectural arrangement of neural cells and neuropils, although it was only about half the thickness of the untreated control. After 2.0 Gy X-irradiation at embryonic day 14, the most of the ventricular cells became necrotic within 6 hours. The paraventricular cell-to-cell interconnections were completely destroyed, and never repaired in subsequent development. Mitosis took place either freely in cell clusters, or in rosettes, which formed randomly in the telencephalic wall between 48 and 72 hours post irradiation. The resulting telencephalic wall of the full-term fetus was also about half the thickness of the control. In the outer part of the tissue, the cortical plate made up of differentiating neuroblasts was hypoplastic, but the inner half was filled with numerous heterotopic masses of pleomorphic cells and bundles of primitive axons. The ependymal layer at the paraventricular surface was never formed. Whether the paraventricular zonulae adherents were destroyed or not by X-irradiation was considered to be an important factor in the determination of the subsequent cytoarchitectural organization of the telecephalic wall.

Takeuchi and Takeuchi[99] reported that a single whole body X-irradiation of pregnant Wistar rats with 1.05 Gy at 10.30, 12.30 or 14.30 hours, on gestation day 10 resulted in significantly high incidences of hydrocephalic offspring. No hydrocephalic offspring were noticed after X-irradiation of pregnant rats with 1.05 Gy at 16.30 hours, whereas, a dose of 1.22 Gy at 16.30 hours resulted in a low but statistically significant incidence of hydrocephalus.

Neither 1.05 nor 1.22 Gy X-irradiation of pregnant rats at 18.30 hours resulted in any hydrocephalic offspring. Dysplasia of the subcommissural organ was noticed in all the hydrocephalic brains on histological examination.

Naturally occurring postnatal cell death in the somatosensory cortex and medial cortical areas was examined in microcephalic rats produced after exposure to 100 cGy X-rays at embryonic day 15. The absolute numbers of dead cells per tissue section were not affected in the future subcortical white matter, but were severely decreased in the cerebral cortex when compared with age-matched controls. Furthermore, the proportion of cortical dead cells per 1000 live neurons in the cerebral cortex and the cumulative cell death across time were significantly reduced in irradiated animals. These results suggested that naturally occurring postnatal cell death in the cerebral cortex was reduced in micrencephaly.

Ferrer[100] reported different cortical malformations in rats after exposure to a single dose of X-rays (200 cGy) given on different days during gestation. These include large cortical ectopic masses after irradiation on day 14; segmentation of the cerebral cortex following irradiation on days 15, 17, 19; and a four-layered "lissencephalic" cortex following irradiation on day 16. Other types of cortical malformation were produced in rates aged 0-2 days by one of the following procedure: focal cortical freezing, focal electrocoagulation, cortical aspiration, and focal brushing of the meanings with a blunt needle covered with cotton. These latter abnormalities include laminar necrosis of layer V, focal cortical dysplasia reminiscent of microgyria, status verrucosus deforms and porencephaly. Experimentally induced cortical malformations in rats can help to increase our understanding of normal and abnormal neurogenesis and organization of the human cerebral cortex.

A four-layered cortex was produced in the dorsal region of the rat brain, after 200 cGy X-irradiation at embryonic day 16.[101] The layer 1 is the molecular layer; layer 2 is the upper cellular layer which is in continuity with normal layers V and VI in the lateral cortex; layer 3 is a sparsely cellular layer; and layer 4 is the inner cellular layer which is composed of neurons normally committed to the upper cortical layers. The failure of neuroblast migration to the upper layers was associated with reduced numbers, and morphological abnormalities, of radial glial cells in the dorsal cortex. This, further supported the idea that abnormal interactions between neuroblastas and radial glial cells result in abnormal migratory patterns. Since a similar four-layered organization was found in human lissencephaly type 1, that can serve better understanding of human cortical malformations.

Caparelli-Daquer and Schmidt[102] reported morphological brain asymmetries using a distinct experimental approach. At embryonic day 16, mice were exposed to a gamma source, receiving a total dose of 2 Gy. AT adulthood 32 irradiated and 28 normal male Swiss mice were evaluated for individual and populational interhemispheric asymmetries of weight, dorsal area, and neocortical volume. All irradiated mice showed callosal defects ranging from total absence to a partial reduction of the midsagittal callosal area. The normal and irradiated groups displayed a pronounced individual asymmetry in all measurements. In contrast, populational asymmetry could be identified only in the normal group. These results suggested that the corpus callosum might play a role in directing interhemispheric morpological brain asymmetries.

The abdominal region of pregnant Swiss mice was exposed to 0.05-0.50 Gy of gamma radiation on day 11.5 p.c. The animals were sacrificed on day 18 of gestation and the fetuses were examined for changes in head size and brain weight. A significant reduction in head size and brain weight was observed at doses above 0.15 Gy. Detectable level of microcephaly was evident even at 0.10 Gy. A linear dose response was seen for these effects in the dose range of 0.05-0.15 Gy. It was concluded that the late period of organogenesis in the mouse, especially between days 10 and 12 p.c., is a particularly sensitive phase in the development of the skull, brain and eye.

The results of Dilmanian *et al.*,[103] indicated that embryonic duck brain has an appreciably higher tolerance to the microbeam modality as compared to the broad beam modality. When the microbeam dose was normalized to the full volume of the irradiated tissue. i.e., the dose averaged over microbeam and the space between the microbeam, brain tolerance was estimated to be about three times higher to microbeam irradiations as compared with broad beam irradiation.

Schindler *et al.*,[104] stated that a first-trimester, neuro-developmental insult in the non-human primate during thalamic neurogenesis produces a complex pattern of thalamic volume loss and shape deformation in adulthood. Low-dose irradiation of the fetal primate may be useful for modelling key features of the pathology described in schizophrenic patients.

Effects on Eye

The developing eye is much more sensitive than the adult eye. A high incidence of ocular anomalies has been observed by Job *et al.*,[1] and Russell[63] in pregnant rats and mice respectively after X-irradiation. Pope[105] and Woods *et al.*,[106] concluded that the actively proliferating cells of anterior epithelium of the lens are the most susceptible to radiation damage. The eye was the organ most frequently rendered anomalous after irradiation day 10 p.c.[56] Anophthalmia or complete absence of the optic cup, optic stalk and lens, occurred rarely after exposure to .5 Gy but it was the most prevalent type of ocular anomaly after the higher doses.[13] Severe eye malformations like anophthalmia and microphthalmia were reported after embryonic exposure on day 9, and especially on day 10 of gestation.[78] The threshold dose for microphthalmia was found in the range of 0.125 to 0.25 Gy of X-rays by Michel *et al.*[107] Further in 1982[108] they also reported cellular changes in developing eye of mice embryo irradiated on 9 p.c. with 0.9 Gy of X-rays. They stated that size of the eye vesicle was significantly reduced in these embryos but it was restored 24 hours later. A decrease in total number of cells and mitotic activities of vesicles was also noticed. Persaud *et al.*,[109] reported a high incidence of developmental defects in the eye and spinal curvature in SD rat fetuses after *in utero* exposure to .5 Gy on embryonic day 9.5.

Friedberg *et al.*,[38] found that in X-irradiated litters, the incidence of anophthalmia was higher in exencephalic than in non-exencephalic embryos and the ratio of these incidences decreased towards 1 with increasing dose. They also noticed higher incidence of bilateral anophthalmia in prenatally X-irradiated embryos than would be expected that if the bilateral form resulted solely from independent injury at each of two equally susceptiable sites.

Pregnant rats were irradiated with soft X-ray at a dose of 6.35 Gy or 12.70 Gy on day 9 or 11 of gestation.[110] The eyes of

F1 offspring were examined by indirect ophthalmoscopy and slitlamp biomicroscopy in postnatal weeks 3 to 7. Ophthalmoscopic examination of offspring irradiated at a dose of 6.35 Gy or 12.70 Gy on gestational day 9 revealed iris and choroidal coloboma together with microphthalmia and anophthalmia at postnatal week 3 or later. Choridal coloboma, an anomaly that cannot be detected by the common external observation method, was found ophthalmoscopically in one rat without any external ocular abnormalities. This fact suggested that ophthalmoscopy in postnatal observation can be useful to detect ocular teratogenic potential of test compounds in preclinical safety studies. Uma Devi *et al.*,[41] also reported micropthalmic fetuses after exposure to 0.1 Gy of gamma rays.

Glineur *et al.*,[111] irradiated two populations of pregnant mice with a single dose of 2.0 Gy on day 8 of gestation for the first population (Po. 1) and on day 9 of gestation for the second population (Po. 2). On gestational days 9 to 17, embryos were submitted to histological and scanning electron microscope examinations. The two populations of embryos presented a high percentage of centro-facial hypoplasia (74.7% for Po. 1 and 75% for Po. 2) which was more pronounced in the latter one. Ocular anomalies were present in 16% of the first population. Cases of anencephay, cleft palate and anomalies of the central nervous system were found in both populations.

Effects on Skeleton

The developing skeletal system of an organism is severely affected after exposure *in utero*. Skull deformities were observed after irradiation on days 10, 12, 13 and in few instance on 14 p.c. The characteristic angulation of the ribs within the osseous parts was noted in the animals exposed on day 15 and 16 of gestation but not in those exposed on previous days. Cleft palate was found after irradiation on day 10, 12, 13, 14, 15 and 16 p.c. but the incidence was particularly higher in the offspring of animals irradiated with both low and high doses on 15 p.c.[112] They also observed newborns with cleft palate and short snout and mandible by irradiating pregnant rats on any one of the several days of gestation with 1.9 R to 19. 5 Gy. Russell and Russell[113, 114] have studied the rodent skeleton most extensively, following X-irradiation and reported that the critical period for posterior shift in the thoraco-lumbar bordered from 7.5 to 9.5 days with the peak of reaction on 8.5 days in the mouse. Exposure to as small a doses as .25 Gy on day 8.5 altered the rib development in BALB/C mice. Prenatally irradiated embryos with doses of 1.0 to 4.0 Gy between 0.5 and 13.5 days of gestation developed variouss skeletal abnormalities: polydactyly, reduced cranium, short snout and limb defect. Similar studies have been made on the digits of mouse embryos[115] and found that exposure to 3 Gy X-rays during the gestation period of 10 to 12 days caused reduction of 1 to 4 digits or increase of 1 or 2 digits. Chang *et al.*,[116] reported production of cleft palate, short tail and fused digits in rabbit and acrania in hamsters after *in utero* irradiation. Murakami *et al.*,[117] also described malformation of the extremities in *in utero* irradiated mouse fetuses.

NMRI mice exposed prenatally in the late organogenetic stages (11-13d.p.c.) to either a single (2.0 Gy) or fractionated (3×1.6 Gy) doses of X-rays developed various types of skull anomalies.[62] Knauss[118] exposed pregnant mice on 7, 10 and 13% to 150 KV X-rays in the range of 0.5-2.5 Gy and day 7 of gestation was reported more sensitive as compared to 13 p.c. for the development of skeletal malformations. Konermann[119] has shown that the spontaneous incidence of 0.6% malformations in the skull increased by a factor of 3 after an exposure to 0.5 Gy on 7%. An exposure to .25 Gy on 10 p.c. doubled the incidence of these malformations.

Radiation effects on collagen and glycosaminoglycan (GAG) synthesis were studied by Cornelissen *et al.*,[120] in non-ossifying embryonic chick cartilage, *in vitro*. Fragments of the non-calcifying region of 15 day old chick sterna were irradiated with 20, 50 and 100 Gy. The counterparts served as controls. Fragments were cultured up to 14 days. Pulse labelling experiments with 3H-proline and 3H-glucosamine started immediately after irradiation and on day 3, 6 and 13 post-irradiation. Fragments for morphological evaluation were fixed on day 1, 4, 7 and 14, simultaneously with the end of the labelling period. The same dose and time dependent inhibition was observed for the collagen and GAG synthesis. Radiation dose of 2.0, 5.0 and 10.0 Gy cause inhibition of about 20, 35 and 60% at the end of the culture period respectively. Morphological alterations were after irradiation with 5.0 and 10.0 Gy. Dose and time dependent necrosis all over the sternum was observed. A restricted number of cells showed an increased GAG synthesis after irradiation (50 Gy). This could explain the levelling off of the radition effect on the incorporation of the labelled precursors towards the end of the culture period for radiation doses up to 50 Gy.

Fetuses of pregnant mice exposed to whole body gamma radiation (4.0 Gy) on the 12th gestational day were compared with control fetuses to assess the effect of radiation on palatogenesis.[121] Animals of experimental and control groups were sacrificed on day 18 of gestations. Histologic examination demonstrated that 100% of the experimental fetuses had palatal clefts. It was concluded that gamma rays have adverse effects on palatal closure of mouse fetuses.

Wang *et al.*,[122] observed that *in utero* irradiation of the fetus during the period of organogenesis induced a dramatic increase in malformation. The correlation between radiation-induced apoptosis and limb malformation was examined in mice. The mice were exposed to X-rays *in utero* on day 11 of gestation during the period of organogenesis of limb buds. A marked increase in the number of apoptotic cells in the predigital region in the forelimb buds was detected 4 hr after irradiation. The pre-interdigital regions of the forelimb buds did not show such an increase at the same time. Aphlangy and ectrodactyly were the main types of anomalies observed on day 19 in the limbs of fetuses irradiated with 5.0 Gy. The increase in prenatal death and teratogenesis in limb digits in living fetuses were dependent on dose.

A decrease in longitudinal growth of bones was observed after exposure to a singel dose of about 5.0 Gy.[123] Riddler *et al.*,[124] reported the complete arrest of metabolism of the cartilage cells as well as longitudinal growth of chick bone after exposure to 2.0 Gy. Yawada *et al.*,[125] reported the increased number of thoracic vertebrae and ribs, after exposure during pre-implantation period only.

Effects of Radioisotopes:

Irradiation by radionuclides poses a special hazard, as the internal emitters continuously emanate radiation to the surrounding tissues of the contaminated animal. Radioactive strontium, phosphorus, iodine, plutonium and tritium can cross the placental barrier and become incorporated in the fetal tissues. However, very few experiments have been made on the effects of radioisotopes on the prenatal mammalian development till date.

Radiostrontium and Plutonium

Finkel[126] studied the transmission of radiostrontium and plutonium from mother to offspring and reported that mice which received radiostrontium or plutonium at the dose of 5.0 or 2.5 µCi/gram body weight before conception produced a smaller number of offspring. Strontium-90 resulted in defects largely of the skeletal system when administered to mice.[127] Long-Evans females were injected with 181 or 382 mCi of strontium-90. Absence of tail, and malformations of eye and skeleton were noticed after exposure at day 2 and 10 of gestation respectively.[128]

Phosphorus-32

Administration of 0.3 to 2.0 millicuries (mCi) of phosphorus-32 in Wistar female rat intraperitoneally on days 7, 9, 10 or 11 of gestation produced malformations and lethality in embryo/ fetus.[129] The frequency and type of malformations were the same, and depend on the day of gestation on which injection was given. Fifty mCi of phosphorus-32 given to a female rat on day 9% gestation produced offspring with gross limb defects.[130]

$Na^{99}mTC_4$

Lathrop *et al.*,[131] administered $Na^{99}mTC_4$ every day to pregnant females throughout gestation, gestation and lactation, or lactation period for three successive generations. At a significant reduction in body weight of offspring was noticed in all the groups. Body growth as determined by weight and body length was found to be greatly reduced in the offspring of mothers who had received I^{-131}, 5500 KBq (150 µCi) at 11.5d.p.c..[132, 133]

Tritiated Thymidine

Greulich[134] administered a total of 245-300 µCi of tritiated thymidine to female mice for 24-32 days before and during gestation, and a decline in fertility was noted after the first litter. Moreover, the offspring showed a reporductive capacity only 70% of the normal. Cleaver[135] reported that the radioactive disintegration of tritium within the nucleus caused a wide range of adverse biological effects, such as delay in growth, chromosomal aberrations and mutation, etc. Fliedner *et al.*,[136] administered tritiated thymidine to rats from 9 to 22d.p.c. and observed malformed fetuses at 8.0 μCi/gram body wt/day dose level. A dose of 3.2, 4.8 or 6.4 μCi/gram body wt/day resulted in marked depression of neonatal growth and bone marrow cells. Fliedner *et al.*,[137] also noted a marked effect of dose relationship in the number of resorptions, still births or neonatal deaths (12 hours after birth) where mother received increasing dose of tritiated thymidine. Chaudhuri *et al.*,[138] studied the radiotoxic effect of continuous infusion of tritiated thymidine (^{3}HTDR) into rats from day 9 of pregnancy till term. All new born rats exposed to ^{3}HTDR infusion had chromosomal abnormalities related to dose. Also there was an increase in number of still births and the offspring suffered from imparied development, decreasing regenerating capacity and malformations.

Excellent reviews[139-141] have appeared on the harmful effects of tritiated thymidine. Vesselinovitch *et al.*,[142] assessed trans-placental neonatal and infant carcinogenic risk by injecting ^{3}H-thymidine (8 μCi/gram) intraperitoneally on day 16 p.c. in mice and noted as accelerated development of hepatomas in addition to epilation and premature graying of hair in both, the male and female offspring. Snow[143] observed the abnormal development of mouse embryos during pre-implantation period, grown *in vitro* with (^{3}H) thymidine. He also reported that mouse embryos treated with two forms of ^{3}H-thymidine methyI-T-thymidine and thymidine-6-T (n) were found to be lethal at concentrations above 0.1 μCi/ml. Both the forms of (^{3}H) thymidine at concentrations between 0.01-0.1 μCi/ml caused a highly significant (P.<001) reduction in blastocyst number.

Schreml *et al.*,[144] and Schremal and Filedner[145] exposed pregnant rats to ^{3}H thymidine and ^{3}H water from day 9 p.c. till term and found that DNA bound tritium appeared three times more effective in causing damage to the somatic development in new-born rats. The development of pre-implantation mouse embryos was prevented by exposure to (35s) methionine, but not to methionine. Such embryos have also been shown to be highly sensitive to (^{3}H) thymidine.[146]

In a multigenerational study Ugarte *et al.*,[147] reported that the reproductive capacity of the two sub-lines treated with HTO and ^{3}HTDR impaired progressively from 19th to 23rd generation. Also, the sub-line exposed to ^{3}HTDR became extinct due to lack of sibling mating. Spindle and Pedersen[148] reported that even 0.1 µCi of (^{3}H) thymidine decreased the number of embryos that could form from the inner cell masse during early post-implantation development. Saito *et al.*,[149] observed offspring which were exposed to tritiated thymidine water during their complete prenatal stage at the dose rate of 185 KBq/ml. The newborn mice were mushed by their own mothers, which were further supplied with tritiated thymidine till 4 weeks after delivery.

Mouse embryos were incubated *in vitro* with (^{3}H) arginine at 2-cell stage at either 370 Bq/ml (10 nCi/ml) or 925 Bq/ml (25 nCi/ml). Development *in vitro* was followed up to the formation of the inner cell mass at 192 hours post conception, the difference in radiotoxicity of the two substances with respect to cell proliferation was not observed. However, formation of blastocyst, trophoblast, outgrowth and inner cell masses were impaired more strongly by (^{3}H) thymidine than arginine when the external exposure concentrations were the same.[150]

Tritium

Studies on the radiotoxicity of tritium date back to the early 1950's after tritiated water became available for tracer work. Investigations of the effects of a continuous application of tritium seem worthwhile from both radiobiological and radioecological point of view because the tritium generated by nucler reactors and fuel reprocessing plants is released into the aquatic environment in low levels as tritiated water.

Besides being of radiological interest, studies of biological hazards of tritium are of practical importance because the animals, which develop in a contaminated aquatic environment are being used for food by man in some areas. Studies of the effects of tritiated water on the embryonic development of mammals have been carried out by a few workers using very high doses of tritium.

Bruses *et al.*,[151] demonstrated that HTO given to mice at the dose rate of 1.0 µCi/gAm body weight caused 50% lethality within 30 days. The animal fell ill 2 day after injection and water consumption dropped drastically. These results were later confirmed by Furchner[152] and for gaseous tritium by Trujillo *et al.*[153] In rats, LD 50/12 and LD 50/30 of tritium was calculated to be 1350.8 and 895.53 Gy, respectively.[154]

Thompson and Ballou[155] reported the uptake of tritium oxide in rates after continuous supply of the tritium isotope through drinking water. Rats were tritiated from conception to 6 months of age at the dose a rate of 3.5 µCi/ml drinking water and were autopsied at various time intervals extending upto 300 days. No adverse effect on the growth was noticed.

Cahill and Yuile[156, 157] maintained pregnant rats throughout pregnancy at constant body activity of HTO ranging from 1-100 mCi/ml body water and microencephaly, sterility, stunting, litter size reduction and resorption of the fetuses were reported a higher dose levels. Decrease in absolute and mean weights of the fetal organs was roughly proportional to the dose of radiation. Further, in 1976, they observed that reproductive capacity of F_1 generation and the effect of life time prenatal exposure on F_2 generation to as low as 3 m rad/day delayed the opening of eye and development of righting reflex in young rats nd 30 rad/day additionally depressed the spontaneous activity of adult male rats.

Hatch *et al.*,[158] studied kangaroo rats living in tritiated environment, (activity ranged from 0.1-0.4 µCi/gram body weight), around the crater of a nuclear detonation in Nevada but did not notice any evidence of harmful effects of radiation on these animals which grew normally and reproduced well in the vicinity of the crater. Effect of continuous *in utero* exposure to HTO was studied by Hass *et al.*,[159] in the development of female germ cells in Wistar arts. The pregnant animals were given daily doses of 290.0-5800.0 µCi HTO intravenously from day 9 of pregnancy to term. During the first 20 days of life, a dose dependent reduction in number of oocytes was observed and a dose of 1450 µCi/day decreased the number of oocytes to 50% of the control, whereas the highest dose of 5800 µCi/day produced total aplasia in the ovary. Laskey *et al.*,[160] exposed the pregnant females to HTO at the dose rate of 0.01, 0.10 and 10.0 µCi/ml of body water and stated that continuous exposure to 10 μCi/ml produced a 30% reduction in the testes weight of F_1 adult males but no impairment of growth or reproductive ability. Statistically significants effects on the F_2 neonates were: reduction in brain weight at 0.1-10.0 μCi/ml; decreased body weight at 1.0 and 10.0 μCi/ml; decreased litter size and increased resorption at 10.0 μCi/ml.

Bursian *et al.*,[161] exposed Sprague Dawley rats continuously to tritiated water (0, 1, 10 and 100 μCi HTO/ml body water) from conception to birth. Cumulative whole body doses to the embryo and fetus were calculated approximately at 0, .066, .66 and 6.60 Gy and intrauterine exposure to as low as .66 Gy produced measurable and persistent decrease in brain weight at 21 and 45 days postnatally. Laskey and Bursian[162] exposed rats for 42 days beginning either from one day of pregnancy, at birth, at 42 or 74 days of post partum age to constant tritium activity of 10 μCi/ml and observed them after either 49, 120 or 300 days of irradiation. They reported a significant reduction in testis weight and sperm contents. The group exposed from the first day of pregnancy was the most sensitive to the effects of chronic irradiation.

Carsten and Cronkite[163] found that continuous ingestion of HTO at a concentration of 3.0 μCi/ml resulted in: (1) reduction in the number of viable embryos present in the female at late pregnancy, from mating when either the female or both the parents have been on HTO and (2) increase in the number of early post-implantation deaths, when both parents were on HTO; but there was no loss in body weight and no apparent effect on breeding efficiency of surviving animals. Bhatia[164] exposed mice to various (acute, fractionated and chronic) doses of tritium at different pre- and postnatal stages of development and observed various histopathological changes in the testes, which were dose dependent. A continuous exposure to HTO from 16d.p.c. caused a gradual reduction in the size of the testes and germ cell population in the offspring.

Torok *et al.*,[165] studied the effects of a single injection of tritiated water on the prenatal and postnatal development of mice given during the organogenetic period to pregnant animals. They observed a significant decrease in weights of brain and genital tract of offspring after injecting 0.07 mCi tritium/gram body weight at 9d.p.c. conception. In 4.5 months old offspring the number of oocyte was substantially reduced but at 2 months, these offspring were fertile. 0.135 µCi/gram body weight affected fertility of females but not of the males. At 18 months the incidence of ovarian tumor in exposed offspring was increased approximately five fold over controls (tritium: 67% control: 14%). At 0.54 µCi/gram body weight at day 7, 9 or 11 of pregnancy, females showed stunted fetuses, but incidence rates of gross malformation as well as of skeletal anomalies were neligible. Retardation of prosencephalon and marked hypoplasy of the gonads were the prominent histological findings in this group. The treatment of dams on day 7 and 9 of pregnancy led to increased resorption of embryos. An injection of 0.81 µCi/gram body weight at 9d.p.c. caused 50% resorption. Tritium administered at higher activity levels than 1 µCi/gram body weight caused maternal mortality and in the uteri of surviving and dead dams all embryos were in a state of resorption.

Jones *et al.*,[166] maintained squirrel monkeys on HTO throughout gestation at 0.05 to 4.0 µCi/ml body water and did not observe discernible effect on new born progeny in terms of body weight, body dimensions and selected organ weights other than ovaries. The number of oocytes in female progeny decreased markedly with the increased dose of tritium in maternal drinking water. Etoh and Hyodo-Taguchi[167] compared the effect of tritiated water on the germ cells of embryo with that of chronic gamma irradiation and reported that number of germ cells in treated groups decreased with increasing concentration of tritium upto 0.2 mCi/ml drinking water. Bhatia and Srivastava[168] subjected mice to continuous HTO treatment at the dose rate of 5.0 µCi/ml drinking water (0.9 rad/day) from 16d.p.c. and various histopathological lesions were reported in the testes. Rochalska *et al.*,[169] measured the amount of tritium incorporated into rat bain. The lipid contents were found to be higher after giving tritiated food than after administration of tritiated water and they were the highest in rats contaminated mainly through milk of their mothers. Dev *et al.*,[170] exposed pregnant Swiss albion mice to tritium at 14¼, 16¼, 18¼d.p.c. at the dose of 5.0 µCi/ml body water and observed reduced body weight of male and female offspring from first day to 12 weeks of postantal age. It was also found that day 14¼ of gestation was the most suceptible stage in respect of postanatal body weight loss.

Bhatia and Sisodia[171] maintained pregnant Swiss albino mice on HTO (2.5 μCi or 92.5 KBq/ml of drinking water; .0045 Gy/day; after a priming injection) at 11, 14 or 18 days of gestation and reported that offspring of both sexes born to HTO irradiated mothers showed various radiopthological changes at different weeks of postnatal age. They also showed a progressive decline in animal and organ weight, body and tail length and a great deviation and organ weight/body weight ratio from the normal values.

The olfactory bulb of Swiss albino mice exhibited a gestational day dependent incidence in severity of damage at 19.25d.p.c. after a continuous HTO exposure at 11.25, 12.25, 14.25, 15.25 and 16.25 days to 19.25d.p.c. at the dose level of 5.5 μCi and 3.5 μCi/ml body water, which varied as the pregnancy advanced. The damage was in the form of various radiopathological lesions; viz. the loss of columnar arrangement of mitral layer, disarray of the tissue with an obliteration of olfactory ventricles, enlarged ventricles, necrosis, pyknosis etc. Blood clots present in some of the fetuses indicated brain hemorrhage. Some fetuses showed foot and tail abnormalities. There was no evidence of microcephaly and enlarged lateral ventricles in fetuses exposed at 11.25d.p.c.. However, 16.25 day old HTO exposed fetuses showed the maximum damage.[172]

One day old mice were continuously exposed to 3μCi HTO/ml of body water and defective movements, hydrocephaly and cerebellar malformations were observed in the offspring at the postnatal age of 1, 2, 3, 4, 5 and 6 weeks. A gradual decline in body and brain weights also observed with a maximum decline in body weight (18%) at 4 weeks of age. The reduction in brain weight was 21, 14 and 19% at 1, 5 and 6 weeks of age respectively.[173] mathur[174] reported that a continuous exposure of mice to 1.0, 2.0 or 5.0 μCi HTO/ml body water at 11.25, 14.25, 16.25 and 18.25d.p.c. resulted in a dose dependent increase in resorption, pre and postnatal mortality and neonatal deaths and decrease in litter size and body weights of surviving fetuses and offspring.

Chromosomal aberrations were reported in germ cells by Mailhes *et al.*,[175] after continuous exposure of female mice to different does (0, 3.0, 7.5, 15.0 or 30.0 μCi/ml) of tritium through their drinking water from 3 to 7 weeks of age. Similar results were also obtained by Pelliccia *et al.*,[176] in their studies on chromosomal aberrations induced by incorporated tritium. Female rats were given tritiated drinking water (3 μCi/ml) from 30 days before mating till delivery and it was found that the course and outcome of pregnancy were normal but the difference between newborn body weight and cerebral weight of the treated rats was on the borderline of significance as compared to control ones.[177]

Sharma and Saini[178] gave a priming injection (im) of tritiated water to pregnant Swiss albino mice (HTO) at the dose rate of 2, 3 and 5 μCi/ml body water (74, 111 or 185 Bq/ml body water) at 0, 6 and 14 day post conception (d.p.c.) and were subsequently maintained on tritiated drinking water *ad libitium* during pre-implatation (0-5d.p.c.), organogenetic (6-12d.p.c.) or fetal (14-18d.p.c.) period, respectively. On day 18 of gestation the females were sacsrificed by cervical dislocation to record the implant sites per dam and embryonic/fetal mortality. Significant reduction was observed in average implant sites per dam when the females were exposed to any of the three doses during the pre-implantation period due to embryonic resorption before implantation. However, the same was found to be within the normal range when mothers were exposed during the organogenetic or fetal period. Prenatal mortality (embryonic resorption/fetal death) was higher after *in utero* exposure to different doses during pre-implantation period as compared to organogenetic period, but mortality did not occur after exposure to any of the doses during the fetal period. Occurrence of mortality was found to be dose dependent.

Effects on Human

In early days, ionizing radiations were widely used to treat many diverse disease. But it was soon realized that irradiaiton leads to severe damage of the fetus and that most, although not all, surviving children showed marked abnormalities such as microcephaly, mental retardation, microphthalmia and cataracts. Maternal illness during pregnancy is not uncommon and sometimes requires radiographic imaging for proper diagnosis and treatment. The patient and her physician may be concerned about potential harm to the fetus from radiation exposure. In reality, however, the risks to the developing fetus are quite small. The accepted cumulative dose of ionizing radiation during pregnancy is .05 Gy, and no single diagnostic study exceeds this maximum. For example, the amount of exposure to the fetus from a two-view chest X-ray of the mother is only 0.00007 rad. The most sensitive time period for central nervous system teratogenesis is between 10 and 17 weeks of gestation. Nonurgent radiologic testing should be avoided during this time. Rare consequences of prenatal radiation exposure include a slight increase in the incidence of childhood leukemia and possibly, a very small change in the frequency of genetic mutations. Such exposure is not an indication for pregnancy termination. Appropriate counseling of patients before radiologic studies are performed and is critical. However, the threshold dose causing the anomaly of a human fetus has not been established, but there are grounds to presume that even diagnostic irradiation of pregnant women (at doses of 0.001-0.2 Gy) can cause substantial malformations, especially if this occurs in the period of the early stage of development of the central nervous system (upto the 38th day after implantation). It is assumed that at least 96% of all deaths caused by congenital malformations can be due to that radiation background, while the 60% increase in malformations observed in the USA during the last 40 years, was a result of the medical use of X-radiation.[12]

Comparison of radiation induced congenital defects in mice and man indicates that rodent data may be extrapolated to man when embryologically comparable gestation ages are considered. Still the literature constains comparatively little information on the action of ionizing radiation on the embryos and fetuses of human. This information is mainly restricted to a description of results of the atomic-bomb survivors at Nagasaki and Hiroshima, Marshall-islanders exposed to fallout from nuclear weapon tests and from persons exposed to therapeutic or diagnostic procedures, from sources that were external or internal. As far as we are aware there is no information from occupational exposures of pregnant women. A general conclusion from these observations is that the embryo is highly radiosensitive. The damaging effect of radiations manifests itself in survived children in the form of various malformations, delays in physical and mental development or combinations thereof.

Irradiation of human fetus during first trimester could lead to gross anomalies, while irradiation during the second and third trimester may lead to subtle functional defects.[26] Implantation in the human occurs at about 11 days after fertilization; the period of major organogenesis extends to about day 38. Irradiation, even with very low doses during the first two weeks of pregnancy results in a high number of spontaneous abortions and in gross abnormalities. Between the third and sixth week of pregnancy, when the major organs systems are being formed, gross abnormalities may be produced by a small dose of radiation. Beyond day 40, the embryo is more radioresistant, although certain radiosensitive neuroblasts and germ cells may be killed with resulting damage to nervous and reporductive systems.

Meyer *et al.*,[179] observed significant alteration in sex-ratio in the offspring of women X-rayed during pregnancy for diagnosis. Among 30 women exposed during pregnancy at Nagasaki, almost half of the pregnancies resulted in fetal, neonatal or infant death.

Lilienfied[180] and Diamond *et al.*,[181] found that intrauterine exposure to diagnostic X-rays was associated with a doubling of the risk of death from all possible causes among exposed children. Dekaban[182] reported that 26 children who received heavy X-irradiation during various stages of gestation showed stunted growth, microcephaly, mental retardation, microphthalmia, pigmentary degeneration of the retina, cataract, genital and skeletal anomalies. Meyer *et al.*,[179] observed that exposed women had almost twice as many fetal deaths and more spontaneous abortions than controls.

Studies of children exposed at Hiroshima and Nagasaki[183-188] or during medical procedures[182] indicated that microcephaly and mental retardation were prominent findings after *in utero* exposure. Some authors further suggested that maldevelopment of the nervous system is a "cardinal manifestation" of exposure of the human conceptus to ionizing radiation.[19, 4] Miller and Mulvihill[51] reported that children born with small head circumference developed severe mental retardation after exposure of women to atomic radiation during pregnancy. They pointed out that mental retardation occurred primarily after exposures at 8-15 weeks of gestational age, with small excess at 16-25 weeks of age. They also reported eye malformation resulting intrauterine radiation effects in the human offspring. A causal relationship between intrauterine exposure to X-rays and increased evidence of leukemia, mangolism, altered sex-ratio, incresed gross death rate and risk of malignant neoplasm among children, was reported by Oppenheim *et al.*,[189] and Shino *et al.*,[190] Harvey *et al.*,[191] reported that when a pregnant woman was irradiated inadvertently, it is the growth, retarded at term and recovers poorly from growth retardation in the post-partum period.

During the last few years there has been a perceptible change in attitudes to the problem of possible irradiation *in utero*. As a consequence of re-examination and re-analysis of observations mostly made many years ago, concern has switched from possible exposure in the first few weeks after conception to consideration of exposures in 3rd to 5th (Lunar) months after the Last Menstrual Period (LMP). The experience of Japanese Bomb survivors irradiated *in utero* has shown clearly that developing brain is most susceptible to damage in the period 10 to 17 weeks after LMP (Otake and Schull[187]) and in contrast, that irradiation at earlier stages in development has not caused a detectable increase in abnormalities of development in other organs or tissues.[4, 7, 9]

Mechanism of Radioembryological Effects

The mechanisms of the radioembryological effect are similar to those considered when describing the basic radiobiological responses of cells. An exclusive feature of an embryo is determined by the presence of a large number of undifferentiated primary cells directed to compensate the cells losses because of which an irradiated organism continues to develop as an entity, although an inevitable result of irradiation is the appearance of a state of insufficiency (defectiveness). The later is to tolerated by an organism depending upon the replacement of the damages cells by ones remaining undifferentiated. The outcome of embryo damage is thus expressed by certain equilibrium between the initial effect of irradiation and the ability of an "organism as a whole" to repair the damage to separate parts thereof recovering by means of the newly oriented viable primary cells. For example, irradiation in a dose of 0.5 Gy may cause the death of practically all the cells of the eye retina in a mouse embryo, but they are replaced by such regeneration, and the infant mouse retains its vision.

Embryos that do not survive after irradiation usually die owing to damage to one or more vital systems, and their damage causes the death of other systems capable of surviving in normal surroundings. The transplantation of organs from irradiated embryos to intact recipients enabled these organs to survive and develop normally.

We must note that the anomalies of discussed are not a specific feature of the action of only ionizing radiation. They are also observed under the action of certain enzyme inhibitors and radiometric alkylating compounds. But owing to the high penetrability of ionizing radiation, the diversity of the post radiation anomalies in comparison with any other agent of the environment is considerably broader. Regardless of the details of the radioembryological effect mechanism, the consequences of irradiating an embryo or fetus are such that they inevitably point to the need of preventing the action of radiation during this period at any cost.

Any action of radiation on a primary germinal cell of an embryo is transmitted to all the cells to its following generations. According to Rugh[192] ionizing radiation is and will remain a very important tool in the flourishing of civilization, as has already been shown in medicine and in other fields of science and engineering. The mean life span of humans has increased considerably owing to the development of radiology.

NOTES

1. Job, T.T., Leibold, G.J. and Fitrmaurice, H.A., Jr. (1935): Biological effects of roentgen rays. Determination of critical periods in mammalian development with X-rays. *Am. J. Anat* 56: 97.
2. UNSCEAR (1977): Sources and effects of ionizing radiation Annex. *J. Developmental effects of irradiation in utero* (New York, U.N. Scientific committee on the effects of atomic radiation, U.N.) p. 656.
3. Beir report (1980): The *effects on populations of exposure to low levels* of *ionizing radiations*. Committee on the biological effects of ionizing radiations. Division of Medical Sciences (Washington, D.C., National Academy Press).
4. *Mole, R.H. (1982):* Consequences of parental radiation exposure for postnotal development. A review *Int. J. Radiat. Biol.* 42: 1.

5. Brent, R.L. (1984): The effect of ionizing radiation microwaves and ultrasound on the developing embryo. Clinical interpretations and applications of the data. *Current problems in pediatrics.* 14: 9.

6. S.S.K. Publications (1985): *Wirkungen nach praenataler bestrahlung,* D. Gumpeclrt and A. Kindt (eds.) (Stutlgart, Gustak Fischer).

7. GSF symposium (1982): *Developmental effects of prenatal irradiation,* H. Kriegel, W. Schmahl, G. Kristner and F.F. Stiene (Eds.), (Stultgert, Gustav Fischer).

8. GSF symposium (1986): *Radiation risk to the developing nervous system.* H. Kriegel, W. Schmahl, G.B. Gerban and F.F. Stieve (eds.), International symposium jointly organized by the commission of European communities and department of nuclear biology, GSF Neuherberyg, 18-20 June, 1985 (Stultgart; Gustav Fischer).

9. EULEP symposium (1984): *Effects of parental irradiation with special emphasis on late effects,* C. Streffer and G. Patrick (eds.), 29 July, 1982 Bordeaux (Franse). Commission of the European Communities Publications. EUR. 8067.

10. Rugh, R. and Wohlfromm, M. (1962): Can the mammalian embryo be killed by X-irradiation? *J. Expt. Zool.* 151: 227.

11. Raugh, R. and Wohlfromm, M. (1965): Parental X-irradiation and postnatal mortality: *Radiat at Res.* 26: 493.

12. Yarmoneko, S.P. (1998): Effect of radiation on embryo and fetus. In: *Radiobiology of humans and animals.* p. 241.

13. Wilson, J.G., Jordon, C. and Brent, R.L. (1953): Effects of irradiation on embryonic development II. X-rays on the ninth day of gestation in the rat. *Am. J. Anat.* 92: 153.

14. Ressell, L.B. (1957): Effects of low dose X-rays on embryonic development, *Proc. Soc. Expt. Biol. Med.* 95: 174.

15. Rugh, R. (1959): Ionizing radiations; their possible relations to etiology of some congenital anomalies and human disorders. *Mol. Med. 124*: 401.

16. Brent, R.L. (1972): *Irradiation in Pregnancy. Gynecol obstetrics.* S. S. Davis (eds.) (New York: Haper and Row). p. 1.

17. Brent, R.L. (1976): *Environmental factors. In: Prevention of embryonic, fetal and perinatal disease,* R.L. Brent and M.I. Harris, Bethescdo (eds.), Maryland, NIH-Dhew Publication No. (NIH). 76-853, p. 179.

18. Brent, R.L. (1977): Radiation and other physical agents. In: *Handbook of Teratology*. J.G. Wilson and F.C. Frazer (eds.), (New York: Plenum Press). p. 153.

19. Brent, R.L. (1980): Radiation teratogenesis. *Teratology*. 21: 282.

20. Kameyama, Y. (1982): Low-dose radiation as an environmental agent affecting intrauterine development. *Environ. Med*. 26: 1.

21. Brent, R.L. (1983): The effects of embryonic and fetal exposure to X-ray, microwaves and ultrasound. *Clin. Obstet. Gynecol*. 26: 484.

22. Russell, L.B. (1980): Radiation damage to the embryo, fetus, and neonates. In: *Biological Risks of Medical irradiations*, G.D. Fullerton, D.T. Kopp, R.G. Waggener, and E.W. Webster, (eds.), American Institute of Physics, New York. pp. 33-54.

23. Houston, C.S. and Shokeir, M.H.K. (1977): Review artical: "Potential hazards of diagnostic radiation". *J. Assoc. Can. Radiologistes*. 28: 62.

24. Brent, R.L. (1984): The effect of ionizing radiation microwaves and ultrasound on the developing embryo: clinical interpretations and applications of the data. *Current problems in Pediatrics* 14: 9.

25. Wilson, J.G. (1973): *Environment and Birth Defects*. (New York, Academic Press). p. 19.

26. Rugh, R. (1971): X-ray induced teratogenesis in the mouse and its possible significance to man. *Radiology* 99: 433.

27. Bithell, J.F. and Stewart, A.M. (1975). Prenatal irradiation and childhood malignancy. A review of British data from the Oxford survey. *Br. J. Cancer*. 31: 271.

28. Hoffman, D.A., Felten, R.P. and Cry. W.H. (1981): Effects of ionizing radiation on developing embryo and fetus. *NH and publication FDA 81-8170*.

29. Michel, C. and Balla, I. (1987): Interaction between cadmium and radiation in mice. *Experientia*. 42: 80.

30. Hande, M.P. and Uma Devi, P. (1995): Teratogenic effects of repeated exposures to X-rays and/or ultrasound in mice. *Neurotoniocol. Teratol*. 17: 179.

31. Jacquet, P., de Saint-Georges, L., Vankerkom, J. and Baugnet-Mahieu, L. (1995): Embryonic death, dwarfism and fetal malformations after irradiation of embryo at the zygate stage: studies on two mouse strains. *Mutat. Res*. 332: 73.

32. Hillcbrandt, S., Streffer, C., Montagutclli, X. and Balling, R. (1998): A locus for radiation induced gastroschisis on mouse chromosome 7. *Mamm. Genome*. 9: 995.

33. Kim, S.H., Lee, J.H., Oh, H., Kim, S.R., Lee, C.S., Jo. S.K., Kim, T.H., and Lee, Y.S. (2001): Dependence of malformation upto gestational age and exposed dose of gamma radition. *Radiat. Res.* 42: 255.

34. Feingold, S. M. and Hahn, E.W. (1972): Post conception development of ova following X-ray induced super-ovulation. *Radiat. Res* 51: 110.

35. Schlesinger, D.M. and Brent, R.L. (1978): Effect of X-irradiation during pre-implantation stage of gestation on cell viability and embryo survival in the mouse. *Radiat. Res.* 75: 202.

36. Zhou, J. and Wang, J. (1982): Studies of radiation effects on mouse embryos in the pre-implantation stage. *Chin, J. Radiol. Med. Prot.* 2: 18 (in Chinese).

37. Roux, L., Harvath, R. and Dupuis, R. (1983): Effects of pre-implantation lowdose radiation on rat embryos. *Health Physic* 45: 993.

38. Fredberg, W., Faulkner, D.N., Neas, B R., Hanneman, G. D., Darden, E.B. Jr., Deal, R.B. Jr. and Parker, D.E. (1987): Dose incidence relationship for exencephalia, anophthalmia and prenatal mortality in mouse embryos irradiated with fission neutrons or 250 KV X-rays. *Int. J. Radiat. Biol.* 52: 223.

39. Angleton, G.M., Benjamin, S.A. and Lee, A.C. (1988): Health effects of low-level irradiation during development. Experimental design and prenatal and early neonatal mortality in beagles exposed to Co^{60} gamma rays. *Radiat. Res.* 115: 70.

40. Uma Devi, P. and Hande, M.P. (1990): Effect of low doses of 70 K Vp X-rays on the intrauterine development of mice. Experientia. 46: 572.

41. Uma Devi, P., Baskar, R. and Hande, M.P. (1994): Effect of exposure to low-dose gamma radiation during the organogenesis in the mouse fetus. *Radiat. Res.* 138, 1: 133-38.

42. Rugh, R. and Grupp, E. (1959): X-irradiation exencephalia. *Am. J. Roentgenol.* 81: 1026.

43. Skreb, N. and Bijelic, N. (1962): Effects of X-rays on the rat embryo during mesoderm formation. *Nature.* 193: 292.

44. Ohzu, E. (1965): Effects of low dose X-irradiation on early mouse embryos *Radiat. Res.* 26: 107.

45. Salko, R.G. (1965): The effect of Co^{60} gamma radiation on development and DNA synthesis in the 11 day rat embryo. *J. Expt. Zool.* 160: 171.

46. Coppenger, C.J. and Brown, S. O. (1976): The gross manifestation of continuous gamma irradiation on the prenatal rat. *Radiat. Res.* 31: 320.
47. Pradhan, D.S. and Sreenivasar, A. (1971): Basic mechanism of radition injury and repair in mammalian system. *J. Sci. and Indus Res.* 30: 703.
48. Brent, R. L. and Gorson, R.O. (1972): Radiation exposure in pregnancy. *Curr. Prob. Radiol.* 2: 1.
49. Rugh, R. (1963): The impact of ionizing radiation on the embryo and fetus. *Amer. J. Roentgenol. Radia. Therapy and nuclear. Med.* 89: 182.
50. Fliedner, T. M., Hass, R.J., Bohne, F. and Harris, E. B. (1969): Radiation effects produced in pregnant rats and their offspring after continuous infusion of tritiated thymidine. In: *Radiation Biol. of fetal and juvenile mammals, Sikov;* M. R. and Mahlum, D.D. (eds.), p. 263, USAEC Oak Ridge.
51. Miller, R.W. and Mulvihill, J.J. (1976): Small head size after atomic irradiation. *Teratology.* 14: 355.
52. Aldeen, K.A.M. and Konermann, G. (1978): Effects of acute X-irradiation on the pre-implantation reaction in the mouse. *Radiat. Environ. Biophys.* 15: 47.
53. Dev, P.K., Pareek, B.P., Goyal, P. K., Mehta, G. and Gupta, S.M. (1983): Effect of prenatal gamma radiations on the development of mice and its modification by MPG *Acta. Anat.* 116: 339.
54. Michel, C. and Fritz-Niggli, H. (1986): Teratogenic interactions between cadmium and radiation in mice. *Experientia.* 42: 80.
55. Kaven, A. (1938): Das auftreten Von gehirnmissbildungen nach rontgenbestrahlung Von mauselembryon: Ztschr. Menschl., *Vererb. U. Konstitutionslebre.* 22: 247.
56. Wilson, J. G. and Karr, J.W. (1951): Effect of irradiation on the embryonic development. 1. X-ray on the 10th day of gestation in the rat. *Am. J. Anat.* 88: 1.
57. Russell, L.B. and Russell, W.L. (1954): An analysis of the changing radiation response of the developing mouse embryo. Symposium on effects of radiation and other deleterious agents on embryonic development. *J. Cell. Comp. Physiol. Suppl.* 1, 43: 103.
58. Rugh, R., Duhamel, L., Osborne, A. W. and Verma, A. (1964): Persistent stunting following X-irradiation of the fetuses *Am. J. Anat.* pp. 115-185.
59. Philipp, J.V. (1975): Influence of ionizing radiation on fetoplacental growth in mice. *Am. J. Obstet. Gynec* 123: 640.

60. Ward, W.F., Aceto, H. and Stallard, D. (1977): Comparison of the teratogeni effectiveness of Co^{60} gamma rays and extended bargg-peak helium ions. *Radiat. Res.* 70: 621.

61. Meyer, M.B., Merz, T. and Diamond, E. L. (1969): Investigation on the effect of prenatal X-ray exposure of human oogonia and oocytes as measured by later reproductive performance. *Am. J. Epidemol* 89: 619.

62. Schmahl, W., Senft, E. and Weber, W. (1979): X-irradiation of mice in early fetal period influences dose dependent sex-ratio of offspring until weaning. *Experientia.* 35: 541.

63. Russell, L.B. (1950): X-ray induced developmental abnormalities in the mouse and their use in the anlaysis of embryologial pattern. 1. External and gross visceral changes. *J. Expt. Zool.* 14: 545.

64. Russell, L.B., Russell, W.L. and Major, M.H. (1951): The effect of hypoxia on the radiation induction of developmental abnormalities in the mouse (abstract) *Acta. Res.* 11: 455.

65. Ershoff, D.P. and Bavetta, L.A. (1958): Potentiating effects of prenatal X-irradiation on dental caries in rat. *Proc. Soc. Expt. Biol. Med.* 97: 202.

66. Nash, D.J. and Gowen, J. W. (1962): Effects of X-irradiation upon post natal growth in the mouse. *Biol. Bull.* 122: 115.

67. Rugh, R., Duhamel, R. and Skaredoff, L. (1966): Relation of embryonic and fetal X-irradiation to life time average weights and tumor incidence in mice. *Proc. Soc. Expt. Biol. Med.* 121: 714.

68. Martin, P.G. and Murphee, R.L. (1969): Growth in prenatally irradiated rats as indicated by change in cell number and cellular materials. *Radiat. Res.* 40: 330.

69. Reindhardt, S. and Kriegal, H. (1969): Effect of fractionated X-irradiation on the development of the mammalian fetus. In: *Radiation biology of the fetal and juvenile mammal* M.R. and Mahlum, D. (eds.), DS, U.S. Atomic Energy Commission, p. 251.

70. Tietz, W. J., Phemister, R.D., Stuart, B. P., Smith, G.E. and Garner, R.J. (1970): Effect of X-irradiation on the implanting canine embryo. *Radiat. Res.* 43: 554.

71. Vogel, H.H. (1977): Neutron irradiation of rat embryos *in utero* *Radiat. Res.* 70: 621.

72. Sasaki, S., Kasuga, T., Saito, F. and Kawashima, N. (1978): Late effects of fetal mice X-irradiated at middle or late intrauterine stage. *Gan.* 69: 167.

73. Jensh, R.P. and Brent, R.L. (1988): Effects of prenatal X-irradiation on the 14-18th days of gestation on postnatal growth and development in the rat. *Teratology.* 38: 431.

74. ICRP Publication (1977): *Recommendation of the international commission on radiological protection.* ICRP Publication 26 (Perganon Press, Oxford).

75. Schull, W.J. and Otake, M. (1986): Learning disabilities in individuals exposed prenatally to ionizing radiation. Hiroshima and Nagasaki experiences. *Adv. Space Res.* 6: 223.

76. Hicks, S.P. (1952): Some effects of ionizing radiations and metabolic inhibition on the developing mammalian nervous system. *J. Pediat* 40: 489.

77. Hicks, S.P. (1953): Developmental malformations produced by radiation. Am. J. Roentgenol. *Radiat. and Nuclear Med.* 69: 272.

78. Hicks, S.P. (1954): Mechanism of radiation aneacephaly, anophthalmia, and pituitary anomalies *Ama. Arch. Pathol.* 54: 363.

79. Hicks, S.P., D'Amato, C.J. and Lowe, M.J. (1959): the development of the mammalian nervous system. I. Malformations of the brain, especially the cerebral cortex, induced in rats by radiation. II. Some mechanism of the malformations of the cortex, *J. comp. Neurol.* 113: 435.

80. Rugh, R. and Grupp. E. (1961): Neuropathological effects of low level X-irradiation of the mammalian embryo. *Mil Med.* 126: 647.

81. Rugh, R. and Grupp, E. (1962): Excephalia Following X-irradiation of the pre-implantation mammalian embryo. *J. Neuropathol. Expt. Neurol.* 18: 468.

82. Konermann, G. (1977): Periodical compensation responses of the mouse brain in the cource of postnatal growth following fractionated irradiation during embryogenesis. *Strahlentherapie.* 153: 399.

83. Heinxmann, U. (1982): X-rayinduced dysplasia in the developing brain. *Science.* 141: 903.

84. Brizzee, K.R., Jacobs, L. and Kharotchko, X. (1961): Effects of total body X-irradiation *in utero* on early postnatal changes in neuron, volumetric relationship and packing density in cerebral cortex. *Radiat. Res.* 14: 96.

85. Brizzee, K.R., Jacobs, L.A. and Bench, C.J. (1967): Histological effects of total body X-irradiation in various dose fractionation patterns on fetal cerebral hemisphere. *Radiat. Res.* 31: 415-429.

86. Brizzee, K. R., Ordy, J.M., Kaack, M.B. and Beavers, T. (1980): Effects of prenatal ionizing radiatin on the visual cortex and hippocampus of new born Squirrel monkeys. *J. Neuropathol. Exp. Neurol* 39: 523.

87. Brizzee, K.R. (1967): Quantitative histological studies on delayed effect of prenatal X-irradiation in rat cerebral cortex. *J. Neuropathol. Exptl. Neurol.* 26: 584.

88. Takeuchi, I. K., Takeuchi, Y.K. and Murakani, U. (1977): Cell surface coat materials on the rasettee formation in the cerebral mantle of the X-irradiated rat embryo. *J. Electron. Microscopy* 26: 151.

89. Das, G.D. (1977a): Experimental analysis of embyogenesis of cerebellum in rat I. Subnormal growth following X-irradition on day 15 of gestation. *J. Comp. Neurol.* 176: 419.

90. Hicks, S.P. and D' Amato, C.J. (1963): Low-dose radiation of developing brain. *Science.* 141: 903.

91. Cowen, D., and Geller, L.M. (1960): Long term pathological effect of prenatal X-irradiation on the cerebral nervous system of the rat. *J. Neuropathol. Expt. Neurol.* 19: 488.

92. Das, G.D. 1978): Premature death of purkinje cells following low level X-ray irradiation during embryonic development. *Acta. Anat.* 101: 225.

93. Hayashi, Y., Hoshino, K. and Kameyama, Y. (1979): Effect of low dose X-irradiation on the developing brain in mice dendritic arburization of pyramidal cells after birth. *J. Radiat. Res.* 20: 16.

94. Norton, S. (1979): Development of rat telencephalic neurons after prenatal X-irradiation. *J. Environ., Sci., and Health* 13: 121.

95. Konermann, G. and Schnald, I. (1980): The formation of trigmoid substance during postnatal, maturation of the brain of mice after pre-and perinatal X-irradiating. *Radiol. Environ. Biophysiol.* 18: 197.

96. Heinzmann, U. (1982): X-ray induced dysplasia in the developing brain. *Science* 141: 903.

97. Yang, T.C. and Tobias, C.A. (1984): Effects of heavy ion radiation on the brain vascular system and embryonic development. *Adv. Space. Res.* 4: 239.

98. Takeuchi, I. K. and Murakami, U. (1979): Two types of congenital hydrocephalus induced in rats by X-irradiation *in utero*: electron microscopic study on the telencephalic wall. *J. Anat.* 128: 693.

99. Takeuchi, Y.K. and Takeuchi, I.K. (1990): Critical period for induction of congenital hydrocephalus and dysplasia of

subcommissural organ by prenatal X-irradiation in rats *Experientia*. 46: 446.

100. Ferrer, I. (1993): Experimentally induced cortical malformations in rats. *Childs. Neru. Syst.* 9: 403.

101. Ferrer, I., Alcantara, S. and Marti, E. (1993): A four layered 'Lissencephalic' cortex induced by prenatal X-irradiation in the rat. *Neuropathol. Appl. Neurobiol.* 19: 74.

102. Caparelli-Daquer, E.M., Schmidt, S.L. (1999): Morphological brain asymmetries in male mice with callosal defects due to prenatal gamma irradiation. *Int. J. Dev. Neuro Sci.* 17: 67.

103. Dilmanian, F.A., Morris G.M., Le Duc, G., Huang, X., Ren, B., Bacarian, T., Allen, J. C., Kalef-Ezra, J., Orion, I., Rosen, E. M., Sandhu, T., Sathe, P., WU, X. Y., Zhong, Z. and Shivaprasad, H.L. (2001): Response of avian embryonic brain to spatially segmented X-ray microbeams. *Cell Mol. Biol.* 47: 485.

104. Schindlcr, M.K., Wang, L., Sclcmon, L.D., Goldman-Rakic, p.c., Rakic, P and Cscnansky, J.G. (2002): Abnormalities of thalamic volume and shape detected in fetally irradiated Rhesus monkeys with high dimensional brain mapping. *Biol. Psychiatry* 51: 827.

105. Poppe, E. (1942): Experimental investigation of the effects of radiation injury and repair in mammalian system. *J. Sci. and industrial Res.* 30: 703.

106. Woods, A.C., Friedwold J.S. and Reese, A.B. (1949): National research council, Washington, D.C. examination by woods, A.C., Friedwold, J.S. and Reese, A.B. *Report of Radioactive Cataract Survey*. Division of Medical Sciences.

107. Michel, C., Fritz-Niggli, H., Blattmann, H. and Cordt-Richle, I. (1977): Effect of Low-dose irradiation with X-rays and pi-mesons on embryos of two different mouse strains. *Strahletherapie*. 153: 674.

108. Michel, C., Fritz-Niggli, I., Cordt-Richle, *et al.*, (1982): *Radiosensitization in embryonio development: the interaction of misonidazole and low radiation doses*. Developmental effects of prenatal irradiation. Gustav Fischer, Stuttgart, p. 287.

109. Persaud, T.V.N., Bruni, J.E. and Froese, G. (1990): CNS development in rat exposed to icnizing radiation during gestation. *Teratology society, Abstracts*. 41: 584.

110. Kuno, H., Inagaki, S., Ban, Y., Konishi, R., Tanaka, K. Fujii, T. and Usui, T. (1992): Usefulness of ophthalimic examination in postnatal observation in teratology studies. *Jikken Dobutsu*. 41: 363.

111. Glineur, R., Louryan, S., Philippson, C., Evrard, L. and de Vos L. (1998): Effects of irradiation on facial development in mouse embryos. *Eur. J. Morphol.* 36: 245.

112. Warkany, J. and Schraffenbergen, E. (1947): Congenital malformations induced in rat by roentgen rays. *Am. J. Roent. Radiat and Nuclear Med.* 92: 153.

113. Russell, L.B. and Russell, W.L. (1950): Changes in the relative proportion of different axial skeletal types within inbred strains of mice, brought about by X-irradiation at critical stages in embryonic development. *Genetics.* 35: 689.

114. Russell, L.B. and Russell, W.L. (1956): "Hazards to the embryo and fetus from ionizing radiation" In: Peaceful use of Atomic Energy. Vol. II. Unted Nations, New York, p. 175.

115. Neyfakh, A.A. (1960): On reduction of the digit in mouse embryos following X-ray irradiation. *Arkh. Anat. Cristol. Embriol* 39: 74.

116. Chang, M.C., Hunt, D.M. and Harvey, E.B. (1963): Effects of radiocobalt irradiation of pregnant rabbits on the development of fetuses. *Anat. Res.* 145: 455.

117. Murakami, V., Kameyama, Y. and Nogami, M. (1963): Malformation of the extremity in the mouse fetus caused by X-irradiation of the mother during pregnancy. *J. Embryol. Expt. Morphol.* 11: 549.

118. Knauss, U. (1980): Dose response relationship of embryonic radiation damage for example in the mouse skeleton after antenatal exposure to X-radiations. Thesis. University of Frieburg Medical Faculty.

119. Keonermann, G. (1982): Consequesces of prenatal radiation exposure on prenatal and postnatal development: morphological, biochemical and histochemical aspects. In Development Effect of Prenatal Irradiation. Gustav fischer, Stuttgart. pp. 237-259.

120. Cornelissen, M., Thierens, H. and De Ridder, L. (1993): Radiation effects on the matrix synthesis in non-ossifying embryonic cartilage *in vitro*: a functional and morphological study. *Tissue Cell.* 25: 343.

121. Saad, A.Y., Abdelazim, A.A. el-khashab, M.M. and mansour, M.A. (1994): Induction of cleft palate by gamma irradiation of prenatal CD-1 mice, *cleft palate craniofac J.* 31: 351.

122. Wang, B., Fujita, K., Ohhira, C., Watanabe, K., Odaka, T., Mitani, H., Hayata, I., Ohyama, H., Yamada, T. and Shima, A. (1999): Radiation induced apoptosis and limb teratogenesis in embryonic mice. *Radiat. Res.* 151: 63.

123. Engstrom, H., Janson, J.O. and Engstrom, C. (1983): Effects of local irradiation on longitudinal bone growth in the rat. A tetracycling labelling investigation. *Acta. Radiologica.* 25: 129.

124. Riddler, L. D., Thierens, M., Cornelissen, M. and Sagaert, O. (1988): Effects of ionizing radiation on the metabolism and longitudinal growth of cartilagenous embryonic chick tibial *in vitro. Int. J. Radiat. Biol.* 53: 965.

125. Yamada, T., USUI, S. and Ohyma, H. (1988): Is malformation induced by X-irradiation at pre-implantation stage of mouse embryo *IIRS Report,* 27: 1987-88.

126. Finkel, M.P. (1974): The transmutation of radiostrontium and plutonium from mother to offspring in laboratory animals. *Physiol. Zool.* 20: 405.

127. Hiraoka, S. (1961): The transplacental effects of radiostrontium-90 upon the mouse embryos. (In Japanese with English Summary). *Acta. Anat. Nippon.* 36: 161.

128. Hopkins, B.J. and Casarett, B.W. (1964): Gross developmental anomalies induced by strontium 90 in the rat embryo. *University of Rochester Atomic Energy Report UR*-643.

129. Sikov, M.R., and Noonan, T.R. (1958): Anomalous development induced in the embryonic rat by the maternal administration of radiophosphorus. *Am. J. Anat.* 103: 137.

130. Arnikar, H. J., Singh, L.M. and Udupa, K.N. (1963): Radiation induced congenital defects in albino rats. *Curr. Sci.* 32: 209.

131. Lathrop, K.A., Gloria, I.V. and Happer, P.V. (1976): Response of mouse fetus to irradiation from Na^{99} $mTcO_4$. In: *Biological and Environmental Effects of Low level Radiation.* Vol. II, IAEA, Vienna, Austria, p. 211.

132. Gupta, S.M. Goyal, P.K. and Dev, P.K. (1981): Weight changes in mice after intrauterine treatment with MPG against-^{131}I irradiation. *Experientia,* 37: 898.

133. Goyal, P.K., Gupta, S.M. and Dev, P.K. (1983): Role of MPG in the postnatal growth of intrauterine ^{131}I irradiated mice. Radiobiol. *Radiobiol Radother.* 24: 243.

134. Greulich, R.C. (1961): Deleterious influences of oral administration of ^{3}HTDR on reporductive capacity of mice. *Radiat. Res.* 14: 83.

135. Cleaver. J.E. (1967): *Thymidine Metabolism and Cell Kinetics* North-Holland, Amstedam.

136. Flieneder, T. M., Hass, R. J., Bohne, F. and Harris, E. B. (1969): Radiation effects produced in pregnant rats and their offspring

after continuous infusion of tritiated thymidine. In: *Radiation biol. of fetal and juvenile mammals* sikov ; M.R. and Mahlum, D.D. (eds.). USAEC Oak Ridge.

137. Fliedner, T.M., Chaudhuri, J., Hass, R.J., Knorr-Gaetrner, H. and Schreml, W. (1971): Radiation effects of tritiated thymidine studies in pregnant rats and their offspring. In: *Biophysical Aspects of radiation Quality,* IAEA, Vienna p. 355.

138. Chaudhuri, J.P., Hass, R.J., Schreml, W., Knor-Gartner, H. and Fliedner, T.M. (1971): Cytogenetic and teratologic Studies on rats continuously infused with various doses of tritiated thymidine during pregnancy. *Teratology.* 4: 395.

139. Bond, V.P. and Feinendegen, L. E. (1966): Intranuclear ^{3}H thymidine: Dosimetric, radiobiological and radiation protection aspects. *Health Physics.* 12: 1007.

140. Cronkite, E.P., Robertson, J.S. and Feinendegen, L.E., (1973): Somatic nd teratogenic effects of tritium. In: *Tritium* Moghissi, A.A. and Carter, M.W. (eds.), Messenger Graphics, Phoenin, Arizona, p. 198.

141. Frinendegen, L.E. (1979): Radiation problems in fusion energy production. *Proceedings of VI International Congress of Radiat. Res.* May 13-19, 1979 Tokyo Okada, S., Lmamura, M., Terastima, T. and Yamaguchi, H. (eds.) p. 32.

142. Vesselinovitch, S.D., Simmons, E., Rao, K.V. N. land Mihailvich, N. (1974): (Abstract) carcinogenic hazards of irradiation during embryonal development in mice, *Radiat. Res.* 59: 101.

143. Snow, M.H.L. (1973): Abnormal development of pre-implantation mouse embryo gown *in vitro* with [^{3}H] thymidine. J. Embryol, Exp. Morph. 29: 601.

144. Schreml, W., Haas, R.J. Planes-Bohne and Steinhardt, B. (1974): Distribution and dosimetry of tritium in new born rats after *in utero* exposure to ^{3}HTDR. *Radiat. Res.* 58: 239.

145. Schreml, W. and Fliedner, T.M. (1977): Distribution of tritiated compounds. Tritiated thymidine and tritiated water in the mother fetus system and its consequences for the radiation effect of tritium. *Curr. Topics, Radiat. Res.* 12: 255.

146. Macqueen, H.A. (1979): Lethality of radioisotopes in early mouse embryos. J. Embryol. *Expt. Morph.* 52: 203.

147. Ugarte A.S., Mewissen, D.J. and Rust. J.H. (1983): *Cumulative genetic effects from exposure of male mice to tritium for twenty three generations.* C4, VII ICRR, Amsterdam.

148. Spindle, A., W.U, K, and Pedersen, R.A. (1982): Sensitivity of early mouse embryos to [^{3}H] thymidine. *Expt. Cell. Res.* 142: 397.

149. Saito, M., Streffer, C. and Molls, M. (1983): Tritium distribution in newborn mice after providing mother mice with drinking water containing tritiated thymidine. *Radiat. Res.* 95: 273.

150. Muller. W.U., Streffer, C., Molls, M. and Gluck, L. (1987): Radiotoxicities of (^{3}H) thymidime and (^{3}H) arginine compared in mouse embryos *in vitro. Radiat. Res.* 110: 192.

151. Brues, A. N., Stroud, A.N. and Rietz, L. (1952): Toxicity of tritium oxide to mice. *Proc. Soc. Expt. Biol. Med.* 79: 174.

152. Furchner, J.E. (1957): Relative biological effectiveness of tritium beta particales and Co^{60} gamma rays measured by lethality in CFI mice. *Radiat. Res.* 6: 483.

153. Trujillo, T.T., Anderson, E.C. and Langham, W. H. (1955): Biological effects of inhalation of high concentrations of tritium gas. *Report No. LA-1986.* Check Alamos Scientific Laboratory. Los Alamos, New Mexico.

154. Moskalev, Y. I., Zhdravie, V.F. and Isotomnia, A.G. (1973): Relative biologicl efficiency of tritium oxide at long period of administration. *Gig. Samitt.* 6: 51.

155. Thompson, R.C. and Ballou, J.E. (1956): Studies of metabolic turnover with tritium as a tracer, V. The predominantly non-dynamic state of body constituents in the rat. *J. Biol. Chem.* 223: 795.

156. Cahill, D.F. and Yuile, C.L. (1969): Some effects of tritiated water on mammalian fetal development. In: *Radiation Biology of fetal and juvenile mammals* M.R. Sikov and D.P. Mahlum (eds.), USAEC, USA. p. 283.

157. Cahill, D.F. and Yuile, C.L. (1970): Tritium: Some effects of continuous exposure *in utero* on mammalian development. *Radiat. Res.* 44: 727.

158. Hatch. F.T., Mazrimas, J.A., Koranda, J.J. and Martin, J.R. (1970): Ecology and radiation exposure of kangaroo rats living in a tritiated environment. *Radiat. Res.* 44: 97.

159. Hass, R. J., Schreml, W., Fliedner, T.M. and Calvo, W. (1973): The effect of tritiated water in the development of the rat oocyte after maternal infusion during pregnancy. *Int. J. Radiat. Biol.* 24: 603.

160. Laskey, J.W., Parrish, J.L. and Cahill, D.F. (1973): Some effects of life time prenatal exposure to low levels of tritium on the F_2 Generation. *Radiat. Res.* 56: 171.

161. Bursian, S.J., Cahill, D. F., Laskey, J. W. and Parker, L.M. (1975): Some aspects of brain neurochemistry after intrauterine exposure to tritium. *Int. J. Radiat. Biol.* 27: 455.

162. Laskey, J.W. and Bursian, S.J. (1976): Some effects of chronic tritium exposure during selected ages in the rat. *Radiat; Res.* 67: 314.

163. Carsten, A.L. and Cronkite, E.P. (1975): The genetic and hematopoietic effects of long term tritiated water (HTO) ingestion in mice. Presented in *'Int. Symp, on Biological Effects of Low level Radiation Pertinent to protection of man and Environment.'* Chicago, U.S.A. IAEA and WHO Sponsored. p. 51.

164. Bhatia, A.L. (1977): Qualitative and Quantitative studies on the effects of tritiated water on testis of mice during prenatal and postnatal development. Ph. D. Thesis, University of Rajasthan, Jaipur, India.

165. Torok, P., Schmahl, W., Meyer, I. And Kistner, G. (1979): *Effect of singel injection of tritiated water during organogenesio on the prenatal and postnatal development of mice.* IAEA-SM-237/24.

166. Jones, D.C. L., Drebs, J.S., Sasmore, D.P. and Mitoma, C. (1980): Evaluation of neonatal squirrel monkeys receiving tritiated water throughout gestation. *Radiat. Res.* 83: 592.

167. Etoh, H. and Hyodo-Taguchi, Y. (1981): Effect of HTO on the germ cells of Oryzias embryos. *J. Radit. Res.* 22: 47.

168. Bhatia and Srivastava, P.N. (1982): Tritium toxicity in mouse testis, effect of continuous exposure during pre and postnatal development. *Strahlenltherapie.* 158: 752.

169. Rochalska, M.R., Van Bruwaene, G.B., Gerber and Kirchman. R. (1982): Organically bound tritium and its distribution in mice fed organically labelled milk powder. *Annales de/Associate Belge de Radioprotect.* 7: 345.

170. Dev, P.K., Mathur, S.P., Patni, N. and Popli, M. K. (1983): Growth response of (2-mercaptopropionylglycine) treated offspring against *in utero* tritium toxicity. *proc. 7 Int. Cong. of Radiat. Res. Amsterdam* July 3-8, C 3-4.

171. Bhatia, A.L. and Sisodia, R. (1983): Morphometric study on neurobiological effect of tritium toxicity in developing mice brain and its modification by MPG. III *Proc. of Int. Symp. on Radiopharmacal.* 21-23 Sept., W. Germany.

172. Bhatia, A. L., and Sisodia, R. (1984): Olfactory bulb vulnerability to *in utero* tritiated exposure in Swiss albino mice. Europ. Develop. Biol. Cong. (Abstract) Southampton UK 2-7 Sep. 1984.

173. Bhatia, A.L. and Sisodia, R. (1984): Neuroteratogenicity in postnatally developing mice brain under HTO-Stress-Europ-Soc. *Radiat. Biol.* 18 Annual meeting. Zurich, Sept. 9-13.

174. Mathur, S.P. (1985): Role of MPG in modifying *in utero* tritium effects on the intestine of F_1 generation mice. Ph. D. Thesis University of Rajasthan, Jaipur, India.

175. Mailhes, J.B., Carsten, A.L. and Benz, D. (1987): Cytogenetic analysis of mouse metaphase II oocytes following exposure to tritiated water. *Radiat. Res.* 3: 438.

176. Pelliccia, F., Belloni, G., Bosi, A., Micheli, A. and Oliveri, G. (1988): Studies on chromosomal aberrations induced by incorporated tritium: effect of post treatment with hydroxyurea and caffeine in G_2. *Mut. Res.* 199: 139.

177. Zamenhoff, S. (1990): Different Effects of chronic ingestion of tritiated water on prenatal brain development. *Radiat. Res.* 122: 101.

178. Sharma, K. and Saini, M.R. (1993): Embryonic/fetal mortality after exposure to triliated water in swiss albino mice during different gestation periods. *Ind. J. Exp. Biol.* 31: 98.

179. Meyer, M. B., Merz, T. and Diamond, E.L., (1969): Investigation on the effects of prenatal X-ray exposure of human oogonia and oocytes as measured by later reproductive performance. *Am. J. Epidemol.* 89: 619.

180. Lilienfeld, A.M. (1966): Epidemiological Studies of the leukemogenic effects of radiation. *Yale J. Biol and Med.* 39: 143.

181. Diamond, E. L., Schmerler, H. and Lilienfeld, A.M. (1973): The relationship of intrauterine radiation to subsequent mortality and development of leukemia in children. *Am. J. Epidemiol.* 97: 283.

182. Dekaban, A.S. (1968): Abnormalities in children exposed to X-radiation during various stages of gestation: Tentative timetable of radiation injury to the human fetus. *J. Nucl. Med.* 9: 471.

183. Woods, J.W., Kuhn, R.J., Kanamoto, S. and Johnson, K.G. (1967): The Growth and development of children exposed *in utero* to the atomic bombs in Hiroshima and Nagasaki. *Am. J. Public. Health.* 57: 1374.

184. Wood, J.W., Johnson, K.G. and Omori, Y. (1976): *In utero* exposure to Hiroshima atomic bomb. An evaluation of head size and mental retardation 20 years later. *Pediartics.* 39: 385.

185. Wood, J.W., Johnson, K.G., Omori, Y., Kawamoto, S. and Kuhn, R.J. (1967): Mental retardation in children exposed *in utero* to

the atomic bombs in Hiroshima and Nagasaki. *Am. J. Publ. Health.* 57: 1381.

186. Blot, W.J. and Miller, R.W. (1973): Mental retardation following *in utero* exposure to the atomic bombs of Hiroshima and Nagasaki. *Radiology* 106: 617.

187. Otake, M. and Schull, W.J. (1984): *In utero* exposure to atomic radiation and mental retardation: a reassessment. *Brit. J. Radiol.,* 57: 409.

188. Schull, W.J. and Otake, M. (1986): Neurological deficit among the survivors exposed *in utero* to the atomic bombing of Hiroshima and Nagasaki: a re-assessment and new directions. In: Radiation Risks to the Developing Nervous System H. Kriegal, W., Schemahl, G.B., Gerber and F.F. Stieve (eds.) Gustry Fischer, Stuttgart New York. p. 399.

189. Oppenheim, B. E., Griem, M.L. and Meier, P. (1976): An investigation of effects of prenatal exposure to diagnostic X-rays. In: *Biological and Environmental Effect of Low Level Radiation* Vol. II, 249 p, IAEA, Vienna.p. 249.

190. Shino, P., Chung, C.S. and Myrianthopoulos, N.C. (1980): Preconception radiation, intrauterine diagnostic radiation and childhood neoplasia. *J. Natl. Cancer. Int.* 65: 681.

191. Harvey, E.B., Boice, J.D. and Honeyman, M. (1985): Prenatal X-ray exposure and childhood cancer in twins. *New engl. J. Med.* 312: 541.

192. Rugh, R. (1962): Response of the nervous system to ionizing radiation. In: *Major Radiobiological Concepts and Ionizing Radiation on the Embryo and Fetus,* Haley and Sinder (eds.) Academic Press, New York. p. 3.

19

On the Nidification and Breeding Biology of Common Babbler *Turdoides caudatus* (Dumont) in Jammu

D.N. Sahi

Department of Zoology, University of Jammu, Jammu-180006

ABSTRACT

Studies on the biology of Common Babbler *(Turdoides caudatus)* were carried out in New University Campus, Jammu University, Jammu. The observations regarding the biological aspects viz., nest construction, nesting tree preferences, the breeding season, clutch size, egg laying pattern, the incubation and nesting period, growth and downing of feathers etc. were made.

The Common Babbler builds its nest on bushes like *Acacia modesta, Zizyplus jujube* and *Capparis sepiaria* in the study area. The material utilized in nest construction was grass roots and grass stems. The breeding period extended from March to mid June in Jammu. The eggs are glossy, turquoise coloured. The time interval between two egg laying was observed to be 24 to 28 hr. Eggs were laid during early hours of the day. The average clutch size 3.75 eggs (x = 30 nests). Both parents shared in nest construction, incubation, brooding and rearing of chicks. The incubation period varied from 13-14 days. The young ones were nidiclolous. The nesting success for the years 1999-2000 was observed to be about 47 percent.

Introduction

The Common Babbler, *Turdoides caudatus* (Fig. 19.1) is a resident bird, widely distributed in the plains and the foothills.

It affects xerophytic thorn scrub jungle, sandy flood plains dotted with clumps of Tamarisk, Sarpat grass (*Sacharum)* and ber bushes (Ali and Ripley, 1983). The species has a preference in the study area for dry open country and semi-desert with thorn scrub and shrubby vegetation. In spite of the fact that this bird has a wide distribution in India, very little is known about its nidification and breeding behaviour except for a few passing references (Baker, 1926; Whistler, 1949; Ali and Fatehally, 1967; Barnes 1981 and Ali and Ripley, 1983), a comprehensive information on the biology of this bird is wanting. Keeping this in view, the present studies on the nidification and breeding ecology of common babbler were undertaken.

Fig 19.1: Adults *Turdoides caudatus* showing morphological features.

Study Area

The study was carried out from January 1999 to September 2000 in New University Campus, Jammu University, Jammu (32.67 N, 74.9 E, 270 M). The area represented a typical dry open, semi-desert with scrub and shrubby vegetation. Predominant plant species in the study area were *Acacia modesta, A. arabica, Zizyphus jujube, Lantana camara, Grevillea robusta. Adhatoda vasica, Cannabis sativa, Doeaona viscose, Dalbergia sissoo, Xanthium strumaruim and Partehnium hysterophorus.*

Material and Methods

Of the 42 nests spotted in the area during the study period, 30 nests were chosen for detailed study. Observations were made with naked eyes and whenever necessary through prismatic field binocular (Bushnell make). The birds were generally watched from a distance in order to avoid any disturbance in their behavioural pattern.

After pair formation, the birds were regularly watched and nesting time and its completion noted. The eggs were marked with Indian ink for recording incubation period and their hatching pattern. For general study, the eggs and chicks were brought in the laboratory at different time intervals. The eggs and chicks were weighed on a chemical balance true to 0.1 gm.

Result and Discussion

Breeding season

The early workers have observed variations in the duration of breeding season of common Babbler. Whistler (1949) observed that breeding season extends throughout the year but majority of the nests are seen from March to May and again from July to September. Ali (1964) recorded it between March to July but irregularly over the entire year. Ali and Ripley (1983) reported that it varies locally to cover almost whole of the year, mostly from March to July and extending upto October. However, in Jammu the common Babbler (*T. caudatus)* does not breed the year around. It has only a single breeding season extending from March to June.

With the commencement of breeding season in March, the pairs separate out temporarily from the flocks to look for nesting sites. The pair stick fairly close together throughout the better part of the day but occasionally join parent flock for foraging or at the alarm calls by the member of flock. At the end of the breeding season, the pairs congregate back into flocks.

Nest construction

(situation and location of the nests):

Ali (1964), Ali and Fatehully (1967), Barnes (1981) and Ali and Ripley (1983) reported that nests are usually made on thorny bushes, at the height ranging from 30 cm to 180 cm. Of the 42 nests, 47.6% nests were recorded between the height range of 91 cm-120 cm, 23.8% at 121-150 cm, 19.1% at 61-90 cm, 7.1% at 31-60 cm and 2.4% at 151-180 cm. Ali and Ripley (op. cit.) observed that *Zizyphus, Prosopics* and *Lantana* are common nesting trees but most favoured situation is hedges of prickly pear (*Opuntia*). Studies on 42 nests showed that 76.2% nests were made in *Acacia modesta,* 14.3% in *Zizyphus jujube* and 7.2% in *Cappris sepiaria* and 2.3% in an unidentified thorny bush.

Some of the newly constructed nests were made on previous nesting trees. Although *Lantana camara* is one of the common bush available in the area, it was not used for nesting. All the plants used for nest construction bear prominent thorns and prickle which act as deterrent to predators. Isolated *Acacia modesta* trees are most preferred host plants. Clumps of *A. modesta* trees are usually avoided and the most plausible reason seems that they act as roosting site for many other bird species that may tract the predators.

The nest is a neat and compact cup, composed of grass stems and roots (Fig. 19.2) as also reported by earlier workers (Ali, 1964; Ali and Fatehally 1967; Barnes, 1981, Ali and Ripley 1983). The walls of the nest are 1-1.5 cm thick. The diameter of cup ranges from 7.5-8 cm and depth 6-8 cm. Both parents participate equally in nesting, site selection and nest building activities as also reported by Ali (1964), Ali and Ripley (1983).

Fig 19.2: Nest of Common Babbler.

Observations made on 18 nests revealed that a pair takes 4-13 days (X=7 days) to complete the nests. Eggs are laid when nests are complete, however, first egg may be laid 2-11 days after completion of nest.

Egg size, weight and colour

The length and breadth measurements of 35 eggs from 10 nests varied from 21.5-23.5 mm × 16.5-8.5 mm (X=22.5×18mm). The average length and breadth recorded by earlier workers were 21.2×16.1 mm (Bakers, 1926) 12×16 mm (Whistler, 1949); 21.2×16.5 mm (Barnes, 1981). Thirty eggs on their first day of laying weighed from 2.000 g to 2.6000 g 9(X=2.400 g). Eggs are oval, mono coloured, turquoise blue having one broader and other pointed end. The colour of eggs gets slightly faded with passage of time.

Egg-laying pattern

Daily observation regarding the egg-laying pattern of this bird revealed that the time interval between the two successive laying was about 24-28 hr. Eggs were generally laid in the morning between 7 and 1 hr of the day. Occasionally eggs were laid before or after this period too.

Clutch size

The no. of eggs laid, as reported by earlier workers is 3 to 4 (Ali, 1964; Ali and Fatehally, 1967; and Barnes, 1981). According to Ali and Ripley (1983), the clutch size is commonly 4, often 3 and exceptionally 5. In the present study, the clutch size varied from 2 to 5 but being 4 the commonest.

Incubation period

The time lag between the laying of eggs and its hatching has been reckoned here as incubation period. Out of 24 nests observed for incubation during the study period, in 16 nests the last egg hatched out after thirteen days of incubation, in 6 nests after fourteen days and in another two, last two eggs did not hatch. The incubation period varied from 13-14 days (X=13.3 days). Ali and Ripley (1983) reported incubation period from 13-15 days in common Babbler.

Hatching pattern

[Numerals in parenthesis ( ) represents the clutch size]. In eight nests 5 (4), 2 (3), 1 (2) chicks hatched out in successive days in the same order in which the eggs were laid. In nine nests 5 (4), 2 (5), 2 (3) all chicks hatched out during the same day but at different timings. Out of the five nests 3 (5), 2 (4), in four, 3 chicks hatched out the same day and the rest on successive day and in fifth nest with 4 eggs, 2 chicks hatched on the same day and other on subsequent days. The two nests where one egg failed to hatch in each is not accounted for hatching pattern. Thus in eight nests incubation sets in with the laying of last egg and in four after third egg and in one when second egg was laid. Thus there appears to be no regular hatching pattern.

Mortality

Out of thirty nests which were chosen for detailed study, 6 nests with eggs were predated upon by some predator. The common predators noticed in the area were garden lizard, white-eye-buzzard, blue winged kite, pariah kite, shikra etc. Out of remaining 24 nests, only 2 eggs in two different nests did not hatch. Thus mortality at the egg level was slightly over twenty percent. Out of 24 nests, in eleven nestling success was cent percent, in seven losses was total as all chicks were predated upon and in six nests the loss was 25 to 50 percent as chicks died of starvation or falling on the ground. The mortality at nestling stage was about thirty eight percent. Unhatched egg remain in the nest even though nestling have left the nest and are not taken care off once all nestling leave nest, however, egg shells and dead chicks are not tolerated in the nest and removed immediately.

Reaction of parents at the removal of eggs or chicks

The birds seem to be more concerned when the chicks were removed out of nests for the data recording than in case of eggs. When the chicks were taken out of nests, the parents nervously twitch their wings and hop from bush to bush, peering down at the intruder, loosely jerking their tails and distressly whistling which-which-which-ri-ri-. . . . ri etc. At times, the sisterhood also joins the chorus.

For observing parental reaction at the removal of chicks, an almost full-grown chick was brought in the laboratory at 10 a.m. and kept there for a day. In the evening the nest was visited, the other two chicks were found missing from the nest, however, parents were observed to visit the nest occasionally. The chick was taken back next day to its nest in a paper bag and kept on the ground. The calls of chick immediately drew the attention of the parents who were nearly ten metres away from nest along with their other two chicks. The parents would reach, at the site and responded to calling chick but on being unable to see the chick, gave several contact calls. The chick also responded to these calls. The parents would then guess the presence of chick in bag. Both the parents fly to spot several times at the bag and unable to release the chick, call distressly. After four min. they made another attempt and in the next thirty min. the parents do not make any attempt but for intermittently calls. The chick is ultimately released from bag and is guided by the parents to a nearby bush.

Nest life

The newly hatched chicks were nidicolous, yellowish pink in colour without any natal down. The beak was yellowish and claws and egg tooth creamish. The mouth cavity and tongue orange coloured. The eyes of chick remain closed up to 2-3 days. The average weight of hatchlings is about 2.200 g (n=40) at the time of emergence from eggs.

Growth rate, weight gain and donning of feathers:

1 Day old chick

Body is naked and only rectrices pierce the skin. Body colour starts turning brown. Chicks weight range from 4-4.500 g ($\overline{X}$=4.00 g).

2 Days old chicks

On second day, the primaries, upper greater primaries, secondaries and alula emerge out of body. The hatchlings weight range from 6.250-7.500 f ($\overline{X}$=6.800).

3 Days old chick

The upper greater secondaries, humerus, spinal tract-cervical and interscapular make their appearance. Head plumage, dorsal, pelvic also get distinct on fourth day. On ventral side the sternal, axillar abdominal pierce the skin. Eyes open in a slit in most of the chicks. The chick weight range 12.500-14.000 g ($\overline{x}$ = 13.200 g, n = 26).

4 Days old chick

The upper middle secondaries, head plumage, femorals, crural, upper tail covert, dorsal and pelvicf also get distinct on fourth day. On ventral side the sternal, axillar, abdominal pierce the skin. Eyes open in a slit in most of the chicks. The chicks weight ranges from 12.500-14.000 g ($\overline{x}$=13.200 g, n=26).

5 Days old chick

The upper middle primaries pierce the skin. The primaries, upper greater primaries, secondaries and upper middle secondaries get very much elongated and some of them sprut in brushes. The fledgling weiged from 15-16.500 g ($\overline{x}$=16.00 g).

6 Days old chick

The anal circlet appears and most of wing quills brust out in brushes. The head plumage spinal tract-cervical region, dorsal and pelvic also sprut to form mini brushes like structures.

7 Days old chick

The eyes open completely and mouth gap turn creamish. The under greater secondaries, middle secondaries and malar pierce the skin on 7th day. The inter-ramals are visible through transparent skin. The feathers grow in size and form fan like structures. The 'rectrices, however, still remain in pins. The fledglings weighed from 18.100-19.800 g ($\overline{x}$ = 19 g, n = 20).

8 Days old chick (Fig.3)

The inter-ramals come out of skin. There is rapid growth of rectrices and remiges and young is almost fledged and may level the nest in many cases. The chick, however, cannot fly at this age but can take a short leap. The fledglings weight ranges from 19.500-22.000g ($\overline{x}$ - 21.200g).

Fig. 19.3: Eight days old chick of Common Babbler

9 Days old chick

Rectrices elongate and body feathers get more denser. The average weight of the chick is 22g. All the chicks were seen to leave the nest by the ninth day. The young one completely resembled its parents, except in the size of the body.

The fledglings leave the nests in 8 to 9 days. After leaving the nests, the fledgling remain for some days on the vicinity of nesting trees but they do not spend the night in the nest. The brood disappears after few days since observer never came across a strongly flying young being fed by parents.

REFERENCES

Ali, S. (1964) *The Book of Indian Birds.* eighth ed. (revised). Bombay Natural History Society, Bombay, p. 162.

Ali, S. and L. Fatehally. (1967), *Common Birds.* National Book Trust, India, New Delhi, p. 118.

Ali, S and S.D. Ripley (1983), *Hand Book of the birds of Indian and Pakistan* (compact edition). Oxford Univ. Press. Delhi, p. 737.

Bakers, E.C.S. (1926), *The Fauna of British India including Ceylon and Burma.* Bird III. London (Taylor and Francis).

Barnes, Edwin, H. (1981), *The Birds of India. A Guide to Indian Ornithology.* Cosmo Publications, New Delhi. p. 245.

Whistler, H. (1949), *Popular Handbook of Indian Birds.* Oliver and Boyd. Edinburgh and London, p. 560.

Index

O

P